Advances in Human Factors/Ergonomics, 15

Workspace, Equipment and Tool Design

Advances in Human Factors/Ergonomics

Series Editor: Gavriel Salvendy, Purdue University, West Lafayette, IN 47907, U.S.A.

Vol. 1.	Human–Computer Interaction (G. Salvendy, Editor)
Vol. 2.	Human–Computer Dialogue Design (R. W. Ehrich and R. C. Williges, Editors)
Vol. 3.	Expertise Transfer for Expert System Design (J. H. Boose)
Vol. 4.	Engineering Physiology: Physiologic Bases of Human Factors/Ergonomics (K. H. E. Kroemer, H. J. Kroemer and K. E. Kroemer-Elbert)
Vol. 5.	Human Factors Testing and Evaluation (D. Meister)
Vol. 6.	Applications of Fuzzy Set Theory in Human Factors (W. Karwowski and A. Mital, Editors)
Vol. 7.	Human Reliability: Analysis, Prediction, and Prevention of Human Errors (K. S. Park)
Vol. 8.	Human Aspects of Occupational Vibration (D. E. Wasserman)
Vol. 9.	Human Factors Research: Methods and Applications for Architects and Interior Designers (J. E. Harrigan)
Vol. 10A.	Social, Ergonomic and Stress Aspects of Work with Computers (G. Salvendy, S. L. Sauter and J. J. Hurrell, Jr., Editors)
Vol. 10B.	Cognitive Engineering in the Design of Human–Computer Interaction and Expert Systems (G. Salvendy, Editor)
Vol. 11.	Occupational Safety and Accident Prevention: Behavioral Strategies and Methods (C. G. Hoyos and B. Zimolong)
Vol. 12A.	Work with Computers: Organizational, Management, Stress and Health Aspects (M. J. Smith and G. Salvendy, Editors)
Vol. 12B.	Designing and Using Human–Computer Interfaces and Knowledge Based Systems (G. Salvendy and M. J. Smith, Editors)
Vol. 13	Designing User Interfaces for International Use (J. Nielsen, Editor)
Vol. 14	Human Factors in Product Design (W. H. Cushman and D. J. Rosenberg)
Vol. 15	Workspace, Equipment and Tool Design (A. Mital and W. Karwowski)

Advances in Human Factors/Ergonomics, 15

Workspace, Equipment and Tool Design

Edited by

Anil Mital PhD PE
Director, Ergonomics & Engineering Controls Research Laboratories
Industrial Engineering, University of Cincinnati
Cincinnati, OH, U.S.A.

Waldemar Karwowski PhD PE
Director, Center for Industrial Ergonomics
Department of Industrial Engineering, University of Louisville
Louisville, KY, U.S.A.

ELSEVIER

Amsterdam–Oxford–New York–Tokyo 1991

ELSEVIER SCIENCE PUBLISHERS B.V.
Sara Burgerhartstraat 25
P.O. Box 211, 1000 AE Amsterdam, The Netherlands

Distributors for the United States and Canada:

ELSEVIER SCIENCE PUBLISHING COMPANY INC.
655 Avenue of the Americas
New York, NY 10010, U.S.A.

ISBN 0-444-87441-0

Printed in The Netherlands

DEDICATED TO OUR
LITTLE DAUGHTERS AASHI AND JESSICA
GROWING SONS ANUBHAV AND MATHEW
AND SUPPORTING WIVES
CHETNA AND BERNARDETTE

Preface

As the ergonomic aspect of many problems facing the industry today attracts more attention from the management, providing scientific knowledge and the know-how to solve such problems is becoming increasingly more important. The impetus for this book originated from our observations of the pressing need to make the state-of-the-art ergonomic information on workspace, equipment and tool design available to practicing ergonomists, safety specialists, engineering designers, and business and technical managers. We want to reinforce the notion that ergonomic data should be explicitly integrated in the design of a system, and should become an indispensable part of the overall design process in production engineering, on an equal basis with such activities as mechanical component design, quality assurance, maintenance, inspection, etc.

In this book, we focus on selected ergonomic data for workspace, equipment and tool design, with special emphasis on the practical aspects of applying the available information to specific problem areas. The book is divided into three parts. Part I deals with workspace design and contains six chapters. The second part is devoted to equipment and tool design and contains four chapters. Three chapters dealing with related issues are included in Part III. In each chapter, authors have attempted to trace the progression of developments in the topical area, critically review the literature, and, wherever possible, summarize the present state of the art in the form of design recommendations, gaps in knowledge, and suggestions for future research.

In the first chapter, Nicholson discusses the applications of anthropometry in workspace design. This is followed by McDaniel's review of two popular computer-aided expert system models (COMBIMAN and CREW CHIEF) for anthropometric accommodation. In the third chapter, Ayoub and Miller present ergonomic guidelines for evaluating and designing workplaces. Kroemer emphasizes furniture design for the computer workstation in the fourth chapter. Particular attention is paid to recording and assessing body postures of seated workers with an aim to alleviating posture-related musculoskeletal disorders.

Design guidelines for workspace safety and maintenance are addressed by Smith and Deivanayagam and Imrhan in the next two chapters. Smith and Deivanayagam discuss the fundamentals of safety analysis and provide specific design guidelines for guards, warnings, and hazard elimination and control. The machine safeguarding issues are discussed in greater detail by Etherton in Chapter 8. The design of warning signs and labels is also addressed separately, by Lehto and Clark, in Chapter 12.

Imrhan, in Chapter 6, outlines the principles of workspace design for maintenance tasks, with special reference to those human activities that call for consideration of such design factors as visual access, mobility, and physical strength and endurance.

Chapter 7 contains design data and guidelines for workspace clearances and access. The considerations include worker performance, freedom of movement in the workplace, proper operation of controls, equipment, and tools and maintainability requirements. In Chapter 9, Mital examins ergonomic problems associated with the design and use of handtools. Human force/torque exertion capabilities with human-powered tools are discussed and specific guidelines for designing handtools are provided.

The ergonomic guidelines for designing and maintaining underground coal mining equipment are presented by Conway and Unger in Chapter 11, the first chapter in Part III of the book. The authors review the impact of equipment design for maintenance on productivity and safety, and present maintainability design guidelines to enhance safety.

In the last chapter, Czaja reviews workplace design issues critical to the elderly adults. Examples are provided to illustrate the role of ergonomic design principles in integrating elderly workers into occupational settings.

The contributors to this book have spent a considerable amount of time and effort in the preparation of their chapters. We consider these efforts very worthwhile since we believe that ergonomic design data and guidelines can have a significant impact on enhancing worker efficiency and safety. The information contained in this volume, however, can be effective only if designers apply it fully and systematically. We hope that users will find the contents of this book relevant to their applications and fruitful. We also hope that this book will assist and encourage designers to incorporate ergonomic principles into their designs.

Anil Mital and Waldemar Karwowski

Cincinnati and Louisville, U.S.A.
August 1990

CONTENTS

PART I
WORKSPACE DESIGN

Workspace, Equipment and Tool Design
edited by A. Mital & W. Karwowski

ANTHROPOMETRY IN WORKSPACE DESIGN

A.S. NICHOLSON
Hu-Tech Associates, Victoria House, 24 York Road,
NORTHAMPTON NN1 5QG, U.K.

ABSTRACT

In this technological age we are surrounded by artifacts and environments which can greatly influence our personal and communal health, safety, effectiveness and well-being. For most of us the majority of our lives are spent in buildings and environments which have been designed and constructed to accommodate us. Without ignoring the impact of the domestic setting, the industrial workplace has the potential to be more hostile to us because of the nature of the very operation and processes being carried out. Within this context it is important that workspace - the area within which we are required to carry out our work - is designed to promote our health and well-being without detriment to our effectiveness.

Anthropometry has been defined as the science of measurement of body size (NASA, 1978). The application of anthropometry is an essential element in the process of designing the workspace to fit the worker. The physical size of a population can be determined by measuring body lengths, breadths and girths, and the data derived can be used to design workplaces, equipment and products which match people's dimensions. In this way the workspace (which includes the equipment, tools, furniture etc) can be fitted to the man or woman's physical dimensions and functional capabilities.

HISTORICAL PERSPECTIVE

In the Roman era anthropometry and design were considered to be related (Pheasant, 1986). The Roman architectural theorist Vitruvius argued that the design of buildings should be based on certain pre-established aesthetic principles of the human body. Simple whole number relationships between the dimensions of the body were considered to be 'harmonious'. It was probably Durer (1471-1528) who marked the beginning of the science of anthropometry. He attempted to categorise the diversity of human physical types according to systematic observation and measurement of a large number of people. Nevertheless, in this Renaissance period the theory of aesthetics remained very influential. The infamous drawing of Leonardo da Vinci (1452-1519) in which a man is shown inscribed within a square and circle is derived directly from Vitruvius.

The science of anthropometry developed in the 19th and early 20th centuries principally because of a close relationship with physical anthropology. This was a time in which attempts were being made to subdivide and classify the human race according to physical dimensions. In the latter part of this century the focus has been on human growth and classification of physiques. Technological advancements such as space travel and military activities have also required more substantial sources of data.

BASICS OF ANTHROPOMETRY

A simple observation of people will show that the human species varies over a range of physical sizes. Indicative of this is the fact that clothes are manufactured to different sizes which not only account for different heights, but also arm or leg lengths and girths, waist and chest circumferences and shoulder breadth. These have been derived from anthropometric data. Such observations however are not sufficient to enable workspaces to be fitted to people's sizes and shapes. Rather, it is necessary to have objective data which has been carefully collected for a large number of people. Examples are the anthropometric surveys of 2000 Royal Air Force Aircrew (Bolton et al, 1973) and of 500 British Royal Armoured Corp Servicemen (Ince et al, 1973) which were used to determine the workspace and clothing requirements of the respective military populations.

The anthropometric characteristics of the user population is distinguished by 2 types of data. First there is the static (also known as structural) anthropometric data concerned with the fixed structural dimensions of the body. Examples include stature, shoulder height (or more correctly acromial height) and eye height (which is commonly used as a reference point for designing visual tasks). The second type of anthropometric data is the dynamic, or functional data. As the title suggests this is different from static data in that the measurement may be of the range of movement of a joint or of the strength of various actions of the joint. It also includes the measurement of reach and clearance made under operational conditions, for example in work clothing. A selection of data from these 2 types of anthropometric characteristics of the user population is presented later in this chapter.

Before considering the ways in which these characteristics might impose constraints upon the design, and how we might determine an effective match, it is necessary to briefly consider the statistical basis of human variability and the main factors of human diversity which have a bearing on workspace design.

The statistical basis of human variability

Most linear dimensions of the human body are normally distributed and a frequency distribution of a particular dimension would reveal a symmetrical bell-shaped curve known as a normal or Gaussian distribution. The example shown in figure 1 is the frequency distribution for the stature of adult British males.

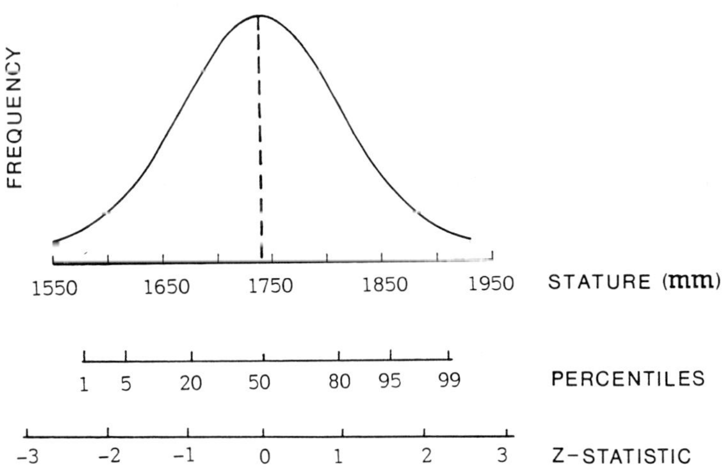

Fig. 1. The frequency distribution for the stature of adult British males showing the percentiles and the z-statistic.

With such a distribution the mathematical equation which describes the curve consists of 2 parameters, the mean and the standard deviation. The mean (μ) indicates where the distribution is located on the horizontal axis (in the case of a Gaussian distribution at the centre of the distribution) and the standard deviation (s) gives a measure of the degree of variability in the population. The standard deviation of a sample of individuals in a population is given by the equation:

$$s = \frac{(x-\mu)}{n-1}$$

where x is any value of the dimension concerned and n is the number of people in the sample.

Since the mean and standard deviation is all that is required to define the Gaussian distribution, it is possible to calculate any percentile value for the population by using the equation:

$$X = \mu + sz$$

where z is a constant for the percentile concerned and can be found in statistical tables.

In figure 1 the percentiles and corresponding z values are indicated for this distribution.

For reasons of practicality, it is necessary to put limits on the range of the population for which the workspace will be designed. To accommodate users at the extreme upper and lower end of the distribution would mean that the degree of variability in the workspace would have to be greatly extended while benefiting relatively few individuals. The constraints on the design would be very limiting. For this reason it is common practice to specify designs to fit 90% of the users, the lower value being defined by the 5th percentile value of a dimension, and the upper value by the 95th percentile value.

HUMAN VARIATION

We have already discussed how people vary in anthropometric characteristics, and that for many linear dimensions of the human body a normal distribution is obtained. However that is not the end of the consideration of human diversity. For example, we have only examined the variance within a general population, but not the variance between populations or groups. The industrial designer, engineer or ergonomist needs to take these and other elements of human diversity into account.

Figure 2 presents the difference in stature between a variety of populations. The 5th and 95th percentile values are indicated for each population. One can see that the range of stature is extremely large, and is dependent upon sex, ethnic origin and selectivity, the latter being seen by comparing the 5th and 95th percentile values between military and civilian populations of the same national group.

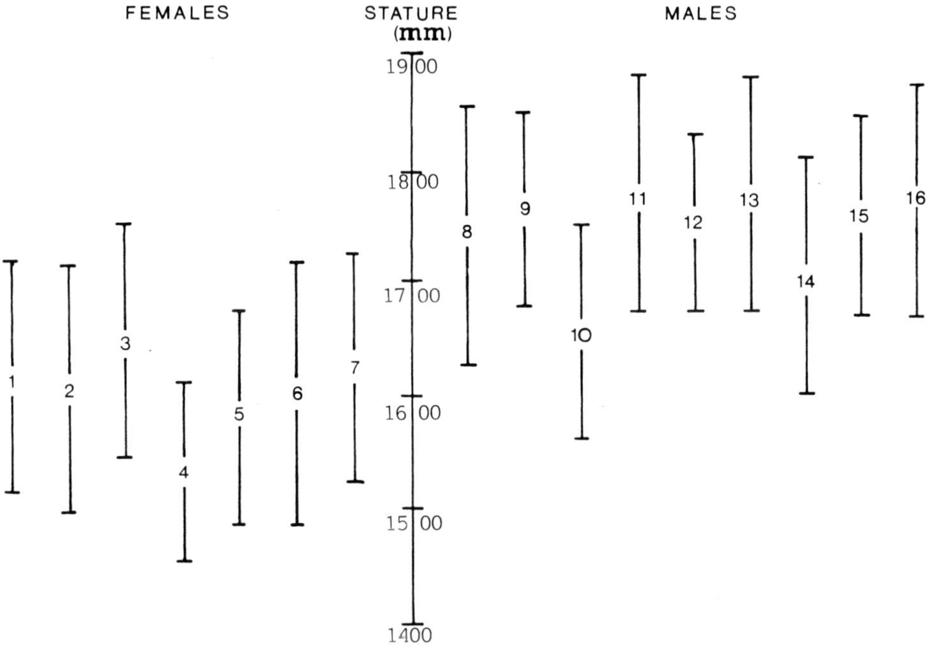

1. US civilians
2. British civilians
3. Swedish civilians
4. Japanese civilians
5. Korean civilians
6. Eygptian civilians
7. US Air Force
8. British civilains
9. British civilian helicopter pilots
10. Japanese civilians
11. US Air Force pilots
12. NASA astronauts
13. Royal Air Force pilots
14. Italian military
15. French Air Force pilots
16. German Air Force pilots

Fig. 2. The range of variability of stature in selected populations.
(Source: Fernandez et al, 1989; Light et al, 1988; Moustafa et al, 1987; NASA Anthropometric Source Book Vol 1, 1978).

Sex differences

An examination of figure 2 shows that the majority of males exceed the stature of most females for the same ethnic origin (compare British civilians for example). Males exceed females for nearly all anthropometric variables. (A careful examination of tables 2 to 5 presented later in this chapter shows this for all the linear bodily dimensions except hip breadth and girth).

In addition to the dimensional anthropometrics described above, men and women differ in their bodily composition. In general, fat represents a greater proportion of body weight in the adult female than in the male. Durnin and Rahaman (1967) found that the percentage which fat contributes to the body to be 13.5% (s = 5.8] for adult men and 24.2% (s = 6.5] for adult women. Subcutaneous fat is distributed differently between the sexes, women accumulating fat in the breasts, hips, thighs and upper arms. Abdominal fat accumulates above the umbilicus in men and below the umbilicus in women).

Pheasant (1983) published an interesting analysis of sex differences in static strength. This showed that the average ratio of female to male strength is 61%, but that the differences range from 37% to 90% for different parts of the body (see table 1).

TABLE 1

Sex differences in strength between males and females.

Part of body	No. of tests	F/M (%) Min.	Ave.	Max.
Lower limb	17	50	66	81
Push/pull/lift	41	38	65	90
Trunk	11	37	62	68
Upper limb	29	44	58	86
Miscellaneous	14	43	53	61
All tests	112	37	61	90

F/M (%) = percentage difference between female and male strength.

Data from Pheasant (1983).

Ethnic differences

A useful starting point when considering ethnic variations is an understanding of what is meant by the phrase 'ethnic group'. Pheasant (1986) defines it as a sample or population of individuals who inhabit a specified geographical distribution and who have certain physical characteristics in common which serve, in statistical terms, to distinguish them from other such groups of people. One assumes that these characteristics are fundamentally hereditary in origin, although the extent to which this is the case is often contentious.

Ethnic groups may or may not be identified with national, linguistic or other boundaries. For example the various ethnic types to be found within the population of Europe are distributed across national boundaries, although the frequency with which a given type is encountered will vary from place to place. To some extent ethnic groups fall into more or less natural clusters, which may be referred to as the Negroid, Caucasoid and Mongoloid 'divisions' or 'major groups' of mankind.

The Negroid division includes most of the dark-skinned peoples of Africa, together with certain minor ethnic groups of Asia and the Pacific islands. The Caucasoid division includes both light- and dark-skinned peoples resident in Europe, North Africa, Asia Minor, the Middle East, India and Polynesia (together with the indigenous population of Australia and some other ethnic groups who form a subdivision of their own). The Mongoloid division comprises a large number of ethnic groups distributed across central, eastern and south-eastern Asia, together with the indigenous populations of the Americas.

There has been, and will no doubt continue to be, a process of migration and intermarriage between the divisions. This means that most samples of national populations will include other ethnic groups besides those indigenous to the nation. Pheasant (1986) states that around 10% of the membership of such samples are of identifiably different ethnic origins, and that this 10% itself is a complex mixture of ethnicity.

Samples of adults may vary from each other either in overall size (as measured by stature or weight) or in bodily proportions. Figure 2 illustrates this feature for stature. Compare for example the stature of Japanese or Korean women with that of US or British women and then with Swedish women. The same trend is observed for Japanese and British men.

By considering the data available (Eveleth and Tanner, 1976 and NASA, 1978) for sitting height and stature, and examining the differences between the ethnic origins, we can see examples of how the divisions vary in linear bodily dimensions.

Black Africans have proportionally longer lower limbs than Europeans; Far Eastern samples have proportionally shorter lower limbs, the difference being most marked in the Japanese, less in the Chinese and Koreans and least in the Thai and Vietnamese. These differences of proportion occur throughout the stature range. If we consider the European data only, there is a tendency for the ratio of sitting height to stature to be slightly greater for short samples than tall ones - suggesting that the lower limbs contribute more to differences in stature than the trunk. The populations of Turkey, the Middle East and India, labelled 'Indo-Mediterranean', have proportions similar to Europeans but, typically, a lesser overall stature.

What are the practical consequences of such differences? How important, for example, is the presence of ethnic minorities in the target population for whom we are designing a certain piece of equipment?

Take for example knee height in a certain design situation, perhaps a VDT workstation. Figure 3 shows what would happen when a target population originally consisting of British men (μ = 545, s = 31) is modified by increasing numbers of a short-legged group (Japanese, μ = 495, s = 21) or a long-legged group (black Americans, μ = 565, s = 32).

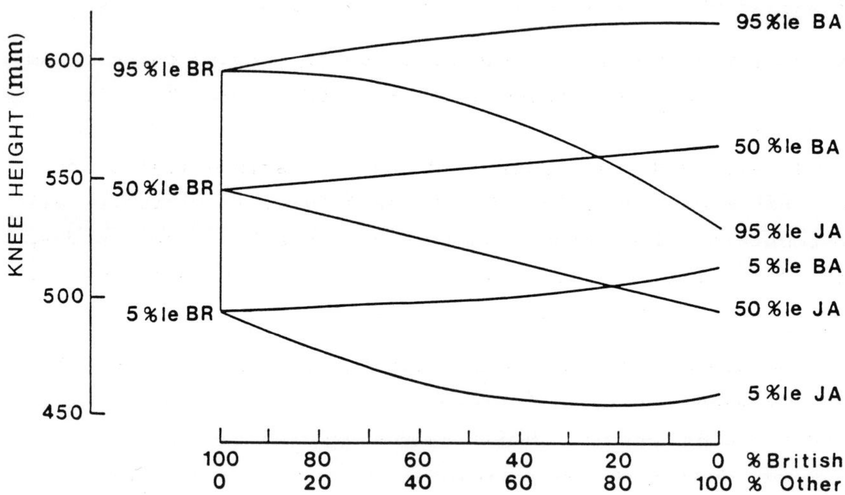

Fig. 3. Effects on the percentile values (%le) of knee height of varying the relative number of British (BR), Japanese (JA) and black American (BA) men in a target population of mixed ethnicity. (From Pheasant, 1986).

The figures suggest that in the case of the Japanese the differences that may interest a furniture designer (25 mm) occur at the level of around 30% membership, whereas for the black Americans the differences are of marginal consequence. There may, of course, be situations in which ethnic differences are much more critical. The research of Thompson and Booth (1982) suggests that there are circumstances in which workers from certain ethnic minorities might be more at risk if industrial safety guarding standards are not modified to take into account their particular anthropometric characteristics.

The secular trend

In this context the term 'secular trend' is often used to describe alterations in the measurable characteristics of a population of human beings occurring over a period of time. Over a period of at least the last century, biosocial changes have been occurring in the population of much of the world which have led to:

i. an increase in the rate of growth of children;
ii. an earlier onset of puberty - as indicated by onset of the menstrual cycle in girls and the adolescent growth spurt in both boys and girls;
iii. an increase in adult stature, with a possible decrease in the age at which adult stature is reached.

The extensive statistical evidence concerning these changes has been reviewed by, amongst others, Tanner (1962, 1978), Meredith (1976) and Roche (1979).

Tanner (1962, 1978) has summarised the available evidence and concluded that from around 1880 to at least 1960, in virtually all European countries (including Scandinavia), together with the USA, Canada and Australia, the magnitude of the trend has been similar. The rate of change has been approximately:

i. 15 mm per decade in stature and 0.5 kg per decade in weight at 5 - 7 years of age;
ii. 25 mm and 2 kg per decade during the time of adolescence;
iii. 10 mm per decade in adult stature.

This has been accompanied by a downward trend of 0.3 years per decade in the age of menarche. Roche (1979) notes that secular changes in size at birth have been small or non-existent.

Japan has shown a particularly dramatic secular trend. The data of Tanner et al. (1972) show that in the decade between 1957 and 1967 Japanese boys increased in stature by:

i. 31 mm at 6 years;
ii. 62 mm at 14 years;
iii. 33 mm at 17 years;

but that in the 1967-1977 period, however, the comparable figures were:

i. 17 mm at 6 years;

ii. 35 mm at 14 years;
iii. 19 mm at 17 years.

In contrast, Roche (1979) cites evidence that in India, and elsewhere in the Third World, there has actually been a secular decrease in adult stature.

What has caused this remarkable secular trend? It seems most likely that genetic inheritance sets a ceiling level to an individual's potential for growth and that environmental circumstances determine whether this ceiling is actually reached. Considerable evidence suggests that the limit to which environmental circumstances can effect any further change in these economically developed countries has indeed been reached. Backwin and McLaughlin (1964) showed that Harvard freshmen from relatively modest social backgrounds increased in stature by around 40 mm from 1930 to 1958, whereas those from wealthy backgrounds showed no change. Cameron (1979) has published data from which it can be concluded that the secular trend for stature had levelled off for children attending schools in the London area, by about 1960. Tanner (1978) also showed that the secular decrease in the age of menarche had come to a halt by about this time both in London and in Oslo. The subsequent national survey of Rona and Altman (1977) confirm the impression that in Great Britain the secular trend has now reached a steady state. Roche (1979) reported that national surveys of US children and youths in 1962 and 1974 show constancy of stature (except at the 5th and 10th percentile levels, where small increases have occurred).

Aging

Figure 4 shows the average heights and weights of the adult civilian populations of Great Britain and the USA plotted against age. A steady decline in stature is apparent, whereas weight increases before subsequently declining at around 50 years in men or 60 years in women.

When considering such features it is important that the combined effects of the aging process and the secular trend are examined. Damon (1973) showed that men of average height and weight had greater longevity than those who deviated strongly in either respect. Other studies have included the Welsh study of Miall et al. (1967) and the comprehensive study of the Veterans Administration (Damon et al. 1972, Friedlander et al. 1977, Borkan et al. 1983). These longitudinal studies show that at around 40 years of age we begin to shrink in stature, that the shrinkage accelerates with age, and that women shrink more than men. The data show a longitudinal increase in weight for height until 55

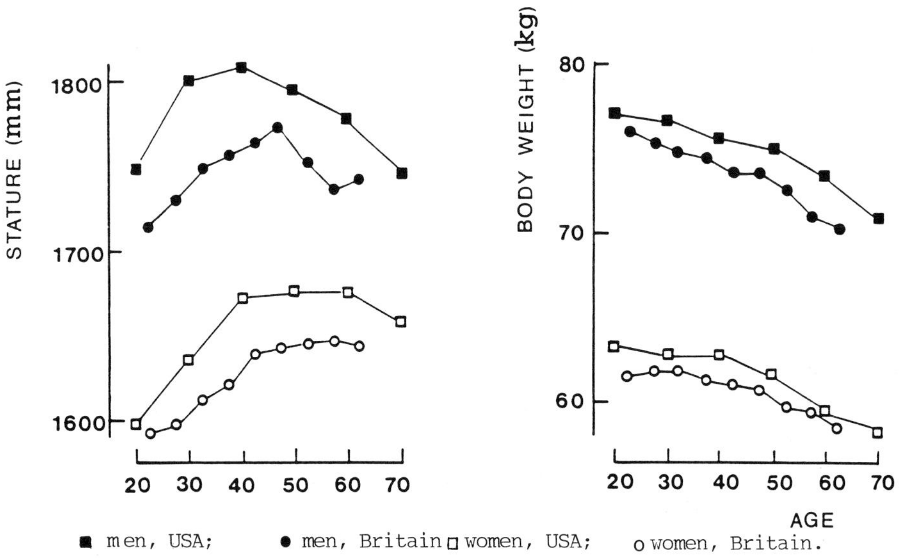

Fig. 4. Average stature and weight in samples of various ages. (From Pheasant, 1986).

years followed by a decline. Friedlander et al. (1977) showed a steady longitudinal increase not only in hip breadth but also in the bi-iliac breadth of the bony pelvis. The mechanism of the latter is not clear but it suggests that 'middle age spread' may not be totally due to the accumulation of fat, but may also involve changes in the bony pelvis.

With increased medical care available to most elderly citizens of developed countries, and increasing in Third World countries, longevity of life is also on the increase. The effect of this on the design of workspace may not be critical since retirement age is commonly 55 to 65 years of age, but the effects of aging on living space is more crucial.

STATIC ANTHROPOMETRIC DATA

Static dimensions are measurements made in fixed bodily positions between anatomical points on the skeleton. The number of possible measurements is enormous. The NASA Anthropometric Source

Book (1978) illustrates 973 such measurements. Many of these are concerned with specific design applications such as helmets, and for the purposes of considering workspace design, the number of variables can be substantially reduced.

Tables 2 to 5 present static anthropometric data for British adults aged 19 to 65 years of age. They are the best estimates from currently available data and have been calculated using a process known as ratio scaling. (For a description of this technique see Pheasant, 1986). The data are for unclothed and barefooted subjects. When designing workspace parameters it is necessary to add to these, or other data, corrections for work clothing. The actual values depend on the type of work carried out and environmental conditions experienced. For normal, light, indoor clothing the effect on the dimensions is so minimal that it can be effectively ignored. For shoes the following data acts as a general guide:

i. For the heels of shoes add 25 mm for men and 45 mm for women;
ii. Add 30 mm to the length of the foot to account for shoe length.

For industrial safety helmets add 40 mm to standing and sitting heights. For heavy outdoor clothing, or similar, add 40 mm to shoulder breadth and other bodily dimensions concerned with clearance and access and egress.

Other sources of static anthropometric data may be found in texts by the Eastman Kodak Company, 1983, NASA Anthropometric Source Book, 1978 and Pheasant, 1986.

These data are used to establish workstation dimensions and workspace requirements. Typical examples would be the dimensions of production/ assembly lines, VDT workstations, office furniture and many other working and domestic situations. The data are used to establish minimum dimensions for such factors as clearance, access and egress and handle size and shape. An example is presented in table 6, quoted from ISO 2860, which gives the minimum dimensions for whole body access. The data are equivalent to the 95th percentile of the worldwide operator population. It should be noted that this means 5% of this population would find it very difficult, if not impossible to gain access!

Such data however has its limitations. For example the practical limit to arm reach is not the shoulder to fingertip length because operators will employ other joint movements to reach

TABLE 2

Anthropometric estimates for British adults (19-65 years)

	Men			Women		
	5th %le	50th %le	95th %le	5th %le	50th %le	95th %le
Stature	1625	1740	1855	1505	1610	1710
Eye height	1515	1630	1745	1405	1505	1610
Neck height	1375	1485	1595	1280	1375	1470
Shoulder height	1315	1425	1535	1215	1310	1405
Chest (bust) height	1175	1270	1365	1080	1170	1255
Elbow height	1005	1090	1180	930	1005	1085
Hig height	840	920	1000	740	810	885
Knuckle height	690	755	820	660	720	780
Fingertip height	590	655	720	560	625	685
Sitting height(1)	850	910	965	795	850	910
Sitting eye height(1)	730	790	845	685	740	795
Sitting shoulder height(1)	540	595	645	505	555	610
Sitting elbow height	190	245	295	185	235	280
Thigh thickness	135	160	185	125	155	180
Knee height	495	545	595	455	500	540
Popliteal height	395	440	490	355	400	445
Buttock-knee length	545	595	645	520	570	620
Buttock-popliteal length	440	495	550	435	480	530
Chest (bust) depth	215	250	285	210	250	295
Abdominal depth	220	270	320	205	255	305
Shoulder breadth (bideltoid)	420	465	510	355	395	435
Shoulder breadth (biacromial)	365	400	430	325	355	385
Elbow-elbow breadth	390	450	510	300	385	475
Hip breadth	310	360	405	310	370	435
Shoulder-fingertip length	720	780	840	655	705	760
Horizontal reach(2)	835	890	945	760	810	860
Shoulder-elbow length	330	365	395	300	330	360
Elbow-fingertip length	440	475	510	400	430	460
Span	1655	1790	1925	1490	1605	1725
Elbow span	865	945	1020	780	850	920
Vertical reach(2)	2040	2170	2300	1895	2010	2125
Sitting vertical reach(2)	1255	1355	1455	1150	1255	1340
Neck girth	345	375	425	335	365	415
Chest (bust) girth	840	950	1110	820	935	1135
Waist girth	705	805	965	565	685	885
Hip girth	890	965	1060	885	985	1165
Crotch height	760	840	920	675	750	825
Maximum body breadth(3)	480	530	580	355	420	485
Maximum body depth(3)	255	290	325	225	275	325
Kneeling height	1210	1295	1380	1130	1205	1280
Buttock-heel length	985	1070	1160	875	965	1055
Body weight (kg)	55	75	94	44	63	81

All dimensions in mm. Percentiles rounded to the nearest 5 mm.

(1) Measurements made in an erect sitting posture. Where appropriate subtract 40 mm to allow for slumping.
(2) These are fingertip measures.
(3) Overall measures taken at the broadest and deepest points wherever they may be.

(Source: Pheasant, 1987)

TABLE 3

Anthropometric estimates for British adults: Hand dimensions

	Men			Women		
	5th %le	50th %le	95th %le	5th %le	50th %le	95th %le
Hand length	173	189	205	159	174	189
Hand breadth (metacarpal)	78	87	95	69	76	83
Hand breadth (including thumb)	97	105	114	84	92	99
Hand girth	201	216	236	169	184	199
Palm length	98	107	116	89	97	105
Index finger length	64	72	79	60	67	74
Index finger breadth	19	21	24	16	18	20
Hand thickness at palm	27	33	38	24	28	33
Hand thickness (including thumb)	44	51	58	40	45	50
Maximum grip diameter	45	52	59	43	48	53

All dimensions in mm. Percentiles rounded to nearest 5 mm.

(Source: Pheasant, 1986)

TABLE 4

Anthropometric estimates for British adults: Foot dimensions

Foot length	240	265	285	215	235	255
Foot breadth	85	95	110	80	90	100
Foot girth	220	245	270	210	225	245
Heel-ball length	175	190	210	160	175	190
Ankle height	60	70	85	55	65	75

All dimensions in mm. Percentiles rounded to nearest 5 mm

(Source: Pheasant, 1986)

TABLE 5

Anthropometric estimates for British adults: Head dimensions

Head length	180	195	205	165	180	190
Head breadth	145	155	165	135	145	155
Head height	205	225	240	200	220	240
Head girth	535	565	590	525	550	575
Ear-ear breadth	125	135	145	120	130	135
Inter-pupillary breadth	55	60	70	55	60	65
Maximum diameter from chin	240	255	265	225	235	245
Nose-back of head	205	220	235	190	205	220

All dimensions in mm. Percentiles rounded to nearest 5 mm

(Source: Pheasant, 1986)

beyond this length. Thus dynamic anthropometric data is also needed to establish such factors as reach.

TABLE 6

Minimum dimensions for whole body access

Minimum dimension in mm	Square	Round	Rectangular	
	W = L	D	W	L
Normal clothing	520	560	330	560
Arctic clothing	600	650	470	650

Note. Maximum 25 radius optional on all corners

L = side to side dimension
W = front to back dimension

From ISO 2860

DYNAMIC (FUNCTIONAL) ANTHROPOMETRY

These dimensions are taken when the body is carrying out some physical activity such as reaching to operate a control device, or when operating a steering wheel or foot control in a vehicle. In such circumstances individual parts, or functional units, of the body are coordinated to act in unison in order to achieve the desired objective. Three concepts of dynamic anthropometry will be examined: the zone of convenient reach, the normal working area, and maximum reach capabilities.

ZONES OF CONVENIENT REACH

This may be defined as the zone or space in which an object may be reached conveniently, that is without undue exertion. It is described by the movements, or sweeps of the upper limb centred upon the joint measured from the shoulder to the fingertip (or to the centre of grip) in a series of arcs for each hand (see figure 5). The volume which is thus defined comprises of two intersecting hemispheres. The radius of each hemisphere is the upper limb length (a) and their centres are a distance (b) equal to biacromial breadth apart.

Many design problems are concerned with the intersection of vertical and horizontal planes with either the volume of the

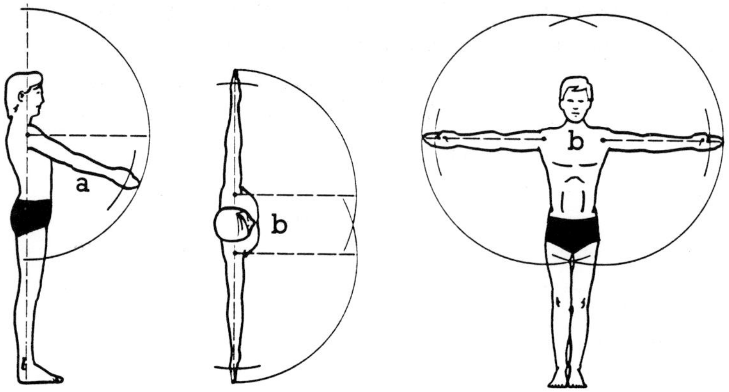

Fig. 5. Zones of convenient reach (ZCR) seen in elevation and plan.

workspace envelope or that of the zone of convenient reach. An example will best serve to illustrate this. In locating a series of controls upon the vertical wall of a control room so that they might be conveniently operated by a standing person, Pheasant (1986) calculated the respective zones of convenient reach for a 5th percentile female and a 95th percentile male.

The intersection of a plane with a sphere produces a circle. The radius of this circle may be calculated by Pythagoras' theorem as:

$$R = \sqrt{a2 - d2}$$

where r is the radius of the circle on the wall, a is the upper limb length (or shoulder grip length) and d is the horizontal distance between the shoulder and the wall. Figure 6 presents the resulting zones for fingertip controls, assuming d = 500.

The zone of convenient reach may be similarly described for any other vertical or horizontal plane parallel to a line joining the shoulders. Such data is presented in table 7. To construct a zone of convenient reach in the vertical plane a distance d in front of the shoulders, draw 2 circles of radius r. The centres of the circles are defined by standing of sitting shoulder height and biacromial breadth (see table 2). For the horizontal plane, draw 2 semi-circles of radius r centred on the position of the shoulders.

The intersection of a horizontal plane, which may be typified in the industrial setting by a bench or production line, with the zone of convenient reach defines an area called the maximum working area. Within this area is a smaller area called the normal working area. These concepts proved particularly useful in developing work study norms and some of the early work was done on the basis of seeking to keep the working area to a minimum, so that the class of shoulder, arm and finger motions be kept to the minimum possible within the context of the work being done.

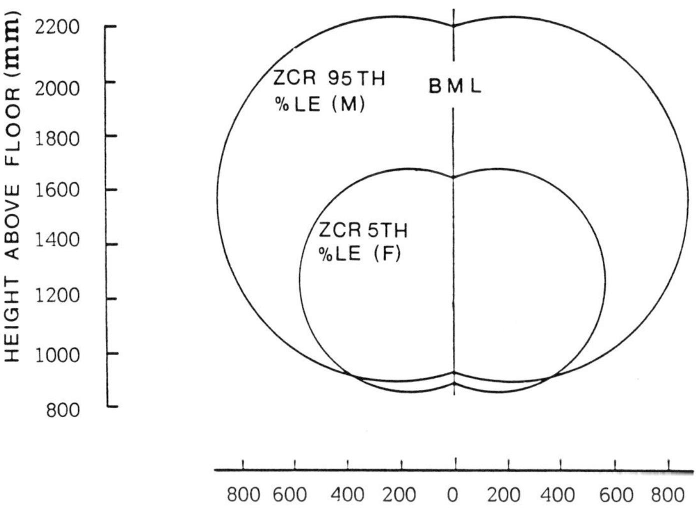

Fig. 6. Zones of convenient reach on a vertical surface 500 mm in front of the shoulders for a 95th percentile (%le) male (M) and 5th percentile female (F).
(BML = body mid-line)
(Adapted from Pheasant, 1986)

NORMAL WORKING AREA

By assuming that the operator was comfortably seated at or standing by the worktable or bench at the correct height, Maynard (1934) determined the normal working area (NWA) in the horizontal plane of the right hand by an arc drawn with a sweep of the arm. Only the forearm was extended, the upper arm hung at the side of the body in a natural position until it tended to swing away as the arm moved toward the outer part of the workplace. Similarly, the normal working area in the horizontal plane for the left hand was determined. The overlapping area in front of the operator constituted a zone for two-handed work.

The normal working area in the vertical plane of the left and right hands included the area made by the lower arm in an upright

TABLE 7

Zones of convenient reach

	Radius (r)					
	Men			Women		
	Percentiles			Percentiles		
d	5th	50th	95th	5th	50th	95th
0	610	665	715	555	600	650
100	600	655	710	545	590	645
200	575	635	685	520	565	620
300	530	595	650	465	520	575
400	460	530	595	385	445	510
500	350	440	510	240	580	415
600	110	285	390			250

(all dimensions in millimetres).

(Adapted from Pheasant, 1986)

position hinged at the elbow moving in an arc. The maximum working area was determined by drawing arcs with the arms fully extended. Both the horizontal and vertical planes were considered, as was the case with the normal working area.

The workplace layouts provided by Maynard were, however, dimensionless and therefore provided negligible guidelines to a designer. Barnes (1940) gave dimensions to Maynard's layouts, and subsequently, Barnes (1958) in citing a study by Farley (1955), gave new dimensions to the layouts developed by Maynard (Figure 7). Farley presented only one work area for male operators and for female operators, based on the average male and female dimensions. He determined the normal working area as being equal to the volume circumscribed by the horizontal forearm pivoting about a relaxed vertical upper arm.

Squires (1956), challenged Farley's work because he recognised that the arm does not move in a regular circular path, since the elbow does not stay at a fixed point, but rather moves out and away from the body as the forearm pivots. Squires' work described only the normal working area in the horizontal plane for the 10th percentile male operator.

Pheasant (1986) discusses the construction of the normal working area, and, based on the concept of Squires, presents the coordinates of the normal working area.

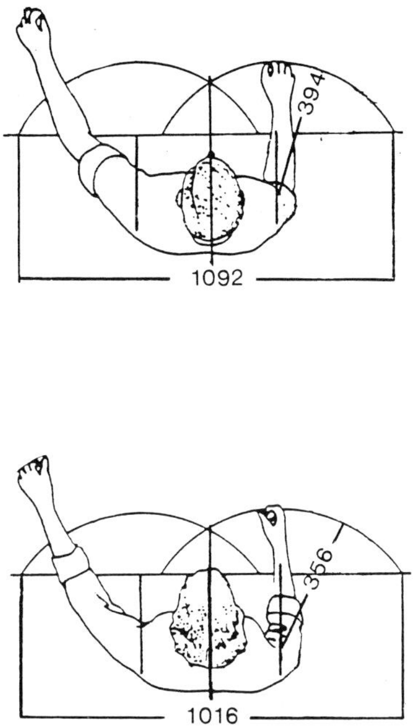

Fig. 7. The normal working area in the horizontal plane for male (top) and female (bottom) operator.

Dimensions in mm.

(After Farley, 1955)

Figure 8 shows the zone of convenient reach and normal working area on a table surface of the right hand of a 5th percentile male and female

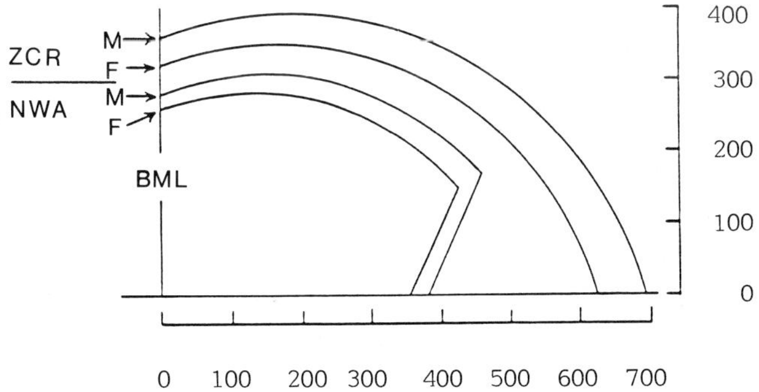

Fig. 8. Zones of convenient reach and normal working area on a table surface for a 5th percentile male (M) and female (F). (Adapted from Pheasant, 1986).

MAXIMUM REACH CAPABILITIES

The zones of convenient reach and the normal working area are necessary criteria in the design of workspaces for normal operations. However, it is at times necessary to know the extent to which operatives can reach to their maximum extent. Most examples are concerned with the safeguarding of equipment to ensure that operatives cannot accidentally make contact with moving parts.

Perhaps the 2 most common examples are arm and leg reach. Consider arm reach. Static anthropometry tells us that the arm length (grip or fingertip to acromial point) is of a certain dimension (or range of dimensions for a population). However the extent to which an individual can reach is also influenced by shoulder movement, trunk rotation, forward flexion of the spine, and hand function.

Such considerations have been embodied in 2 recent Standards developed to promote personal safety when operating machinery or technical equipment. DIN 31 001 (1984) was designed to protect 95% of the German workforce. BS 5304 (1988) is identical, but relates only to machinery.

Figure 9 shows the concept of reaching over barriers that is used in both these standards. It also illustrates the hazard (or danger) point and presents the various distances which are used to define the safety distances from the hazard point. Table 8 presents the data from the Standards (in this case related to the German Standard).

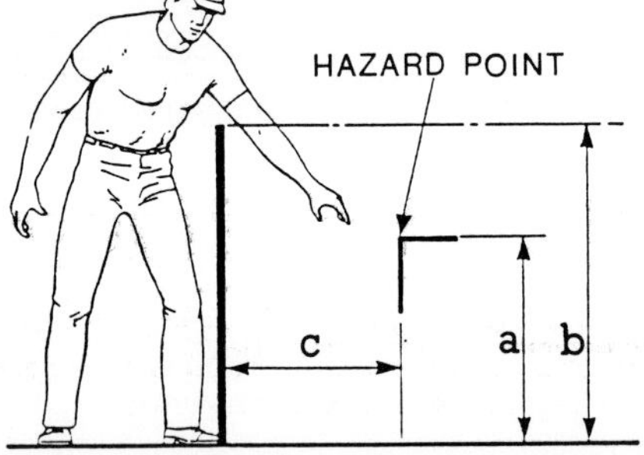

Fig. 9. Hazard point reaching down or over.

Thompson and Booth (1982) found that the reach distances quoted in the DIN Standard would not protect 95% of the British population. Thompson (1989) has subsequently presented data which has

TABLE 8

Safety distances from hazard point

Distance of hazard point from floor a	Height of edge of safety device b							
	2400	2200	2000	1800	1600	1400	1200	1000
	Horizontal distance c from hazard point							
2400	–	100	100	100	100	100	100	100
2200	–	250	350	400	500	500	600	600
2000	–	–	350	500	600	700	900	1100
1800	–	–	–	600	900	900	1000	1100
1600	–	–	–	500	900	900	1000	1300
1400	–	–	–	100	800	900	1000	1300
1200	–	–	–	–	500	900	1000	1400
1000	–	–	–	–	300	900	1000	1400
800	–	–	–	–	–	600	900	1300
600	–	–	–	–	–	–	500	1200
400	–	–	–	–	–	–	300	1200
200	–	–	–	–	–	–	200	1100

All dimensions in mm

Values for edge b under 1000 mm have not been included because the reach does not increase any further. In addition there is the risk of falling into the hazard area.

(From DIN 31 001, 1984)

confirmed this. He has also published comparable data which is designed to protect 99% of the British workforce.

Figure 10 presents the safety distances for the arm according to different combinations of horizontal barriers. DIN 31 001 (1984) presents further data relating to the safeguarding of machines and similar equipment.

Part of the body	Safety distance r mm	
Hand from root of finger to fingertip	≥120	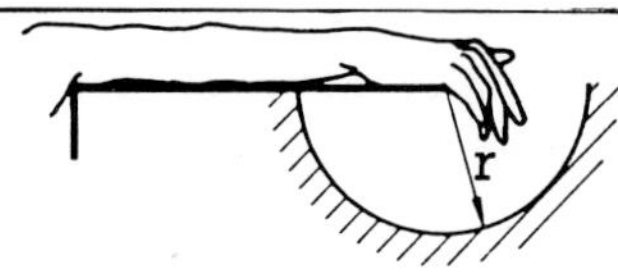
Hand from wrist to fingertip	≥230	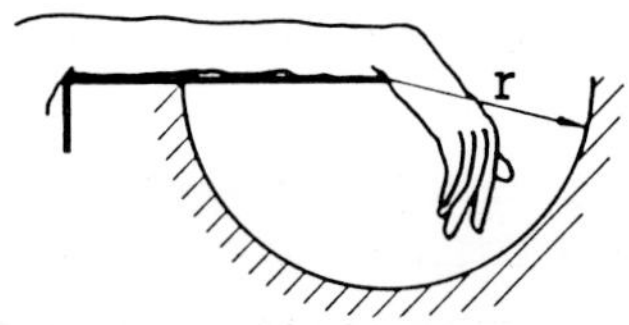
Arm from elbow to fingertip	≥550	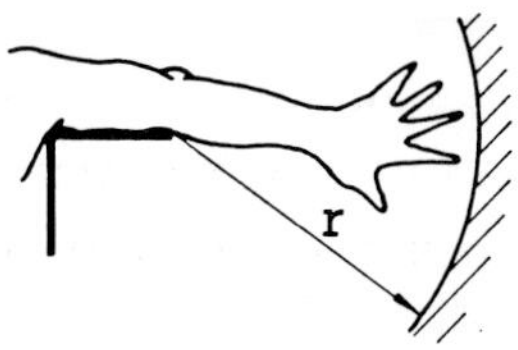
Arm from armpit to fingertip	≥850	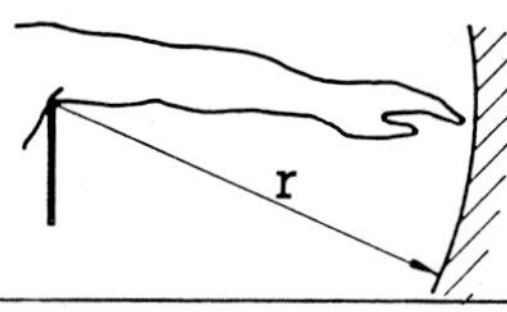

Fig 10. Safety reach distances for the arm according to DIN 31 001 (1984)

Similar data has also been presented for foot reach under guard rails on agricultural machinery (Bottoms and Butterworth, 1990). The same considerations apply as with arm reach. They conclude that the height of a single bar above the ground should be such as to limit the operator's leg reach to about 300 mm.

AN EXAMPLE OF THE USE OF ANTHROPOMETRY IN WORKSPACE DESIGN

Anthropometric selection criteria for airline cabin crew

Aircraft cabin crew are required to work in the confined physical environments of aisles and gallies and other miscellaneous areas. A study was carried out by Buckle and David (1989) in order to determine stature criteria for cabin crew based on the functional requirements of the tasks they perform.

As a first step they identified difficult or awkward situations with respect to reach, clearance or strength for cabin crew staff, and also where anthropometric limitations were critical. They then measured the dimensions of the workplace and equipment used in these "worst case" situations on training simulators and aircraft. These were compared with the appropriate anthropometric dimensions for both male and female populations in the age range 19 - 45 years (Pheasant, 1986). A shortlist of 11 worst case situations were then compiled based on relevant observations and data. Three of these were further investigated by mock-up simulation, and 20 individuals (12 female, 8 male) of measured anthropometric dimensions within the 4th - 93rd percentile range for female stature, and 3rd - 99th percentile range for male stature attempted to perform each task. The results of this trial were used to confirm the estimated dimensions which had been calculated from scaled diagrams.

The 11 worst case situations are presented in table 9 along with the limiting dimension and the selection limit. The results were used to propose the following stature limits for cabin crew selection:

Female staff: stature between 1605 and 1875 mm (equivalent to the 45th to >99th percentile British female stature)

Male staff: stature between 1605 and 1895 mm (equivalent to the 5th to 98th percentile British male stature).

These limits were based on the need for the full range of duties to be performed by an all female staff, and for individuals to carry them out in a composed, skilled and dignified manner. They were also based upon the understanding that the current uniform design imposes only very minor restrictions to body movements required to perform the tasks. It was assumed that female footwear adds 45 mm to the height of female staff, and male footwear to add 25 mm to the height of male staff.

These recommended selection criteria would exclude from potential recruitment, 45% of the British female population, and 7% of the male population. The situation would be even more limiting for other ethnic groups operating the same aircraft types.

TABLE 9

Summary of Results for Worst Case Activities

Worst Case Activity	Limiting Dimension	Selection Limit (*)
Walking upright in cabin	Height in shoes	Stature Females should be under 1875mm (>99) Males should be under 1895mm (98)
Walking along aisles between seats	Hip width between arm rests	Hip width Females and males should be less than 425mm (95,99)
Reaching to close overhead lockers	Overhead functional reach	Stature Females should be over 1615mm (50) Males should be over 1635mm(5)
Reaching to open overhead lockers	Combined overhead and forward functional reach	Stature Females should be over 1605mm (45) Males should be over 1635mm (5)
Standing preparing food without stooping in the galley interior	Height in shoes	Stature Females should be under 1705mm (93) Males should be under 1725mm (40)
Reaching for carry-cot in overhead stowage compartment	Overhead functional reach with wrist flexed at right angle	Stature Females should be over 1725mm (96) Males should be over 1745mm (50)
Removing trays of glassware from stowage in galley	Overhead functional reach	Stature Females should be over 1615mm (50) Males should be over 1635mm (5)
	Upper body strength	Force application Lift 45N (5, 1)
Reaching to far side of trolley	Forward reach when squatting or stooping	Stature Females should be over 1600mm (40) Males should be over 1635mm (5)
Pushing and pulling trolleys	Physical strength	Force application More than 180N (10, 1)
Reaching to open lockers which contain safety/emergency equipment	Overhead functional reach	Stature Females should be over 1545mm (12) Males should be over 1550mm (1)
Opening cabin doors	Upper body strength	Force application More than 200N (10, 1)

* The figures in brackets refer to the corresponding female and male percentile values

(Source: Buckle and David, 1989)

It may be concluded from this study that the safety of the smaller in stature female members of cabin crew staff could be compromised by the design of these aircraft interiors. Aircraft travellers will not be unused to seeing shorter female cabin crew standing on the edge of seats to gain access to overhead lockers, or, having to ask for the assistance of taller passengers to help them with their duties.

Such a study shows that designers need to consider the target working population when designing workplaces and work spaces. It also highlights the difficulties employers may face in recruitment procedures if such measures are not taken.

REFERENCES

Backwin, H. and McLaughlin, S.D., 1964. Increase in stature - is the end in sight? Lancet, ii: 1195-1197.

Barnes, R.M., 1940. Motion and Time Study, 2nd Edition. John Wiley and Sons, New York.

Barnes, R.M., 1958. Motion and Time Study, 4th Edition. John Wiley and Sons, New York.

Bolton, C.B., Kenwood, M., Simpson, R.E. and Turner, G.M., 1973. An anthropometric survey of 2000 RAF Aircrew 1970/71. RAF Institute of Aviation Medicine Report No. 531, Ministry of Defence, UK.

Borkan, G.A., Hults, D.E. and Glynn, R.J., 1983. Role of longitudinal change and secular trend in age differences in male body dimensions. Human Biology, 55: 629-641.

Bottoms, D.J. and Butterworth, D.J., 1990. Foot reach under guard rails on agricultural machinery. Applied Ergonomics, 21(3): 179-186.

British Standard Code of Practice, BS 5304, 1988. For Safety of Machinery. British Satndards Institution, London.

Buckle, P.W. and David, G.C., 1987. Development of Anthropometric selection criteria for airline cabin crew. In Contemporary Ergonomics (Editor: E.D. Megaw). Taylor and Francis, London, pp. 320-325.

Cameron, N., 1979. The growth of London schoolchildren 1904-1966: an analysis of secular trend and intra-county variation. Annals of Human Biology, 6: 505-525.

Damon, A., 1973. Ongoing human evolution. In Heredity and Society (Editors: I.H. Porter and R.E. Skalko), Academic Press, New York, pp 45-74.

Damon, A., Seltzer, C.C., Stoudt, H.W. and Bell, B., 1972. Age and physique in healthy white veterans at Boston. Journal of Gerontology, 27: 202-208.

Deutsches Institut fur Normung, DIN 331 001 Part 1 (draft), 1984. Safety design of technical equipment, safety devices, definitions, safety distances for adults and children. DIN, Berlin.

Durnin, J.V.G.A. and Rahaman, M.M., 1967. The assessment of the amount of fat in the human body from measurements of skinfold thickness. British Journal of Nutrition, 21: 681-689.

Eastman Kodak Company, 1983. Ergonomic design for people at work Vol 1. Van Nostrand Reinhold, New York.

Eveleth, P.B. and Tanner, J.M., 1976. Worldwidde variation in human growth. Cambridge University Press.

Farley, R.R., 1955. Some principles of methods and motion study as used in development work. General Motors Engineering J., 2(6): 20-25.
Fernandez, J.E., Malzahn, D.E., Eyada, O.K. and Kim, C.H., 1989. Anthropometry of Korean female industrial workers. Ergonomics, 32(5): 491-495.
Friedlander, J.S., Costa, P.T., Bosse, R., Ellis, E., Rhoads, J.G. and Stoudt, H.W., 1977. Longitudinal physique changes among healthy white veterans at Boston. Human Biology, 49: 541-558.
Ince, N.E., Redcup, S. and Piper, J., 1973. Anthropometry of 500 Royal Armoured Corp Servicemen 1972. Army Personnel Research Establishment, Ministry of Defence, UK.
International Organisation for Standardization, ISO 2860, 1983. Earth moving machinery - minimum access dimensions. ISO, Geneva.
Light, I.M., Masson, J. and Ross, H., 1988. Anthropometry of North Sea helicopter pilots. Ergomomics, 31(6): 905-913.
Maynard, H.B., 1934. Workplace layouts that save time, effort and money. Iron Age, 134(23): 28-30, 92.
Meredith, H.W., 1976. Findings from Asia, Australia, Europe and North America on secular change in mean height of children, youths and young adults. American Journal of Physical Anthropology, 44: 315-326.
Miall, W.E., Ashcroft, M.T., Lovell, H.G. and Moore, F., 1967. A longitudinal study of the decline of adult height with age in two Welsh commumities. Human Biology, 39: 445-454.
Moustafa, A.W., Davies, B.T., Darwich, M.S. and Ibraheem, M.A., 1987. Anthropometric study of Eygptian women. Ergonomics, 30 (7): 1089-1098).
NASA, 1978. Anthropometric Sorce Book. (NASA Publication No. 1024) US Natioanl Aeronautics and Space Administration, Ohio, US.
Pheasant, S.T., 1983. Sex differences in strength - some observations on their variability. Applied Ergonomics, 14: 205-211.
Pheasant, S.T., 1986. Bodyspace: Anthropometry, Ergonomics and Design. Taylor and Francis, Lodon, UK.
Pheasant, S.T., 1987. Ergonomics - standards and guidelines for designers. British Standards Institution, Milton Keynes, UK.
Roche, A.F., 1979. Secular trends in stature, weight and maturation. Monographs of the Society for Research in Child Development, Serial No. 179, 44(3-4): 3-27.
Rona, R.J. and Altman, D.G., 1977. National study of health and growth: standards of attained height, weight and triceps skinfold in English children 5-11 years old. Annals of Human Biology, 4: 501-523.
Squires, P.C., 1956. The shape of the normal working area. Report No 275. US Navy Department, Bureau of Medicine and Surgery, Medical Research Laboratories, New London, CT, USA.
Tanner, J.M., 1962. Growth at adolescence. Blackwell, Oxford, UK.
Tanner, J.M., 1978. Foetus in man. Open Books, London.
Thompson, D., 1989. Reach distance and safety standards. Ergonomics. 32(9): 1061-1076.
Thompson, D. and Booth, R.T., 1982. The collection and application of anthropometric data for domestic and industrial applications. In Anthropometry and biomechanics: Theory and Application (Editors: R. Easterby, K.H.E. Kroemer, and D.B. Chaffin). Plenum Press, New York, New York, pp. 279-291.

Workspace, Equipment and Tool Design
edited by A. Mital & W. Karwowski

THE DEVELOPMENT OF COMPUTER MODELS FOR ERGONOMIC ACCOMMODATION

Joe W. McDaniel, Ph.D.

Armstrong Aerospace Medical Research Laboratory, Wright-Patterson AFB, Ohio 45433, U.S.A.

ABSTRACT

Tools for supporting ergonomic analyses continue to improve and are adopting the latest in computer technology. It is important to understand the evolution of development of these tools for two reasons. First, it helps potential users understand what they are, and what they are not. Second, it helps those who aspire to develop similar tools to understand the computer-aided engineering tool technology and perhaps avoid some of the pitfalls.

The chapter focus is on the evolutionary development of two such computer-aided engineering (CAE) tools developed by the Air Force Human Systems Division: COMBIMAN (Computerized Biomechanical Man-model, a model of an aircraft pilot) and CREW CHIEF (a model of an aircraft maintenance technician) which was derived from COMBIMAN. Functioning as expert systems, both models allow the designer to perform the functions of an expert ergonomist. The designer may simulate operations or maintenance activities on the computer-aided design image to determine if the activity is feasible.

These software models automatically create the correct body size and proportions for males and females, the encumbrance of clothing, personal protective equipment, joint mobility, physical access for reaching into confined areas (with hands, tools, and objects), visual access, and strength.

This chapter describes these models and their utility to show how they relate back to the technology base research into the characteristics and capabilities of the human operator and maintainer. A computer model is a simulation of something, in these cases, ergonomic data. Having the model, analyses can be done in minutes which would require weeks or months to accomplish by experimentation with real human subjects.

INTRODUCTION

The availability of modern computer-aided design (CAD) systems provides a new environment for implementing and transferring knowledge of ergonomics and its applications to the designers and evaluators of systems and equipment. Modern CAD systems create conceptual designs that can be used as "electronic mockups." New ergonomic design tools are being developed which provide a quick and accurate means of evaluating the interaction between an operator or maintenance technician and a system design. Without such tools, these types of evaluations are rarely cost effective, and often of insufficient quality.

The expanding use of computers to support the design industry has become so great that the field is beginning to subdivide. What started out as CAD (Computer-Aided design) has split many times, producing these distinctive categories:

- CAD (Computer-Aided Design or Computer-Aided Drafting) refers to the creation of a product design on a 2-D or 3-D computer-graphics workstation, replacing the traditional drawing boards. The term Computer-Aided "Design" is more descriptive than "Drafting" because traditional drafting was formerly done by a technician who merely put the designer's ideas on paper. Modern designers often operate the CAD systems themselves, eliminating most of the drafting effort just as modern authors type their work directly on a word processor, reducing the need for most typing by secretaries.

- CAM (Computer-Aided Manufacturing) refers to programming digital controlled manufacturing machines directly from the CAD system software and the digital design, without additional human programmer support.

- CAE (Computer-Aided Engineering) refers to computer tools which analyze the CAD design. These include Finite Element Analysis modeling tools for stress analysis. More properly, design evaluation tools such as COMBIMAN and CREW CHIEF fall into this new category.

- CIM (Computer Integrated Manufacturing) is industry's new name for an integrated set of tools which intercommunicate to perform the CAD, CAM, and CAE functions. The Department of Defense uses additional terms CALS (Computer-aided Acquisition and Logistics Support) and PDD (Product Definition Data) to refer to the deliverable output files of the CIM tools.

This chapter will discuss development two computer-graphics tools for ergonomic analysis: COMBIMAN (Computerized Biomechanical Man-model, a model of a seated vehicle operator) and CREW CHIEF (a model of a maintenance technician). These two models are actually expert CAE tools which allow the user to perform the functions of an ergonomist by simply defining the task to be performed by the operator or maintainer. Both models perform four types of accommodation analyses: fit, reach, strength, and vision. Extensive empirical data bases were used to develop these models. The models are, in effect, a way to access and use the complicated and interactive data describing the physical capabilities and characteristics of the human operator and maintainer.

Traditional anthropometric and biomechanical accommodation analyses require a system prototype or mockup which has a mature design as a prerequisite. However, even a hardware mockup has a significant cost and effort to change. Analyzing designs on a CAD system, on the other hand, have little or no additional cost because modern designs have been created on CAD systems. The CAD design serves as a 3-D "electronic" mockup.

Another major drawback associated with using a hardware mockup or actual hardware to evaluate a new design is the difficulty of finding and testing human subjects. Even in major systems development programs, only four or five subjects have typically been used to evaluate ergonomic accommodation, too few to represent the user population. Because of variation in proportionality among body dimensions, there is a significant possibility of erroneous conclusions when such a small number of subjects is tested.

Considering the world community, the range of adult human stature (standing height) is from two to eight feet and adult body weight ranges from approximately 20 to over 1000 pounds. Fortunately for the designer, these extreme individuals are so rare, they need not be considered in general purpose designs. This great range of body sizes is not found in the military populations because the military has admission standards that screen out the rare individuals. Typically, military populations have statures ranging from 4.5 to 6.5 feet, and weights from just below 100 pounds to slightly over 250 pounds. But even these ranges are still too large for some equipment to accommodate. For example, more stringent body size requirements are necessary for aircraft pilots, and many of the physically demanding jobs now have minimum strength requirements as well. Because the body size and strength characteristics of military populations has been studied more thoroughly than civilian population, designers sometimes erroneously apply these data to civilian designs. Military populations, then, represent a subset of the general population which is young and fit, having a limited range of body size, as compared to the civil sector.

Accommodation analysis with a few human subjects placed in an actual mockup is risky because of the lack of proportionality among people. Two people with exactly the same stature will have different length arms, legs, etc. Typical design specifications may require accommodation of a size range from a 5th percentile (small) female to a 95th percentile (large) male. There are, however, no 5th percentile women and there are no 95th percentile men. The term "percentile" refers to a statistical property of a single measure. Fifth percentile arm length, for example, refers to that measure of arm length which five percent of the population fall at or below and 95 percent fall above. People near the extremes of a distribution in one dimension are usually nearer the average in other dimensions. So, if one selects a person with a 5th percentile stature (short), that person's arms (and therefore arm-reach characteristics) are likely to be much longer than a 5th percentile arm reach, and can even be longer than average arm reach. If one evaluates reach to controls with such a subject, one will likely over-estimate the arm reach and arrive at an erroneous conclusion.

Demands for improved quality, reliability, and performance have placed additional requirements upon the design process itself. Early in the design of a new system, decisions are made which commit the project to an approach with implicit limitations in quality, reliability, and performance. If meaningful changes to a design are going to be made, the changes must be made while the design is still on the drawing board. CAE tools are best used in this situation. COMBIMAN and CREW CHIEF are actually expert systems which allow the system designer to perform the functions of an expert ergonomist. These models also make it easy to evaluate the accommodation of different population subgroups. Other 3-D models provide complex animation with multiple human figures to aid in task visualization.

COMBIMAN and CREW CHIEF are not just 3-D anthropometric manikins, although they are also that. They have unique capabilities for automatically simulating the performance of task, while accounting for the complex and interacting effects of body size, joint mobility, clothing restraints, harnessing restraints, type of grip, type of control, type of tool, type of object, obstacles, posture, and the operation performed. Because of the large number of interacting variables to be considered, more than 100 person years of effort was required to gather the empirical data which was modeled to create these programs.

Both of these models use a new enfleshment technique to represent the irregular surface necessary to depict clothing and personal protective equipment. A wire-frame mesh surface is created by an array of small, interconnected triangles, similar to the technique used in finite element analysis. An algorithm determines which lines comprise the profile view (from any view direction) and also which lines are essential (such as facial features, etc.) and eliminates all other lines before generating the displayed image. The result is a profile view of a high fidelity depiction of the irregular human figure. Although presented on a two-dimensional display, the crewstation and human-model are three-dimensional and can be rotated for viewing at any angle.

Description of COMBIMAN

Beginning in 1972, COMBIMAN was developed by the Armstrong Aerospace Medical Research Laboratory (AAMRL) as a computer-aided Engineering (CAE) model of an aircraft pilot (Kroemer, 1973; McDaniel, 1976; Krauskopf, et al., 1989). COMBIMAN has been in distribution since 1978, and has been used by industry, the U.S. Army and the U.S. Air Force to evaluate design changes, saving the costs associated with hardware mockups and prototypes. COMBIMAN is the predecessor of CREW CHIEF (a model of a maintenance technician), and much of the COMBIMAN technology was the basis of the CREW CHIEF program. The link system, joint system, reach algorithms, vision analysis, strength algorithms, and other features originated in COMBIMAN. COMBIMAN gives the user more control over body size definition, since that is most critical in the limited volume of the cockpit. CREW CHIEF gives the user more control over task definition, since the maintenance tasks are highly variable in nature.

COMBIMAN is a 3-D (three-dimensional) interactive computer-graphics model of an aircraft pilot used to evaluate the physical accommodation of the pilot in existing or conceptual crewstation designs. It is an expert system which performs four types of analyses: fit analysis, visibility analysis, reach analysis, and strength for operating controls with the arms and legs. The user of COMBIMAN does not have to be an expert in ergonomics or anthropometric accommodation because that expertise is automated in the software. The user defines and directs the task to be performed, just as one would a human subject in a mockup review. The program generates accurate body sizes, proportions, strength, clothing restraints, harness restraints, and vision obscuration. Postures can be manipulated automatically by task-driven commands, or manually by positioning individual joints.

COMBIMAN gives the user complete control over sizing the human model and provides several alternatives for dimensioning and changing the dimensions of the model. The user can define the model's dimensions either by entering actual measures or selecting percentiles from a list (in the range of 1st through 99th percentiles of a known population). The typical evaluation session begins with dimensioning the human model by one of the following techniques:

- Computer-Aided Dimensioning - statistically correct human models can be generated from stored anthropometric data. After the user selects the single body dimension which is most relevant to the task being simulated, the program constructs a proportioned human-model using regression equations. For example, arm length is used in evaluating reach capabilities. Sitting height is used for ceiling clearance. Leg length is used to reach to pedals. The model's size and proportions are computed from both the selected measure and body weight, providing a statistically accurate human-model for task analyses.

- Direct Measure - twelve measurements of an individual are entered from the keyboard to size the model to represent a specific individual. In addition to its use in validation of the model, this method can be used to enter a set of individuals with extreme combinations of dimensions for multi-variate accommodation.

- Changing Individual Dimensions - beginning with a predefined model, the user can change any dimension or joint angle by selecting the old value in a table and typing in its replacement value.

- Data-base Summary Statistics - Percentiles computed from large samples are stored in the model and used to define the man model. Individual segments may be set to the desired percentile of the selected population. Selecting percentiles is available with both the two- or twelve-variable method. Currently, there are six populations modeled in COMBIMAN: USAF male pilots, USAF women, USAF female pilots, Army male pilots, Army women, and Navy male pilots. These surveys have been edited to reflect recent changes in body size and weight standards for the military. As newer surveys become available, they are added or substituted, as appropriate. The user is provided a utility program which allows adding new surveys.

Description of Crew Chief

Beginning in 1984, CREW CHIEF, a computer graphics model of a maintenance technician, was developed by the Armstrong Aerospace Medical Research

Laboratory (AAMRL) and the Air Force Human Resources Laboratory (HRL), (McDaniel & Askren, 1985; Korna, et al., 1990). This model is interfaced to several commercial CAD systems used by aerospace designers and may be used to evaluate the maintainability of aircraft and other complex systems. The CREW CHIEF model allows the designer to simulate a maintenance activity on the computer-generated image of the design and to determine if the required maintenance activities are feasible. As with COMBIMAN, the user of CREW CHIEF need only define the task to be performed, just as one would define a task for a human subject in a mockup review.

CREW CHIEF generates ten sizes of human models (five male and five female) with the encumbrance of four types of clothing and personal protective equipment, mobility limitations for twelve postures, physical access for reaching into confined areas (with hands, tools, and objects), automated obstacle avoidance, visual accessibility, and strength analysis (for using hand tools and manual materials handling of objects).

The skeletal link system used in CREW CHIEF is the same as COMBIMAN. These links connect the major points of rotation of body segments and are used to graphically display the model in different postures. The lengths of the links are calculated by regression from 13 (the 12 used in COMBIMAN, plus stature) readily measured anthropometric variables (e.g., stature, sitting height, weight, etc.). To portray clothing bulk, subjects were measured both nude and in the four clothing configurations (fatigues, fatigues with jacket, arctic, and chemical defense).

Approximately 35 percent of the lifetime cost of a military system is spent for maintenance. Excessive repair time is caused by failure of the system design to adequately consider maintenance. The maintenance technician will spend hours making a repair which could have been completed in minutes if accessibility had been adequate. Ultimately, development costs and acquisition time as well as life cycle costs and maintenance time will be reduced. Designers, human factors engineers, and maintainability engineers may use CREW CHIEF as an electronic mockup. Analyses, which would have required weeks or months to perform by conventional methods, can be performed in minutes. Distribution of CREW CHIEF Version 1 began in May 1988, with Version 2, about a year later.

HISTORY OF ANTHROPOMETRIC COMPUTER MODELS

As soon as graphics displays became available for CAD use in the late 1960s, engineers and computer scientists began to dabble in modeling humans on the computer. Most of these human-computer models were abandoned at some point in their development, and do not warrant discussion. Of the many early attempts at graphical human models, only COMBIMAN and two others are still under development: SAMMIE (Bonney, et al., 1969) and Torque Man (Chaffin, 1969). All of the current sophisticated models have been around for many years and can trace their origin to these earliest developments. NRC (1988) describes many of the current models.

Modern engineering graphics workstations and CAD software have 3-D modeling features which make it very easy to create and assemble the geometric shapes which are human-like in appearance. There are even some graphics software created for animation and robotics which make it easy to make these manikins dynamic. Because this CAD technology is widely available from a number of vendors, there are an unprecedented number of human modeling efforts in progress. Body size and joint mobility data are readily available, and human-computer models progress rapidly to stage of a 3-D manikin. The arm can now be extended to simulate a reach. This pseudo-reach capability provides the first functionality. Vision capability is also easy add, since most 3-D CAD systems allow the user to change the view location and distance. Even crude animation is possible with some of the new CAD software. However, meaningful data on segment trajectories and times is not available.

At this point in the model evolution, further development becomes expensive. Enfleshing over the joint segments is very difficult. Most models either let the segments overlap or leave a gap between the segments. Useful data on arm and leg kinematics is not found. Modelers now notice the disparity between where the model's joints articulate and where people articulate. The problem is that joint centers are usually not at the geometric centers of the segments. Some joints are neither hinge or ball joints, for example, the shoulder joint with its six-degree-of-freedom mobility is particularly troublesome to define and model. What other capabilities should be added: strength? endurance? dynamics? intelligence?

At each stage of model progression, the model becomes increasingly demanding for data. After the modeler has tapped all the sources of available data, development can only continue with additional and costly research. At some point in this evolution, the development ceases, and the model is judged to be "good enough" as it is. It is not an underestimate to say that this human-model development scenario is now being played out at a dozens locations. While user-friendly technology draws developers to begin, computer graphics will take you only so far. True engineering functionality is not just a picture on a computer screen. A useful computer tool must first have the function, then the picture which represents the function.

The COMBIMAN and CREW CHIEF models differ from other developments, then, in that they are implementations of ergonomics data and ergonomics models which either existed before the computer model, or else was gathered specifically to meet the needs of the model. These programs started as data and understanding about the physical characteristics and capabilities of the human worker, including body size, body proportion, skeletal links, strength, reach, vision, etc. COMBIMAN and CREW CHIEF models are more products of research than development. First, there was the research and then, the development.

Another distinction between models is whether they are 2-D or 3-D. The answer is not always apparent because the computer display is 2-D. If a model is true 3-D, however, it can be rotated and viewed from any angle. In fact, some of the earliest human models were 2-D drawing board manikins consisting of cutouts of the body segments with rivets for joints. Figure 1 has examples of a small and large quarter-scale manikins from Dempster, 1955. The segments are cut out, then assembled by riveting the joints. The oversized hole at the shoulder is an attempt to provide for the proper range of mobility, but a skillful user is required for accurate positioning. The arcs around the joint centers have additional rivets, and provide limits of mobility. These scale manikins were laid atop a paper drawing, positioned, and traced around with a pencil. Some of the early computer models used this same technique to create a 2-D computerized manikin.

The evolution of COMBIMAN generally followed the development described above. However, since the development began in the early 70's, COMBIMAN did not have any established CAD technology to build on. Version 1 started, in fact, as a stick figure which represented the 3-D skeletal links of the body. Later, in Version 2, the stick figure was "enfleshed" by attaching wireframe rectangular solids about the links, which gave the model's segments a box-like appearance.

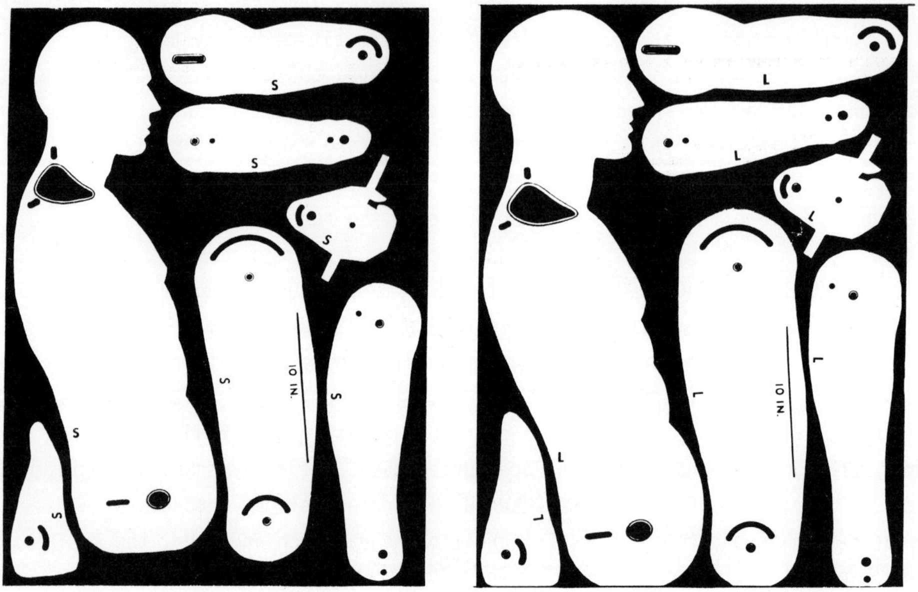

Figure 1. Dempster's (1955) 2-D drafting board manikins (Small and Large).

To give COMBIMAN a more human-like appearance, Version 3 positioned ellipsoids (3-D ellipses) around the link joints, then projected these ellipsoids onto the 2-D viewing plane, and finally connected the ellipses with tangent lines to define the exterior profile of each body segment. The resulting enfleshment is shown in Figure 2. The advantages of this technique were:

- a minimum number of lines were used to depict the enfleshment, because only the profile was shown. It was easy to distinguish which lines belonged to the human model and which to the workplace.

- the enfleshment was 3-D, and could be viewed from any orientation, retaining its accuracy while keeping the number of lines to a minimum.

- the enfleshment was defined parametrically, that is, as equations, so the size and proportions were under complete control of the program and the user. Sizes and proportions could be changed by substituting different values in the equations. Many other models have fixed-shaped segments which are scaled to show various sizes.

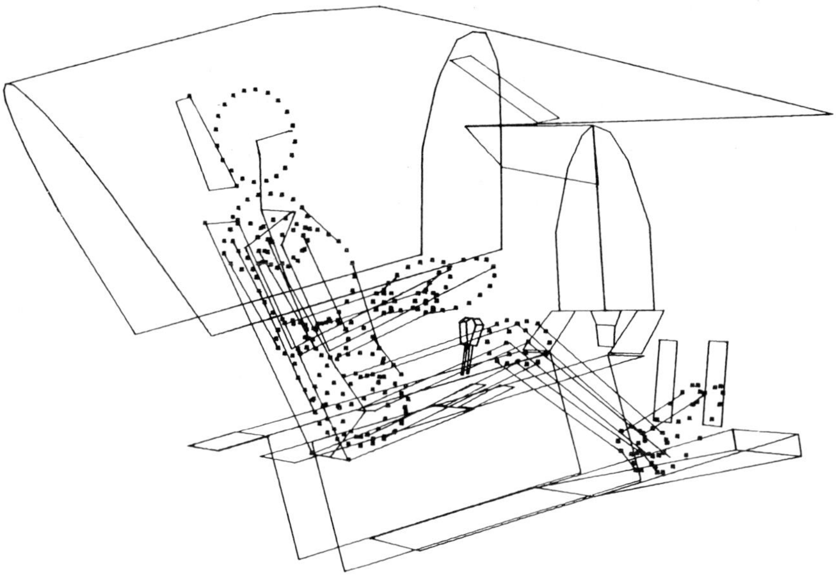

Figure 2. COMBIMAN enfleshment using ellipsoids to enflesh over the joints.

- the ellipsoids could be sized to represent the length, breadth, and depth of each joint using equations, so that the curvature over the joints was realistic. The centers of the ellipsoids are offset form the joint centers to produce accurate relationship between the skeletal links and the external enfleshment. The general outline of the human-model was depicted on the screen by a series of dots and line segments.

Beginning with Version 7, a finite element mesh (FEM) technique was used for the COMBIMAN enfleshment. These FEM techniques, which were not available at the beginning of the development, allowed representation of irregular details such as clothing, helmets, boots, facial features, etc. The finite element technique requires representing the 3-D surface of the body as a wire-frame covered with triangular facets, with the vertices and edges of the triangle identified. A sonic digitizer, laser scanner, etc. is used to define the local X,Y and Z coordinates of the triangle vertices. An algorithm determines which lines comprise the profile view (from any view direction) and also which lines are essential (such as facial features, etc.), and eliminates all other lines before generating the displayed image. The new enfleshment is shown in Figure 6 (later in the chapter). This same enfleshment technique is used in the CREW CHIEF model.

One unique feature of COMBIMAN is the clothing model. For each clothing configuration the program internally limits the range of joint mobility to account for encumbrances of the bulk of the clothing configuration. There are 7 clothing configurations in COMBIMAN:

- Semiclad - the base line for validating the human model's dimensions.
- Summer suit - the bulk of the summer flying suit with flight helmet.
- Summer suit w/vest - the summer suit, helmet, and survival vest.
- Winter suit - the bulk of the winter flying suit, helmet.
- Winter suit w/vest - the bulk of the winter flying suit, helmet, and survival vest.
- Winter suit w/vest, jacket, and life preserver - the winter flying suit, helmet, survival vest, jacket and life preserver. This is the bulkiest clothing combination.
- Chemical definse - the bulk of the chemical defense clothing ensemble.

Because COMBIMAN began before CAD systems were available, it had to include its own CAD features. These CAD features, retained through Version 8, include: Change View, Identify Object, Omit Object, Retrieve Crewstation, Display Anthropometry, Display Links, Design Panel, Modify Posture, Seat Adjust, Zoom, Paper Plots, and Add Crewstation.

Beginning with Version 9, COMBIMAN was reconfigured to operate in conjunction with existing CAD systems, which have provisions for designing the work environment and even drawing the human model as part of the workplace. In the new configuration, the CAD capabilities described above are accomplished by the host CAD software, not COMBIMAN. This technique was used from the beginning in the development of CREW CHIEF. The advantage of this method is that the user does not have to translate and/or transfer computer files to use the biomechanical model, which is always available, and can be called into the user's own design with a simple command. The disadvantage of this approach is the dependency on specific CAD software. Someday, if CAD systems develop a common format, this dependency may be removed.

PROBLEMS AND DIFFICULTIES IN DEVELOPING MODELS

<u>Problems in modeling body size</u>. The major obstacle to anthropometric accommodation has been finding a sample of subjects which represents the exact measures on one or more dimensions which are pertinent to the design. The

major advantage of using computer models of the human body to evaluate accommodation is that all of the models dimensions can be adjusted to the required values. Thus, the computer-generated subjects chosen to evaluate the design are as good as one's ability to define the needed values.

Another reason anthropometric evaluation has been difficult is because the body-size measures are uni-dimensional segment lengths and circumferences, not the 3-D motion envelopes required by workplace designers. Each is the measure of an important body segment or characteristic, but the measures are not attached to each other nor to any common coordinate system. For example, the measure "shoulder-elbow length" is measured to the bottom surface of the bent elbow while "elbow-wrist length" is measured to the back surface of the elbow. Neither of these measures are referenced to the center of rotation of the elbow joint. Even if a design engineer were given these measures, they would not be sufficient to assess arm reach capability.

Anthropometric measures are made on the surface of the body, while the engineer usually needs internal dimensions associated with the skeletal link system, that is, from one joint center to another. The science of relating external and internal body dimensions is too complicated for a designer to apply routinely, but it is not too difficult to incorporate into a computer model and calibrate against empirical task analyses. Development of task-related data is usually very difficult because it is also task-specific. While computer models are based on more general data, the user must beware not to apply them to unusual designs which fall outside the region of the model's data bases. For this reason alone, there will continue to be a need for some accommodation analyses with human subjects.

Although considerable data are found in published studies, the data were gathered for specific applications and can rarely be generalized to other applications. COMBIMAN and CREW CHIEF model developers have developed a technique for using benchmark tests performed on large samples of the population to adjust the data gathered on small samples, but with large numbers of interacting variables. However, all of the task-related data bases in the three-dimensional CAE models COMBIMAN and CREW CHIEF were gathered especially for these models. The modelers have compiled and developed the most accurate representations of the link system and task representations available. The engineer can now use the model without having to develop the model.

The military aircraft cockpit is a particularly difficult design problem because the need to save weight and volume require the cockpit to have the minimum acceptable size and adjustment. When designing a complex workplace such as an aircraft cockpit, several dimensions must be accommodated simultaneously. To be specific, pilots must be able to see out the window and to reach the hand controls and the foot controls simultaneously. This is referred to as "multivariate accommodation". While the concept of multivariate accommodation is not new (see Moroney & Smith, 1972), the procedures for accomplishing it have been costly and laborious. Computer models can facilitate multivariate accommodation by allowing a designer to enter several specific individuals whose dimensions are at the limits of a multidimensional distribution of critical body dimensions. Each individual in this set represents the worst case of several critical dimensions. For example, one model may have the shortest arms and legs appropriate for a person with a tall sitting height. Another model in the set may have short sitting height as well as short arms and legs, etc. By defining a population with combinations of realistic extremes, a multivariate accommodation can be verified if all the members of the set are accommodated.

CAE models are the only practical way to represent the unusual dimensions needed for multivariate analyses. To achieve a similar quality of evaluation with human subjects, several hundred have to be sampled to find the critical ones. Computer models can do this by searching existing surveys for the required values, then dimensioning the model with those values.

The link system of a human model is its skeleton. The link system is made up of three-dimensional vectors, attached head to tail, just as the bones of the human body attach to make the skeleton. These vectors represent the distances and angular relation from one joint center to another. The link system is the sum of a series of vectors. Each link vector has a local coordinate system with its origin at the distal end. The orientation of the next link is defined in this local coordinate system. Phi, Theta, and Psi correspond to Euler angles as shown in Figure 3. These local coordinate systems are usually not aligned with the base system which has its origin at the Seat Reference Point (SRP).

These vector links more represent the function of the human skeleton, because they are always straight, while the bones of the skeleton are often curved. These links then, represent internal dimensions of the human body. The relationship of internal-to-external body dimensions was first defined by

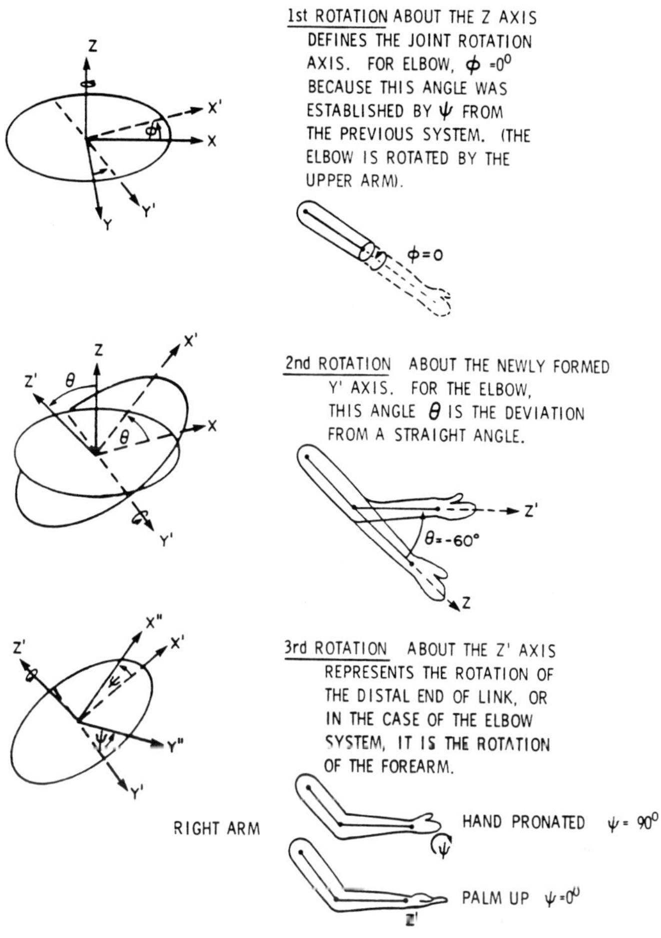

Figure 3. Euler angles use to define motion of the elbow joint.

Dempster (1955). Dempster's data have since been supplemented by making adjustments to joint centers based on externally observable kinematics, and additional data on individual joints. Over the years, Dempster's data has passed through so many generations of reports, today's modelers do not even realize they are using his basic link system.

Problems in modeling reach operation. Reach to hand controls must consider the type of grip involved. Vehicles, including aircraft, have only three types of control couplings: hand grips (levers, wheels, etc.), rotary knobs, and push buttons. COMBIMAN has three different hand grips for evaluating reach to crewstation controls: grip center (whole-hand grip), functional grip (for operating knobs), and fingertip (for operating push buttons and switches). The program evaluates reach capability as a function of body size, clothing, and restraints (harness) in two ways. First, the user can select a control handle, or simply key in the 3-D coordinates, and COMBIMAN will simulate a reach to that point. Second, the user can select a control panel, and the model will compute the maximum reach envelope in the plane of that panel. The reach analysis can be performed with the arms or legs, and movements are automatically limited by the restraints of lap belts and shoulder harnesses.

The reach models for COMBIMAN are based on the most complex single ergonomic study in any of these models. The study's variables included the three types of grip (grip center, functional grip, and fingertip), three types of harness restraint (lap belt, shoulder unlocked, and shoulder locked), six combinations of clothing, three planes of reach (forward, above, and side). The data collection on the twenty-six male and female subjects required 2.5 years, with the analysis requiring another two years. Reach capability was not only measured mechanically at the grip, but also photographically to establish the posture of the body's links while reaching.

Joint mobility is more complicated to model than most suspect. Although most joints are hinge joints (elbow, knee, phalanges, etc) or ball joints (shoulder, hip, ankle, etc), the muscles and ligaments attached across the joints usually cause interactions in mobility limits. In other words, the position of one joint affects the mobility of an adjacent joint. For example, if the upper arm is abducted (raised from the side of the body, outward to the side, and finally up above the shoulder), the upper arm can be raised higher if the elbow is straight than if it is bent 90 degrees with the lower arm pointing forward.

Joint mobility data used to construct mobility equations (Figure 3) involving the Euler transformation variables (phi), (theta) and (psi) were gathered specifically for the COMBIMAN development. The mobility limits are used to place realistic limits on human-model movement when using the analysis functions (Reach, Reach Curve and Strength) which automatically reposition the human-model body and body segments. The joint mobility limits are modified by the application of constraint factors caused by the bulk of programmed clothing types and mobility restraints such as lap belts and shoulder harnesses. The equations are internal to the program and cannot be changed by the user.

Visibility Analysis. The visual field analysis plots the crewstation as seen by the human-model. For COMBIMAN, if the visibility plots are made in the forward-looking direction, they satisfy the requirements of MIL-STD-850 (DoD, 1970). Unlike manual techniques which are limited to forward-looking from the fixed reference eye position, COMBIMAN's mobility allows a range of head and eye positions, as occur with a real pilot. The size of the operator, seat adjustment, head position, and visual restrictions may all be quickly varied for a realistic evaluation of the visual field of the pilot. Obscuration templates for peripheral vision, field of resolution, and helmet/mask assemblies may be superimposed upon the visibility plot. If the user has data describing a restriction to vision, a new template can be constructed (up to a total of 15) and added to the data base using the procedure described below.

Data describing a restriction to vision would be collected with a peripheral vision testing device. There are a variety of commercial and non-commercial devices available. Figure 4 and 5 describe two devices commonly used to measure field of vision limits. A smooth curve will be generated to fit the points provided. It is recommended that no fewer than eight coordinate pairs be used. Although 100 pairs may be entered, increasing the set beyond 36 pairs does not cause a significant difference in the shape of the vision envelope.

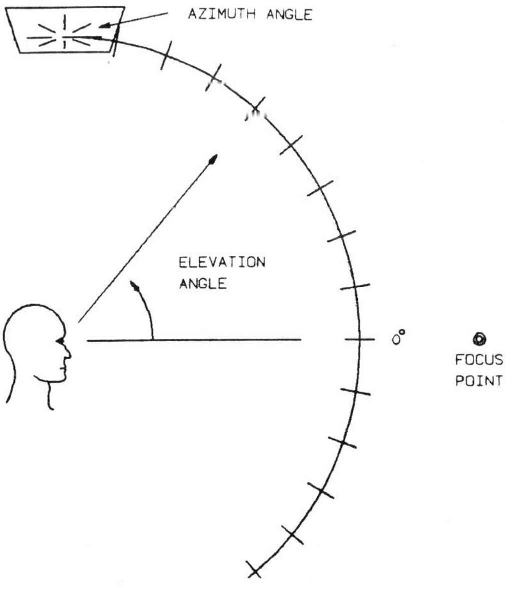

Figure 4. Vision limit measuring apparatus with azimuth outer gimbal.

The data collected with the device shown in Figure 4 are in azimuth/elevation angles can be entered directly into the data base. Using the azimuth outer gimbal device, subjects sit beneath the vertical pivot of a semicircular protractor. Data describe (first) the azimuth angle (left or right of forward) and the elevation angle (above or below horizontal). This azimuth/elevation data can be entered directly into the data base without transformation. The data collected from the device in Figure 5 need to be converted into azimuth/elevation angles. With this device, the subject faces the horizontal pivot of a semicircular protractor which pivots about the longitude axis. Data describe (first) the angle of the protractor plane (Angle A) and (second) the angle of vision limit in that plane (Angle B). The angle of the protractor plane is between 0 and 360 degrees, and the angle from forward is between 0 and 180 degrees. The following equations will convert the spherical angles to rectangular (azimuth, elevation) angles:

$$\text{Elevation angle} = \sin(\sin A \sin B)$$

$$\text{Azimuth angle} = \cos(\cos B / \cos(\sin(\sin A \sin B)))$$

A = angle of the protractor plane.

B = angle from forward.

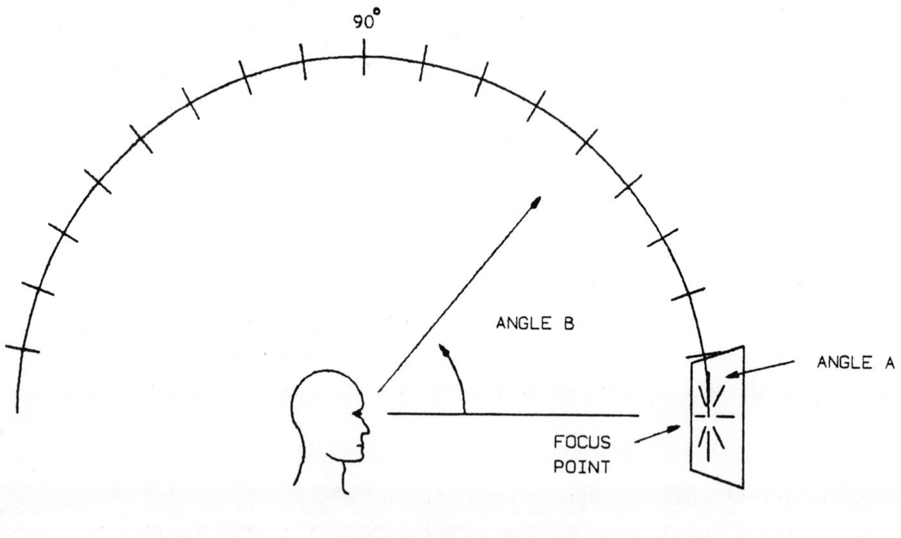

Figure 5. Vision limit measuring apparatus forward outer gimbal.

Problems with modeling human strength

Human strength is task specific. In other words, if you change the nature of the task by moving the point of application of force, changing the direction of force, using a different type of hand grip, orientation of handgrip, etc., the available strength will also change. The reason is that many different combinations of muscles are used for any exertion of strength. The amount of force produced by muscular exertion also depends on the duration of the exertion, that is, endurance. The duration of exertion reduces the strength by an exponential of the time the force must be sustained. Because of the task specific nature of strength, very large data bases are required to develop a general model, such as COMBIMAN and CREW CHIEF. The strength research which preceded the CREW CHIEF model, for example, involved over 100,000 strength measures, and more will be made in the future (Gibbons, 1990).

Table 1 shows combinations of variables and types of strength measured for the CREW CHIEF model development. An "X" indicates that a particular variable and type of strength variable was researched. In most cases the "X" represents a number of individual studies. For example, for the first combination of standing and tool torque, seven separate studies were performed with different combinations of other variables. Other combinations of variables included different sizes and lengths of wrench handles, different orientations of wrench handles and bolt heads, loosening and tightening exertions, different combinations of hands (right, left, or both), different types and sizes of wrenches, with and without gloves, different types and degrees of obstructing barriers, and extensions and U-Joint sockets. All studies used both male and female subjects.

Effects of posture. A sonic digitizer was used to measure body posture in many of the strength studies. Posture is an important variable when shifting the center of body mass effects the force generated. For example, in pushing or pulling, the body mass may be shifted by bending or straightening the elbows. In one study of pushing strength, for example, men averaged 48 percent more and women 30 percent more when pushing with bent elbows versus straight elbows.

The sonic digitizer employs an array of microphones surrounding the subject. Electric spark gaps are taped to the subject's joint centers or other anatomical features useful in tracking posture. The sonic digitizer

TABLE 1
Variables Tested for CREW CHIEF Development

VARIABLE	TOOL TORQUE	LIFT	PUSH & PULL	CARRY	HOLD & POSITION	CONNECTOR
GENDER	X	X	X	X	X	X
OBJECT HEIGHT	X	X	X		X	X
ORIENTATION	X	X	X		X	X
BARRIERS	X			X	X	X
HANDLE SIZE	X					X
ONE HANDED	X	X	X	X	X	X
TWO HANDED	X	X	X	X		
POSTURE						
STAND	X	X	X		X	X
SIT	X	X	X		X	
BEND	X	X	X	X	X	X
SUPINE	X	X	X			
PRONE	X	X	X			
SIDE	X	X	X		X	
KNEEL	X	X	X		X	
SQUAT	X	X	X	X	X	
WALK				X		
CRAWL			X	X		

measures the time delay between the generation of the spark and when each of the microphones detect the "popping" sound of the spark. The delay is translated into a slant range distance, then the 3-D coordinates are computed. The sonic digitizer can locate points in 3-D space at the rate of 48 Hz. By surrounding the subject with an array of 8 microphones, masking by the body parts is eliminated, since only three of the eight data points are needed.

Posture is critically important in all types of physical activities, especially important in maintenance tasks where the objects being maintained often create obstacles and forces the worker into restricted postures. When carrying under a low ceiling in a passage way or under the wing or fuselage of an aircraft, the available strength is reduced, as shown in Table 2 below. Maximum weight was carried in a box 18 inch wide box with no handles while using two hands. The ceiling height was set to a percentage of stature.

TABLE 2
Maximum weight carried in a box under a low ceiling.

CEILING HEIGHT	MALE	FEMALE
UNLIMITED	153	79
80%	146	73
60%	113	54
40%	64	41

Lowering the ceiling forces the worker into a bent posture, then a semi-squat, and finally crawling (40% of stature). At each progressive level, the amount of weight that an individual can carry is reduced, until, in a crawling posture, it averages only 45 percent of the unrestricted condition.

The effect of posture on weight lift capability can be seen in Table 3. The values shown are the maximum weight of a 24 inch wide box with no handles people were able to lift from the floor and place on a shelf. The maximum weight was determined using the incremental technique, increasing the amount of weight lifted until the subjects were no longer able to lift the box onto the shelf. Five postures were used: standing, kneeling, sitting, squatting, and lying on the side. All lifts used two hands, and the height of the shelf was adjusted to 35 percent of the subject's reach height in that posture, except for lying, where the shelf was ten inches high. The values are in pounds and averaged over each group of subjects.

TABLE 3
Maximum Weight Lift (Pounds) Capability for Different Postures.

POSTURE	50 MALES	50 FEMALES
Standing	118	58
Kneeling	99	53
Sitting	92	49
Squatting	79	43
Lying, side	42	21

The table shows that the effective strength decreases when the body support becomes less stable. The kneeling posture allows more mobility of the lower torso in adjusting the posture toward the load while still providing a stable support. Sitting provides a stable support but reduces the mobility of the lower torso, forcing the reach to the shelf to be farther. The squatting posture has little support, as the subject is only supported by the balls of the feet and must exert some effort to maintain balance. The lying-on-the-side posture has little stability in the axis of load and requires the exertion of lateral forces to raise the weight. A maintenance technician would prefer to work standing up. Often this is not possible because of constraints in the workplace. Obstructions limit the access, forcing a less than desirable posture.

Point of force application. The location and direction of a force greatly affects the magnitude of strength available. An example of this interaction of posture and direction of force is found in measures of torque produced with a socket wrench. In a study of isometric torque using a 1/2-inch square drive ratchet to turn a bolt with a 3/4 inch head, it was found that the least favorable location/orientation of a bolt head allowed only fifteen percent of the torque produced in the most favorable location/orientation.

Of course a maintenance technician would not intentionally choose the less favorable configuration, but may be forced to do so by obstacles. In another study measuring wrench torque, but where the subject had to reach over or around obstacles in the workplace, the available torque was reduced up to 80 percent due to the obstacles.

Benchmarking Procedures. Many previous ergonomics models have failed to achieve their goal simply because model developers incorrectly assumed that all required data was available. There is a vast quantity of data available in the ergonomics literature, but most are not suitable for development of a general purpose model. Most data are limited in the range of variables, the sample size, the applicability of the subjects to military populations, and non-availability of raw data for modeling. Developers of the COMBIMAN and CREW CHIEF models have programmed a large portion of their resources for ergonomics research.

To overcome the limitations described above, a seven step testing and modeling procedure was developed. Because of the complexity and the amount of research data needed for the human-computer models, it was not possible to gather all data on a large sample of operational personnel. Rather, a benchmarking technique was developed to allow laboratory research to represent the population of workers. This seven step process, described for the CREW CHIEF application, insures that research data is representative of the population of Air Force maintenance personnel:

First, laboratory subjects were screened to represents the size and age of Air Force maintenance personnel. Since more than 99 percent of personnel doing manual work are age 30 or younger, research subjects were limited to the range of 18 to 30 years. The Air Force also has strict height and weight allowances defined by Air Force Regulation 160-43 which were also applied to research subjects. These restrictions may limit the utility of CREW CHIEF to represent older civilian populations, but Army and Navy personnel have almost identical characteristics.

Second, subjects were given benchmark strength tests. This battery of tests has been given to large samples of Air Force maintenance personnel over the years. One of these tests is the Maximum Incremental Weight Lift to Six Feet which is given to all Air Force recruits and has been demonstrated to be highly correlated with manual materials handling tasks. Three static (isometric) strength tests are also given: the one-arm pull, which involves bracing with the straightened left arm while pulling on a vertical handle with the right; the elbow height lift, which involves lifting against vertical handles positioned at elbow height; and the 38cm lift, which involves lifting with two hands against a horizontal handle 38 cm above the floor. These tests have also been given to several thousand military personnel.

Third, the subject's body size is measured. For subjects tested at AAMRL, 69 measures were taken, for some of the tests made off-site, 20 measures were taken. These measures were made on several thousand military personnel. Only subjects meeting the height/weight standards were used.

Fourth, the subjects participated in simulated working tasks wherein their strength was measured. In most of these simulated work tasks, from 40 to 100 subjects were tested in each combination of variables. Treatments were randomized with suitable rest periods between all strength measures. Some treatment conditions were repeated at both the beginning and the end of each test session to verify the reliability of the subject's performance. The tool torque and push/pull tests were static (isometric) while the lifting, carrying, and holding were dynamic. Static measures were gathered with a computerized data collection system which evaluated the data against goodness criteria as it was collected and identified exertions to be repeated.

Fifth, the data were sorted, collated and edited. This process used both within subject and between subject relationships to identify outlying data values.

Sixth, the data were adjusted to represent the population of workers. This was accomplished using regression equations developed on large samples performing both the benchmark tests and some of the work tasks.

Seventh, the adjusted data were converted to algorithmic models for CREW CHIEF. When user of CREW CHIEF defines a task to be performed, the model determines which conditions apply, and select the appropriate strength models of male or female data. Predicted strengths for the 1st, 5th, 50th, 95th, and 99th percentiles are displayed on the workstation. The procedures for benchmarking the COMBIMAN relationships used a similar seven step procedure.

Strength for control operations. If a point or control in the forward hemisphere can be reached, the analyst can evaluate the force (application of strength) the COMBIMAN human-model can exert as a function of control location, direction of force, and type of control. Available strength is computed by COMBIMAN for the 1st, 5th, 50th, 95th, and 99th percentiles of the strength population for both male and female pilots as a function of control location, control type, and direction of force. Years of strength research on more than 1,000 subjects have been modeled to provide accurate computations of available strength for operating controls.

One of the strength studies for COMBIMAN required measurement of over 500 maximum isometric exertions on each of 40 male and 40 female subjects. Each subject exerted force in six different directions at each of the 76 handle locations used in the study. The six directions were forward, back, right, left, up, and down. The 76 handgrip locations form a three dimensional grid with 10 inch spacing in the X, Y or Z directions. These data were normalized to the total population by adjustments based on benchmark exertions measured on a large sample of Air Force personnel.

The strength data for aircraft controls (stick, wheel, and rudder pedals) was compiled on more then 400 female and 600 male subjects. Many of these were students from the Air Force Academy and Officers Training School. The large data base was used to calculate strength percentiles for each direction of force. Ejection control forces were measured on an additional 100 male and 100 female subjects in the lab. Another study determined the relationship between leg strength, knee angle and vertical location of the rudder pedal. The computer model predicts the 1st, 5th, 50th, 95th and 99th percentiles available strengths.

Strength for maintenance operations. Most of the textbook strength information describes the standing or seated posture. In fact, most previous strength studies have used the standing posture, working in front of the body. We have all seen the promotions for the correct methods of lifting (keep the back straight, lift with the legs, etc.). However, maintainers must work in all postures. Much of the strength research needed required postures which all safety engineers and most human factors engineers considered to be hazardous. One of the most difficult problems associated with gathering the strength data for the maintenance tasks was the variety of postures. That we were able to gather so much data in these unusual postures without injuring any of the subjects is testament to the strength measurement technology developed for these research programs.

CREW CHIEF has models of strength for torquing with all types of wrenches, even torquing the lock ring on electrical connectors. In developing the CREW CHIEF model, strength measures with all types of wrenches were made with different combinations of barriers. The many variables in the wrench torque studies included posture; bolt elevation as a percentage of the finger tip vertical reach for a given posture; distance of the body from the bolt head; wrench handle orientation (selected position around the 360-degree circle of the bolt axis); bolt orientation (relative to the mid-sagittal plane of the body); type wrench (ratchet with socket, open end, box end, etc.); and the reach around and over obstacles to the wrench handle.

Most maintenance work with tools involves working around barriers or obstacles. The work environment has three types of barriers which influence tool selection: (1) Opening barriers - access opening through which the maintainer must reach with arms and tools to do the job; (2) Intermediate barriers - barriers which occur between the opening barrier and the work. Intermediate barriers are usually obstacles created by other components, internal structure, internal walls, wiring, piping, etc.; (3) Clearance barriers - barriers around the work itself. Clearance barriers limit access to fasteners with tools and reduce tool/hand movement. These barriers can be features of the airframe, other components, or features of the component being removed. Clearance barriers may obstruct vision or two hand operation. The object being removed/replaced may itself be a clearance barrier.

A series of experiments were conducted to develop the Materials Handling data bases. The lifting phases of the experiments were done in various postures (kneeling, standing, sitting, supine and prone). Lifting tasks are one- or two-handed. Various size boxes are used for the different tasks, and the weight can be adjusted. Materials handling strength used the psychophysical method. The subject could request the experimenter to add or remove weight until the subject reached his or her maximum effort without risking injury.

A common maintenance task involves lifting an object into position and holding it in place with one hand while installing a fastener with the other hand. A series of Hold/Position studies were conducted with variations in posture, target height (expressed as a percentage of the subject's finger-tip vertical reach for the posture used in the exertion), barriers (used above and in front of the target to simulate working conditions), and the weight of the object (variable from 10 to 110 pounds). The object was positioned with both hands and, once positioned, was held in place with only the left hand. The time the object is held within the confines of the target is measured for different weights to compute the maximum weight one can hold for 60 seconds.

Another series of experiments were performed to collect data for Push/Pull capabilities. Variables for the exertions included five postures, direction of force, handle altitude, foot distance from the handle, and angle of the elbow (straight or bent). Exertions were also performed in two methods (freestyle and controlled). In "freestyle", the subject was allowed to push or pull to attain the maximum resultant force on a fixed handle. In "controlled", only the horizontal component of the force was possible. A sonic digitizer was used to record the posture of the body during each exertion and to relate body angles to force data.

CURRENT MODEL USAGE

In describing the development of the COMBIMAN and CREW CHIEF models above, the stated purpose was to evaluate the physical accommodation of the operator or maintainer with respect to fit, reach, strength, and vision. It has also been said that this will aid in the early identification of potential design-induced maintainability problems, which is essential to correct a problem before mock-up, fabrication, or production. The designer may simulate operations or maintenance activities on the computer-aided design image to determine if the activity is feasible.

These models use expert system software to automatically create the correct body size and proportions for males and females, the encumbrance of clothing, personal protective equipment, and mobility, as well as physical access for reaching into confined areas (with hands, tools, and objects), visual access, and strength. The usage of the models is best illustrated with a few examples.

Analyzing outside vision. One of the challenges of vehicle design, whether the vehicle is an automobile or aircraft, is outside vision. The range in sitting height of the operator must be accommodated by seat adjustment, or making the windows large enough so that operators of all sitting heights can see out. Figure 6 shows one of the most difficult vision problems found in aircraft, which is the tandem seating arrangement, that is,

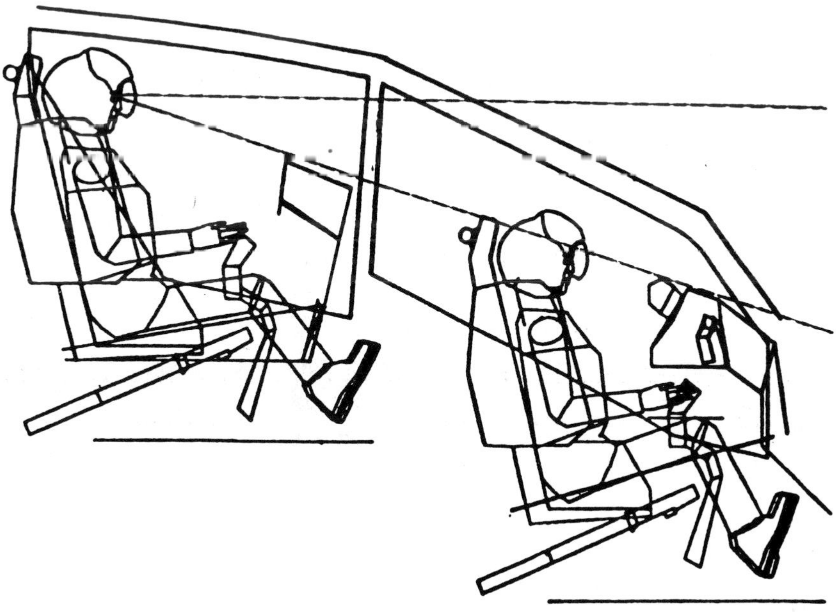

Figure 6. COMBIMAN Electronic Mockup of Helicopter Crewstation.

with the pilot sitting behind the copilot. In the tandem helicopter cockpit is shown in this figure, the pilot's downward vision is limited by the copilot's helmet which is approximately 6.5 feet forward of the pilot's eye position and the nose of the helicopter which is about 12 feet forward.

The COMBIMAN program has a function which plots the crew station from the pilot's view point. MIL-STD-850 requires such visibility plots to ensure adequate vision over the nose and around the window frames. While MIL-STD-850 requires only plots from the arbitrary "Design Eye Position", COMBIMAN vision analysis additionally considers realistic combinations of body size and seat adjustment, as well as realistic offset of the eye position due to head motion and obscuration overlays which represent the masking of helmets, respirators, night vision goggles, and other head gear. COMBIMAN's vision plots can also show objects outside the aircraft, such as other aircraft, runways, landing pads, tanker aircraft, etc. Notice that for the purpose of this type of analysis, the graphics representation is quite simple, containing only the relevant features, such as instrument panels, controls, and windows. Once the crew station model is created and entered into the program. The visibility plots in Figure 7 shows the view of the pilot depicted in Figure 6. The Head

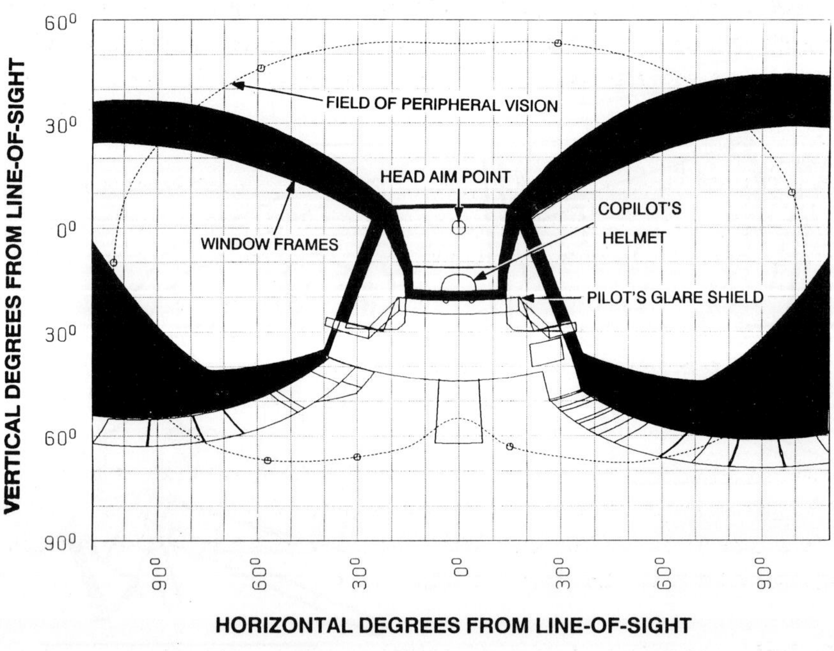

Figure 7. Visibility Plot from COMBIMAN with peripheral vision envelope.

Aim point is where the pilot's head is pointing. Cockpit details include the window frames, front and side instrument panels. The oval-shaped Field of Peripheral Vision indicates the limits of the pilot's field of view. COMBIMAN allows the user to install additional obscuration envelopes.

Analyzing reach with a shoulder harness. Today, most vehicle operators must wear a seat belt, but this does not significantly restrict reach. In many aircraft, pilots must also wear a shoulder harness. When practicing acrobatics and other violent maneuvers, the shoulder harness must be locked. Also, aircraft ejection seats adjust vertically, but usually not horizontally, so the pilot's knees or feet will not strike the instrument panel or canopy frame when ejecting. This would happen if a long-legged pilot adjusted the seat too far forward. Since the result of a knee strike is catastrophic, ejection seats adjust only vertically. The "ejection clearance line" is based on the large pilot's buttock-knee length. However, when flying with the shoulder harness locked, as often happens during maneuvers, the small pilot's arms may not be long enough to reach controls on the forward instrument panel.

The seat and human model in COMBIMAN can be adjusted without limit in three dimensions. The Reach Analysis Model is based on data gathered with three types of controls, six types of clothing, three types of harnessing, and seven reaching planes. The data collection required over two years of full-time effort, providing the most comprehensive description of arm reach available anywhere.

The COMBIMAN program evaluates arm reach with three different harness configurations: (1) lap belt only, representative of some non-pilot crew members, (2) shoulder harness on but not locked, representing the non-critical situation, and (3) the shoulder harness on and locked, representing the situation where maneuvering or inverted flight is anticipated. The harness is usually locked when practicing for combat. When the shoulder harness is locked, arm reach envelopes are limited to arm and shoulder movement.

Figure 8 shows an analysis of arm reach with COMBIMAN in the summer-weight flight suit and 50th percentile arm length. When COMBIMAN is asked to reach to a control on the forward instrument panel with the shoulder harness locked, the pilot fails the reach by 4.83 inches.

Analyzing the effect of seat adjustment on reach. Variability in body size can be accommodated by making the seats adjustable. This simple solution

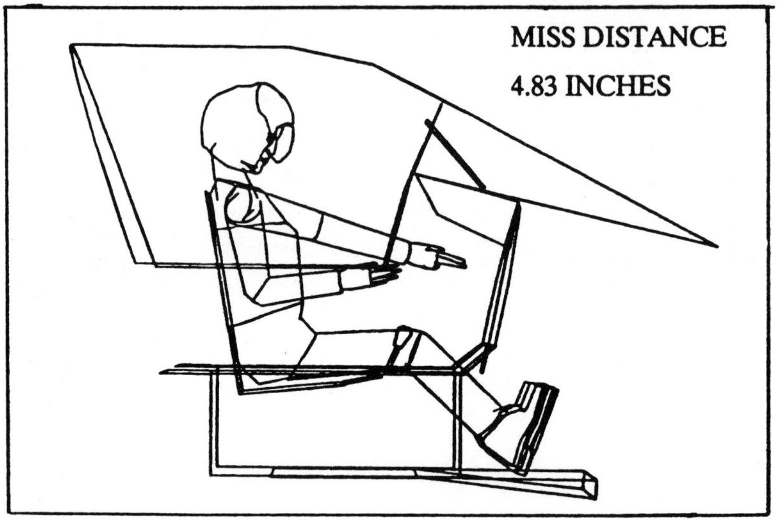

Figure 8. COMBIMAN fails Reach Analysis due to restrictive shoulder harness.

can have some unsuspected pitfalls, however, because seat adjustment interacts with reach to controls. A large pilot may have to adjust the seat down to achieve head clearance beneath the canopy. For the large pilot, reach and vision are seldom problems; however, the fit may be cramped in extreme cases. For the small pilot, everything is a problem. If the small pilot adjusts the seat up to get adequate vision through the windows, he or she is moving farther from the controls and side consoles.

COMBIMAN can compute and display a 3-D reach envelope, which is the intersection of the reach envelope and any surface of the crew station. This is called the Reach Envelope Analysis and it considers body size, the type of clothing worn by the pilot (more layers and bulk reduce the mobility of the joints), and the type of harness (lap belt, shoulder locked, and shoulder unlocked). Figure 9A shows the result of a reach envelope analysis to the right side console of a helicopter for a male pilot with a 50th percentile arm length and the seat adjusted to the full down position. The reach curve superimposed on the side console defines the area that can be reached with a functional grip. By contrast, Figure 9B shows the same conditions for the seat adjusted to the full up position. The difference in the area of the side console that can be reached without leaning is very dramatic.

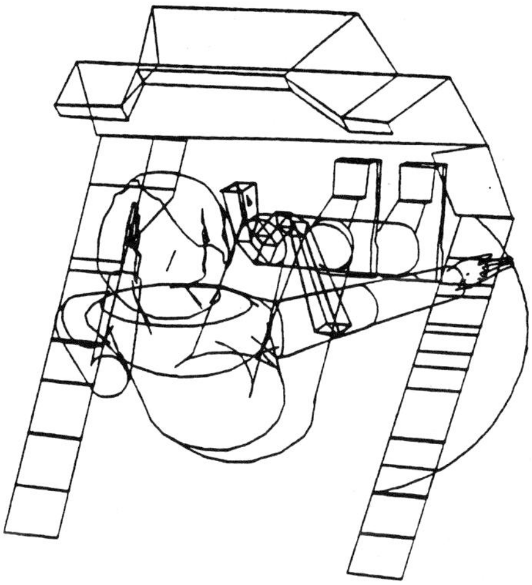

Figure 9A. Seat Adjusted Full Down.

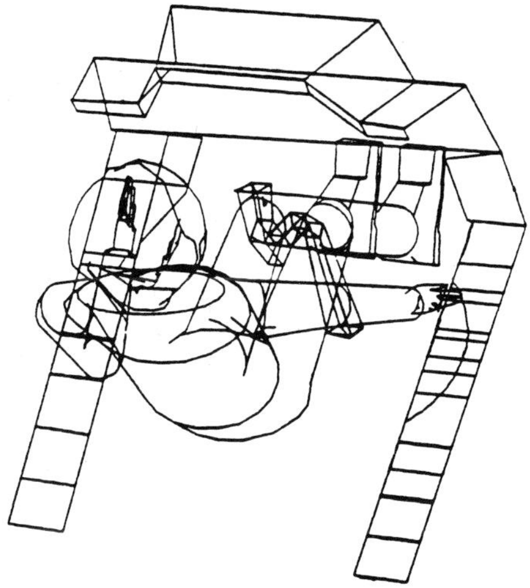

Figure 9B. Seat Adjusted Full Up.

Analyzing strength for control operation. Vehicle operators should have sufficient strength to operate all the controls. From a designer's perspective, this involves designing controls with a resistance that is acceptable to all the operators. The COMBIMAN program has models to predict the strength a seated operator can exert on eight types of controls: stick-type aileron and elevator, wheel-type aileron and elevator, general-type hand grip (any location in the forward hemisphere reach envelope), side-mounted ejection handles, center-mounted ejection handles, and rudder pedals. The controls can be operated by left, right, or both hands, and the pedals with left or right foot. The strength models are based on measurements of over 1,000 subjects. The strength analysis models provide 1st, 5th, 50th, 95th, and 99th percentile forces for the task defined.

The strength available for operating a control depends on the type of control, location of control, orientation of the control, and the direction of force applied. These control characteristics are important because different joint angles and combinations of muscles are involved in exerting forces in different directions and locations. Also, the orientation of the handgrip may produce awkward postures which can reduce the available strength.

In the case of the ejection handles, the side-mounted handles are located in line with the shoulders and the direction of force is toward the shoulders. Pilots can exert a larger force on side-mounted handles than on a center-mounted handle, which is not in line with the shoulders and requires a force at an angle to the line between the hand and shoulder.

The results of the strength analysis for the center pull ejection handle are shown in Figure 10. The analysis simulates left hand operation of a center-mounted ejection handle by a female pilot. The results show that 1st percentile female strength value is 29.48 pounds, less than the 49-pound resistance of the control. The 5th percentile force is 49.25 pounds, which is the value of the control resistance. This indicates that almost 5 percent of the women would not be able to operate the center-mounted handle with one hand.

If the center-mounted handle were analyzed with a simultaneous two-hand pull, the 1st percentile force would have been 62.35 pounds, well above the control resistance. The reason the two-hand force of 62 pounds is more than double the one-hand force of 29 pounds is because the symmetry of the two hands pulling outward balance each other, causing the sum of the two forces to be aligned with the direction of control movement.

ALL MAGNITUDES IN LBS FORCE
EJECT - CENTER

PERC	FMAG
1ST	29.48
5TH	49.25
50TH	64.00
95TH	93.75
99TH	101.60

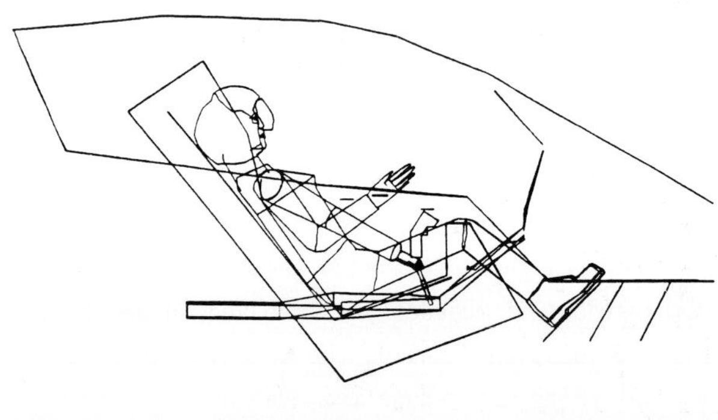

Figure 10. COMBIMAN strength analysis for the center pull ejection handle.

Analyzing carrying under a low ceiling. The CREW CHIEF illustrates the performance of this task in Figure 11, where a small female model (5th percentile) is the most limiting condition. (Remember, in CREW CHIEF, a 5th percentile female model has only stature, arm reach, and weight parameters set at 5th percentile population values; the other dimensions are regressed from these three.) If the population of workers includes both men and women, MIL-STD-1472 requires accommodation of a 5th percentile female strength, as a minimum. The box carried is medium size, approximately 10 inches high and deep and about 18 inches wide. The 30-inch passage ceiling is denoted by a horizontal line above the floor. To perform the analysis, the designer identifies the location of the work (in this case, the box to be carried), the task (CARRY), and the location of the worker from the work. The CREW CHIEF program will assemble the human model automatically, posture it, and place it in the workplace in the location specified. The available strength for those conditions is then computed and displayed as shown in Figure 11. Note the 5th percentile strength value is pounds. If the object weighs this amount or less, the design criterion is satisfied.

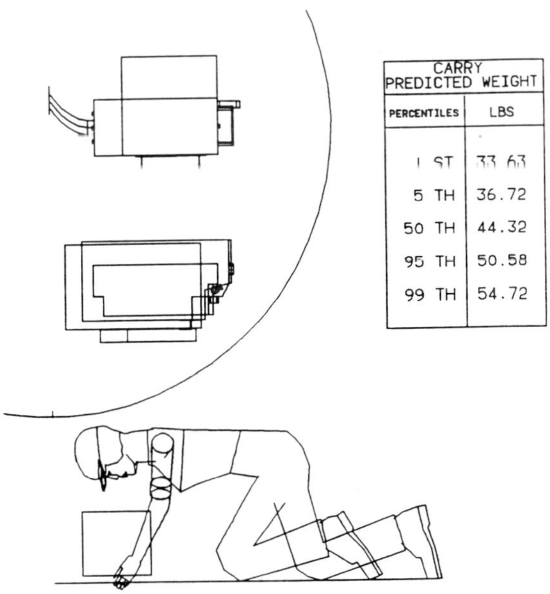

Figure 11. CREW CHIEF strength analysis for carry under a low ceiling.

Analyzing tool clearance. A maintainer selects a tool for a particular task depending on the tools efficiency in the situation. Without regard for environment, the typical efficiency (ordered from more to less efficient) for tools to remove/replace hex-head fasteners is: speedhandle, nut driver (small bolts only), ratchet, ratcheting box-end, box-end, open-end, breaker-bar, and torque. When removing several fasteners, a change of tool may be less time consuming than continuing with a less efficient tool.

With socket wrenches, extensions and universal sockets will be used whenever possible as a means of improving accessibility. Note that some torque is lost with a universal socket. The torque transmitted is equal to the cosine of the angle of flexion of the joint. If the universal joint is flexed 90 degrees, no torque is transmitted. Typically, the deflection angle should be less than 45 degrees.

CREW CHIEF can be used to verify that a particular tool can fit into a tight clearance, that there is sufficient range of movement of the tool to remove or replace a fastener, and that the maintainer wearing even arctic or chemical defense clothing has the accessibility to perform the operation. CREW CHIEF has a 222 piece tool kit, containing all the types of common hand tools, each accurately dimensioned so that precise fit can be evaluated. The software also allows users to add other tools, if desired.

Analyzing tool envelope. In cases in which a barrier or other component interferes with the movement of the tool handle, the tool envelope analysis function of CREW CHIEF automatically shows the arc of free range of movement of the tool. If, for example, there were obstructions on a box limiting the rotation of a ratchet wrench, the Tool Envelope analysis would show the limited rotation arc. If an extension rod is then added between the ratchet and the socket, so that the rotation is not obstructed, the Tool Envelope analysis shows a full 360 degrees of tool handle rotation is possible. The tool envelope function works for all of the tools in CREW CHIEF's tool box. The designer could have selected chemical defense clothing or arctic clothing for analysis if hand clearance had been limited.

WHERE TO WE GO FROM HERE

In the immediate future, there are two advances in computer technology that will influence the direction of CAE tools. The most obvious is the increased capability of low cost engineering workstations. The new graphics workstations are approaching capabilities assigned only to super computers a few years ago. This combination of declining cost and great capability are tipping the balance toward people, that is, the labor costs are relatively higher, so each designer will have a separate computer as opposed to time-sharing with a terminal on a large mainframe computer where the user's pace is limited by the machine. Most of the major CAD systems have or are developing versions of their product which operate on these workstations. There is a significant move toward standardizing on an operating system, so that exchange of graphics data among different CAD vendors may soon be possible. Desk-top, or personal computers already have the capability to effectively host CAE tools, and will do so as soon as the support software is available. In the near future, CREW CHIEF will be re-hosted on several other CAD systems to make it available to a greater number of users. A personal computer version of CREW CHIEF is also in the planning.

The second technology push will come from 3-D stereo displays which allow the user to see a 3-D image in true 3-D. These displays are viewed with special glasses which alternately occlude one eye's view of the display. At the same frequency, the display alternates the two stereo images, so that the result appears to be a single 3-D image. Very high resolution displays were needed to bring this about, but the technology is already moving out of the laboratory, and into the market.

Beyond the immediate future we can predict that the virtual display technology will be applied to CAD/CAE tools. The virtual display is a 3-D stereo display which is mounted on the head, so that each eye sees a separate display, not just an alternating image. Since each image is dedicate to a single eye, the perceived resolution and picture quality will increase. The largest improvement will result from the wide field of views achieved by having the displays located near the eye. Not only will the design appear in 3-D, but the designer will appear to be inside the design. By panning the image with head movement, the perception of presence will be complete. In such a virtual display environment, the workstation controls should also be virtual. Instead of selecting a line to be changed or moved with a cursor, you can reach out and grab it with your hand, or rather an virtual image reproduction of your hand.

CAE models must not only use new technology, but should also anticipate the design problems of the future. Planned next enhancements for the CREW

CHIEF model will include a Task Time Estimator (TTE). Already CREW CHIEF can tell the user if a maintenance task is physically possible. With the TTE, the model will be able to tell how long it will take.

Development of Predetermined Time Systems (PTS) for the factory environment began in the early 1900s, and is a mature technology. In the factory, however, jobs are designed for efficient operation, free of accessibility problems. This is not the case in aircraft maintenance, where limited accessibility is the rule rather than the exception. The Predetermined Time Systems do a very poor job of predicting mean-time-to-repair on the flightline. The increasing cost of maintenance is attracting new attention to this aspect of design. Designers need better tools for determining time to repair. The future CREW CHIEF system of programs will have a task-time estimating capability for remove/replace tasks involving very limited accessibility, as is common in aircraft maintenance.

New technology is even making inroads in body size measurement, an art that has changed little in a hundred years. Someday, we may be able to measure the human body in three dimensions. Already, new technology uses a laser scanner to gather data on head and face shapes (Beecher, 1986; Zehner, 1986). In the near term, such data may be useful for head gear design, such as helmets and protective masks. High density face data are currently being used for face contour comparisons, landmark spatial distribution analyses, and subregion delineations. The volume scanned by current laser mappers is too small to scan an entire body at one time, but work is in progress to develop a system which will.

REFERENCES

Beecher, R.M., 1986. Computer Graphics and Diagnostics. Proceedings of the Human Factors Society - 30th Annual Meeting - 1986, Human Factors Society, pp 211-215.

Bonney, M.C., Evershed, D.G., and Roberts, E.A., 1969. SAMMIE - A computer model of man and his environment. Proceedings of the 1969 Annual Scientific Meeting of the Ergonomics Research Society, Bristol, England, March, 1969.

Chaffin, D.B., 1969. A computerized biomechanical model - development of and use in studying cross body actions, J. of Biomechanics, 2:4, 429-441.

Dempster, W.T., 1955. Space requirements of the seated operator. WADC TR-55-159, Wright Air Development Center, Wright-Patterson AFB, Ohio.

DoD (Department of Defense), 1989. MIL-STD-1472D, Human Engineering Design Criteria for Military Systems, Equipment and Facilities.

DoD (Department of Defense), 1980. DOD-HDBK-743, Anthropometry of U.S. Military Personnel, Military Handbook.

DoD (Department of Defense), 1970. MIL-STD-850: Aircrew Station Vision Requirements for Military Aircraft.

Gibbons, L.E. (editor) 1990. Summary of Ergonomics Research for the CREW CHIEF model Development; Armstrong Aerospace Medical Research Laboratory Technical Report AAMRL-TR-90-038, Wright-Patterson AFB, Ohio.

Korna, M., Krauskopf, P., Haddox, D., Hardyal, S., Jones, M. Polzinetti, J., and McDaniel, J.W., 1990. User's Guide for CREW CHIEF: A Computer Graphics Simulation of an Aircraft Maintenance Technician (Version 2-CD21). Armstrong Aerospace Medical Research Laboratory Technical Report AAMRL-TR-90-014, Wright-Patterson AFB, Ohio.

Krauskopf, P., Quinn, J., Berlin, R., Stump, W., Gibbons, L., and McDaniel, J., 1989. User's Guide for the COMBIMAN Programs, Version 8, AAMRL-TR-89-024, Armstrong Aerospace Medical Research Laboratory, Wright-Patterson Air Force Base, Ohio.

Kroemer, K.H.E., 1973. COMBIMAN (Computerized Biomechanical Man-Model), Aerospace Medical Research Laboratory Technical Report AMRL-TR-72-16. Wright-Patterson AFB, Ohio.

McDaniel, J.W. and Askren, W.B., 1985. Computer-aided Design Models to Support Ergonomics, Armstrong Aerospace Medical Research Laboratory Technical Report AAMRL-TR-85-075, Wright-Patterson AFB, Ohio.

Moroney, W.F. and Smith, M.J., 1972. Empirical Reduction of Potential User Population as the Result of Imposed Multivariate Anthropometric Limits. Naval Aerospace Medical Research Laboratory, Pensacola, Florida.

NRC (National Research Council), 1988. Committee on Human Factors, Ergonomic Models of Anthropometry, Human Biomechanics, and Operator-Equipment Interfaces, Proceedings of a Workshop, National Academy Press, Washington, D.C.

Zehner, G.F., 1986. Three Dimensional Summarization of Face Shape. Proceedings of the Human Factors Society - 30th Annual Meeting, Human Factors Society, pp 206-210.

BIBLIOGRAPHY

Ayoub, M.M., Smith, J.L., McDaniel, J.W., and Selan, J.L., 1985. Job Demands in Some Maintenance Operations, Proceedings of the 9th Congress of the International Ergonomics Association, Bournemouth, England, 2-6 September.

Ayoub, M.M., Denardo, J.D., Smith, J.L., Bethea, N.J., Lambert, B.K., Alley, L.R., & Duran, B.S., 1982. Establishing Physical Criteria for Assigning Personnel to Air Force Jobs. Institute for Ergonomic Research, Texas Tech University, Lubbock, Texas.

Ayoub, M.M., Mital, A., Asfour, S., and Bethea, N., 1980. Review, evaluation and comparison of models for predicting lifting capacity. Human Factors, 22(3), pages 257-270.

Badler, N.I., O'Rourke, J., and Kaufman, B., 1980. Special problems in human movement simulation. Computer Graphic, Proceedings, Special Interest Group on Computer Graphics, Association for Computing Machinery (SIGGRAPH 80), Vol. 14, pages 189-197.

Bapu, P., Korna, M. and McDaniel, J., 1983. User's Guide for COMBIMAN Programs, Version 6, AFAMRL-TR-83-097, Air Force Aerospace Medical Research Laboratory, Wright-Patterson Air Force Base, Ohio.

Bates, F.J., Evans, S.M., Krause, H.E. and Luming, H., 1974. Three Dimensional Display of the COMBIMAN Man-Model and Workspace, AMRL-TR-74-15, Air Force Aerospace Medical Research Laboratory, Wright-Patterson Air Force Base, Ohio.

Bittner, A.C., 1976. Computerized accommodated percentage evaluation (CAPE): Review and prospectus. Proceedings of the 6th Congress of the International Ergonomics Association. University of Maryland, College Park Maryland.

Bonney, M.C. and Case, K., 1976. The development of SAMMIE for computer aided work place and work task design. Proceeding of the 6th Congress of the International Ergonomics Association. University of Maryland, College Park Maryland.

Chaffin, D.B., Herrin, G.D., Keyserling, W.M., and Garg, A., 1977. A method for evaluating the biomechanical stresses from manual materials handling jobs. American Industrial Hygiene Association Journal, 38:662-675.

Evans, S.M., 1978. Updated User's Guide for the COMBIMAN Programs, AMRL-TR-78-31, Air Force Aerospace Medical Research Laboratory, Wright-Patterson Air Force Base, Ohio.

Fortin, C., Gilbert, R., Beuter, A., Laurent, F., Schiettekatte, J., Carrier, R., and DeChamplain, B., 1988. SAFEWORK: Micro Computer-Aided Workstation Design and Analysis, New Advances, Directions, and Future. Genicom Consultants, Inc., Montreal, Quebec.

Harris, R.M. and Iavecchia, H.T., 1984. Crewstation Assessment of Reach Revision IV (CAR IV) User's Guide. Analytics, Inc. Technical Report TR1800.10A, Willow Grove, Pa.

Leetch, B.D. and Bowman, W.L., 1983. Articulated Total Body (ATB) "View" Program Software Report, Part II, User's Guide, Volume 2. AFAMRL-TR-81-111. Air Force Aerospace Medical Research Laboratory, Wright-Patterson AFB, Ohio.

Karwowski, W., Genaidy, A.M., and Asfour, S.S. (editors) 1990. Computer-Aided Ergonomics: A Researcher's Guide. Taylor & Francis, London.

Robinson, J.C., Robinette, K.M., and Zehner, G.F., 1988. User's Guide to Accessing the Anthropometric Data Base at the Center for Anthropometric Research Data (CARD). Armstrong Aerospace Medical Research Laboratory Technical Report AAMRL-TR-88-012, Wright-Patterson AFB, Ohio.

Woolford, B.J. and Lewis, J.L., 1981. Applications of digital image acquisition in anthropometry. Proceedings of the SPIE Technical Symposium East, Volume 283, 3-D Machine Perception, Washington, D.C.

Workspace, Equipment and Tool Design
edited by A. Mital & W. Karwowski

INDUSTRIAL WORKPLACE DESIGN

M.M. Ayoub[1] and Marta Miller[2]
[1]Department of Industrial Engineering, Texas Tech University, Lubbock, Texas 79409 USA
[2]Department of Psychology, Texas Tech University, Lubbock, Texas 79409 USA

ABSTRACT

A workplace is the place which contains the space, tools, and equipment where workers perform the functions and tasks of their jobs. The area of workplace design is a rich field with many facets. Historically, the study of workplace design has developed out of many different disciplines, including industrial engineering, human factors, and industrial psychology. Generally, a good workplace design is one where a worker can easily change positions to relieve the stresses and pressures of a posture while still performing job tasks efficiently and easily. Specific considerations for the design of the workplace include the type of workplace (sitting vs standing), work envelopes, and postural concerns are discussed. Finally, guidelines for the evaluation of existing and proposed workplaces are presented.

THE HISTORICAL PERSPECTIVES

The study of the workplace design emerged out of the industrial revolution and its accompanying social changes which brought about a concern for worker welfare. One of the earliest indications of consideration for the condition of worker well-being came in the early 1700's from a man named Bernardino Ramazzini. The concern arose out of an interest in the health of workers who often suffered detrimental effects from work stress caused by physical labor (in Tichauer, 1978). Ramazzini identified two different types of hazards to workers. The first was the harmful characteristics of the materials used and the second was "certain violent and irregular motions and unnatural postures of the body" (in Anderson, 1988). Both of these hazards are still valid concerns in industrial settings today, and the second hazard is directly related to the focus of this chapter.

The early concerns, however, seem to be more of a concern for the worker without any of the focus on the actual design of the workplace as a solution to worker problems. Not until almost 200 years later did techniques, the time and motion studies, begin to develop which laid the ground work for scientific study of the workplace. Consequently, the history of workplace design is tied up in the history of motion and time study. The time studies of the turn of the century, which looked at the economy of motion, proved important to the area of workplace design because the techniques relevant to the design of the workplace began to be developed (Chaffin, 1987).

Two of the early founders of the motion and time studies where Frederick W. Taylor and the Gilbreths, Frank and Lillian. The work of Taylor, which began in the 1880's, was the foundation of modern time analysis. Taylor, without contest, is given the title of the "Father of modern time study." Taylor's work included studies done in several industrial settings, the development of efficient work schedules for workers, and shop management techniques to name a few (Niebel, 1988). In one of his more remembered efforts, Taylor studied the ability of workers to shovel various materials in an industrial yard. From his study, he was able to devise methods for 140 men to complete the work that was previously only accomplished by 400 or more men (Barnes, 1980).

A more important contribution than the actual findings of his studies are the time analysis techniques (Barnes, 1980). Taylor's methods utilized a stop watch to create analyzable segments of work for the systematic and scientific study of industrial work. It is for these pioneering efforts that Taylor should most be remembered.

In the early 1900's, the Gilbreths, a husband and wife team, carried out more of the early studies using techniques which lent themselves to the study of work stress (Barnes, 1989; Chaffin, 1987). The Gilbreths had an interesting perspective on the workers because the husband was an engineer and the wife a psychologist. The combination gave the couple unique insight into the worker psyche as well as the masses, weights, and forces of the relationship between worker, tools, and materials (Barnes, 1980).

Frank Gilbreth quite often employed photographs in his studies, but it was the use of the motion picture camera that is one of the Gilbreths most significant contributions to the study of worker performance. Their technique became known as "micromotion analysis," one of their lasting contributions to the study of the workplace. As Barnes (1980, p. 18) describes it, "micromotion study is the study of the fundamental element or subdivisions of an operation by means of a motion picture camera and a timing device which accurately indicates the time intervals on the motion picture film."

In addition to micromotion analysis, the Gilbreths developed the cyclegraphic and chronocyclegraphic analysis techniques. These two techniques were employed to study the motion path of the worker. Both techniques involved filming the movements of a worker. In the cyclegraphic technique, a small light bulb was attached to a moving part of the workers body. In the chronocyclegraphic technique, the current to the light was systematically interrupted so the light would blink. Both techniques recorded the motion paths which could later be analyzed for efficiency (Niebel, 1988).

It was the Gilbreths who gained the acceptance of industry for the detailed study of motion of the body of workers as a tool to develop methods to reduce worker stress and increase productivity (Niebel, 1988). Their success does not stop there either. The Gilbreths' techniques and principals have proved enduring. They are still being adopted by industry today (Barnes, 1980).

These early pioneers, along with others not mentioned here, helped bring about the conditions for the scientific design of the workplace. Nevertheless, it can be seen that the emphasis in development was not yet on workplace design but on the broader techniques which lend themselves to the design of workplace. It was not until World War I that there was any real effort or interest in designing the workplace for the advantage of the workers (Tichauer, 1978).

Once the interest began in workplace design, the idea was picked up by many different disciplines. The disciplines from which workplace design can claim its roots are diverse and plentiful. Among the major disciplines which are responsible for many of the workplace design concepts are: 1) industrial engineering, 2) ergonomics or human factors engineering, 3) industrial hygiene and safety engineering, 4) industrial relations, 5) behavioral science, and 6) industrial psychology (Das & Grandy, 1983). The diverse nature of the roots of workplace design is a good indication that the area is a rich and multi-faceted discipline. This is indeed the case. The field of workplace design has a wealth of knowledge for those who wish to explore the area.

WHAT DEFINES A WORKPLACE?

Before addressing any concerns in the area of workplace design, it is important to know exactly what a workplace is. The workplace has been defined as "a location where a person or persons perform tasks for a relatively long period of time" (Cushman, Nielsen, & Pugsley, 1983, p. 14). Others have defined the workplace as "that small area of an industrial plant which workers are generally confined when on the job" (Das & Grandy, 1983, p. 105). Such definitions of the workplace are somewhat vague and leave what a workplace may actually encompass wide open. For our purposes, the workplace is the place which contains the space, tools and equipment where workers perform the functions and tasks of their jobs.

Similar terms which are often confused or used interchangeably with workplace include work station and work environment. A work station has been defined as "one of a series of workplaces that may be occupied or used by the same person sequentially when performing his or her job" (Cushman, et al., 1983). Several work stations are often encompassed by one workplace. Work stations are subparts of the work area.

Work environment (sometimes called work area) on the other hand is a more expansive term which includes the workplace, but also such things as climate, lighting, and work clothing (Das & Grandy, 1983). Work stations and work areas are ultimately important to worker comfort and performance, but these are not the focus of this chapter.

FINDING THE OPTIMUM WORKPLACE LAYOUT

The optimum industrial workplace would be the quest of this chapter if it were a journey. Such a workplace has several characteristics important to the layout of the workplace. Das and Grandy (1983) listed several necessary features for the optimum workplace. The optimum industrial workplace must include: 1) adequate postural support, 2) proper body and limb weight distribution, 3) natural limb positions, and 4) positions should require little demand for maximum reach or force in the performance of job. These are the ultimate goals a workplace design hopes to attain.

The concept of work tolerance discussed by Tichauer (1978) is another useful view of the optimum workplace from a slightly different angle. "Work tolerance is a state in which the individual worker performs at economically acceptable rates, while enjoying high levels of emotional and physiological well-being" (Tichauer, 1978, p. 32). Tichauer lists "important prerequisites" to biomechanical work tolerance which compliment the features mentioned previously by Das and Grandy (see Table 1).

GENERAL FACTORS/COMPONENTS IMPORTANT TO THE DESIGN OF WORKPLACES

Some overall, general considerations important to workplace design must be addressed before looking at specific concerns in the workplace. In 1934, Maynard proposed two general principles of industrial workplace layout that were historically important and are still valid today (in Das & Grandy, 1983). His first concept says that all motions of the worker should be reduced to the lowest class. The second concept is in reference to normal and maximum work areas in the horizontal and vertical planes. The main thrust of the two concepts is to minimize the effort the worker must exert to perform any given task. While Maynard's principles are a bit general for present standards, his concepts provided a foundation that others have built on.

Principles presented by Cushman, et al. in 1983 provide a good starting place for the development of an adequate workplace design. According to Cushman, et al. (p.14), the following

TABLE 1

Prerequisites of Biomechanical Work Tolerance
(from Tichauer, 1978, p. 33)

Postural

1. Keep elbows down.
2. Minimize moments on spine.
3. Consider sex differences.
4. Optimize skeletal configuration.
5. Avoid head movement.

Engineering

1. Avoid compression ischemia.
2. Avoid critical vibrations.
3. Individualize chair design.
4. Avoid stress concentration.
5. Keep wrist straight.

Kinesiological

1. Keep forward reaches short.
2. Avoid muscular insufficiency.
3. Avoid straight-line motions.
4. Consider working gloves.
5. Avoid antagonist fatigue.

---

principles should be utilized in workplace design: 1) services needed by several workers should be located in a central place, 2) after assessing communication needs, workers and workplaces should be located to maximize communication, 3) operator lines of sight need to be kept clear, 4) distracting environmental factors, such as noise, odor, and heat, should be minimized, 5) layout should be such that the production moves in one direction with a minimum of rehandling, and 6) workers must be allowed postural flexibility and the ability to change positions.

The last principle is a theme which will reoccur. Workplace design should be such that, when the design is complete, the worker can easily change positions through out the course of the workday to relieve the pressures and stresses of a posture. The ability to change postures will reduce fatigue and stress that accumulate when one position is maintained too long.

Two final concerns for workplace design are 1) what percent of the population will fit in the completed workplace, and 2) economics. First, fitting the majority of the population of workers with a single design is important in the design of the workplace. Obviously, not all workers are the same size. Some workers are smaller and some are larger. Even those workers of average size are seldom average in all measurements. A woman of average height may have longer than average legs and a shorter than average torso. Designing for a wide range of workers may be difficult but it is necessary to a functional workplace design. At the outset of the design, the percent of the worker population, usually 80 or 90 percent, should be selected to fit in the completed workplace.

Second, while it would be nice if it did not play a role in the design of the workplace, economics cannot be ignored. Quite often the restrictions of finances is the determining factor in the design. A good design will then have a small cost for building, little maintenance costs and, finally, a long life expectancy for the design.

SPECIFIC CONSIDERATIONS FOR THE DESIGN OF THE WORKPLACE

Sitting vs standing

It is generally agreed workplaces can be divided into three different types. There is the sitting workplace where most of the work is carried on while the worker is in a seated position. In the standing workplace the operator performs the majority of the work from a standing position. Finally, there is the sit/stand workplace where the worker varies regularly from the sitting to the standing positions. There are advantages and disadvantages to each type of workplace.

Worker characteristics and the task characteristics play the major role in determining which type of workplace is used. Operator-related dimensional factors that must be taken into consideration are: 1) postural control and distribution of body weight, 2) reach envelopes, and 3) eye position with regard to display areas (Hertzberg, 1972; in Das & Grandy, 1983). Task characteristics include size and weight of materials, forces generated by the worker, reaches employed to perform the task, and the frequency and duration of the work cycle.

The seated workplace

There are situations where sitting is the best option for the workplace. According to Cushman, et al. (1983, p.15), these situations are when:

1) all items needed in the short-term task cycle can be easily supplied and handled within the seated work space;

2) the items being handled do not require the hands to work at an average level of more than 15 cm (6 in) above the work surface;

3) no large forces are required such as handling weights greater than 4.5 kg (10 lbs) (adapted from Rehnlund, 1973). These large forces may be eliminated by using mechanical assists; and

4) fine assembly or writing tasks are done for a majority of the shift.

Additionally, a fifth condition should be added to Cushman's list. The seated workplace is desirable when the worker must operate foot controls. The standing worker is too susceptible to injury from falling due to a loss of balance when switching from a two-footed stance to a one-foot balance in order to operate the foot control with the other foot. Loss of balance in an industrial setting is all too likely to result in the operator falling into moving parts of a machine and receiving serious injuries. When these conditions exist, it is desireable to place workers in seated workplaces.

There are several advantages of sitting over standing in the workplace. Fatigue is reduced because less muscular activity is needed to maintain the posture. Sitting is better for doing tasks requiring fine motor activity because of the stability the seated posture provides. Sitting also reduces the intervascular pressure in the lower legs and feet, which is especially desirable for female workers. Finally, foot controls can be better utilized and still maintain good working posture (Ayoub, 1973).

While the seated workplace is the most desireable type of workplace, a good design is not easily developed without taking into consideration many factors. To develop an adequate seated workplace, it is necessary to explore the design of the chair, the work surface, and their inter-relationship.

First, let's examine the seat. Chairs are extremely important to the industrial worker. A chair developed with a sound design can add as much as an additional 40 minutes of work to each day (Tichauer, 1973). In selecting a chair for the workplace the seat pan, the back rest, and the seat height with any subsequent foot rests should be examined.

(i) The seat pan:

There are several shapes the seat pan design can take. Some of the major shapes are saddle, waterfall, flat, crown, and concave (see Figure 1; Ayoub, 1971). There is disagreement about which design is best for which application.

It has been recommended that the flat seat pan is good in industrial settings due to the fact change of position is best achieved from this seat pan (Ayoub, et al., 1987). Remembering that the ability for the worker to change position is highly desirable quality in the workplace.

Once the seat pan shape is chosen, the angle of the seat pan is the next important consideration. The seat pan of a chair with a backrest should be tiltable (Bendix, 1986). The flat seat pan chair with a backrest will eject the worker from his or her seat when the worker leans back if the seat pan is level. For this reason, it is desirable to angle the seat pan slightly backwards if the chair is to have a backrest.

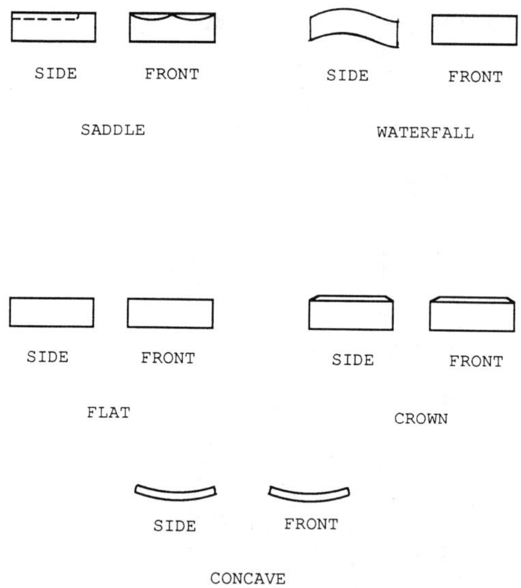

FIGURE 1. SHAPES OF SEAT PANS (AYOUB, 1971)

The actual slope of the seat pan should have is subject to some debate as well. Most theories suggest the angle should be between three and five degrees backwards (Ayoub, et al., 1987). A backwards tilt of this degree will be enough to keep the worker in his or her seat without leaning so far as to interfere with their work.

However, Mandal's theory (1976), one of the more popular opposing views, suggests that the angle of the seat pan should be from 10 to 15 degrees forward to maintain spinal curvature (see Figure 2). Mandal's ideas may have possibilities for application in the industrial work environment, but up to this point has mostly been applied to the seats of school children.

It is also desirable to have the seat pan upholstered. The upholstery should be such that it helps the worker avoid slipping from the seat (Bendix, 1986). Upholstered seats have the added advantage of reducing the pressure on the buttocks. The compressive forces exerted on some areas of the buttocks can be quite high. It has been estimated that the pressure under the ischial tuberosities is between 85 and 100 psi (Tichauer, 1978).

While some upholstery is good, it should not be too elaborate. Upholstery should be of "breathable" material to ensure air circulation and reduce worker discomfort created by sweating. Furthermore, upholstery that is too soft will reduce the ability of the worker to change position and should be avoided. The cushion should not give way more than one inch or

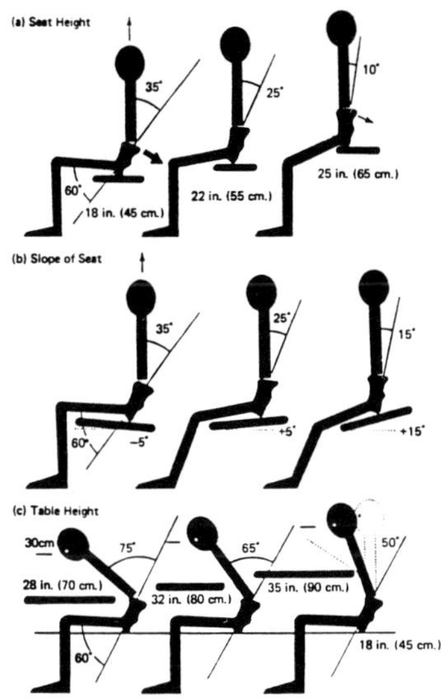

FIGURE 2. EFFECTS OF SEAT HEIGHT, SLOPE OF SEAT, AND TABLE HEIGHT ON THE LUMBAR FLEXION OF THE LUMBAR (LOWER) SECTION OF THE SPINE (From Mandal, 1982).

it will create a situation where it is difficult for the worker to shift positions. The cushion should be stiff and flat for best results (Ayoub, et al., 1987).

The seat pan width and depth are another set of important characteristics. The width is the dimension from side to side. The depth is the dimension from front to back. The width should be designed to fit the 90th or 95th percentile person of the worker population (Ayoub, et al., 1987) and to fit any workbench or spatial constraints in the workplace. Tichauer (1978) recommends that the breadth of the seat pan be 25% wider than the width of the buttocks of the user.

Ultimately, depth is more important than width. The depth needs to accommodate the length of the thighs, a difficult task considering the amount of human variation on this dimension. It is important that there is enough room in the popliteal areas. Excessive pressure to the back of the knees can lead to discomfort because of the number of nerves and blood vessel in the popliteal area. There should be 10 cm (4 in) clearance between the back of the knees and the front of the seat pan (Ayoub, et al., 1987). The front of the seat pan should also be rounded to give more comfort in the popliteal area.

The depth is most easily made adjustable if there is a backrest present. With a backrest that can be moved forward or backwards and a good seat pan, the depth can be either reduced or increased. The adjustability of this dimension will be a factor in accommodating the majority of the user population. (See Figure 2 for adjustable dimensions in chair design specifications; Anderson, 1988).

(ii) The backrest:

The size and placement of the backrest are most critical to good design. The purpose of the backrest is lumbar support which will reduce stress and fatigue. The size of the backrest and the height of the backrest from the seat pan are the major characteristics of importance in the backrest.

The ideal size of the backrest is dependent on activity the task requires to perform (Sanders & McCormick, 1987). The seats of workers performing tasks requiring more activity and a greater range of motion should have a smaller backrest. A small backrest will not interfere with the movements of the person working but still provide lumbar support. Due to the types of tasks being done in the industrial workplace, it is most likely the small backrest will be called for. On the other hand, backrests for chairs assigned to workers who have less of a need for motion in the performance of their tasks can be larger. Larger backrests will provide more stability to tasks requiring lots of writing and similar fine activities.

Finding a good height for the backrest is a bit more complex. The backrest should be positioned so that it provides good lumbar support without interfering with the worker's ribs or pelvis. In industrial settings, the backrest should be placed so that it is between the iliac crest and the 12th thoracic vertebra (Ayoub, et al., 1987). The range of human variation on this dimension will dictate up and down adjustability of the backrest to accommodate the majority of workers. The backrest should have a lumbar pad (be upholstered) and swivel about a horizontal axis for better fit to the curvature of the spine which also varies greatly from worker to worker. (See Figure 3).

Furthermore, to maintain spinal curvature, the backrest should create a small, forward force exerted on the lumbar region. Bendix (1986, p. 361) suggests "the backrest should be positioned against the back while the ischial tuberosities are placed about four to six centimeters behind the axis for tilting the seat."

(iii) The seat height:

Seat height needs to be such that pressure is taken off the thighs. Excessive pressure on the thighs will reduce blood flow to the lower legs and feet increasing fatigue. Ideally, the seat height should be such that the workers lower legs are vertical to the floor, the feet can be placed flat on the floor, and the thighs are horizontal. To relieve pressure on the back

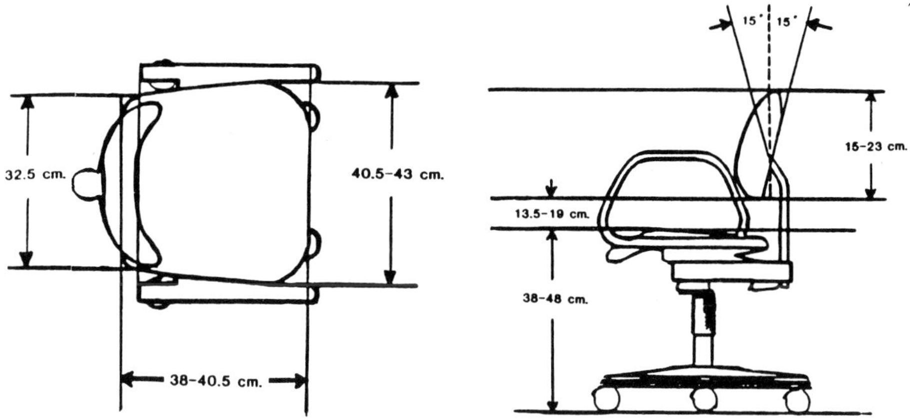

FIGURE 3. EXAMPLE OF AN ADJUSTABLE CHAIR WITH DESIGN SPECIFICATIONS THAT PROVIDES BOTH ARM AND LUMBAR SUPPORT. (From Anderson, 1988)

of the thighs, it is most often suggested that the seat height should be slightly less than the popliteal height. Most recommendations suggest seat height be anywhere from 1 cm to 3-5 cm below popliteal height (Ayoub, et al., 1987; Bendix, 1986).

If for some reason the seat height must be such that the feet cannot be flat on the floor, a foot rest should be designed into the chair or on the workbench where the chair will be used, preferably the workbench. The location of the foot rest should be positioned such that the relation to the seat maintains the relief from pressure on the thighs.

There are several important characteristics to a footrest. A footrest should provide the ability for the worker to change position of the foot and legs. For this reason, small footrests should be avoided. A bar footrest is especially undesirable because of the constant effort to maintain the foot on it. A good footrest should be wide and concave to accommodate the normal movements of the foot about the knee (Grandjean, 1988; Ayoub, et al., 1987). A good footrest should adjust up and down, forwards and backwards, and tilt to provide the best fit to most workers. (See Figure 4 - Work space design specifications).

<u>(iv) The work surface height</u>:

The height of the workbench is often mistakenly considered to be the work surface height. The work surface height may or may not be the actual workbench height. If the materials the worker is using are small, the workbench height is the work surface height. However, if the materials the worker is using are of any significant dimensions, the work surface height becomes higher than the workbench. The worksurface height, in this case, is the level the hands are actually required to work at with the material on the bench. (See Figure 5).

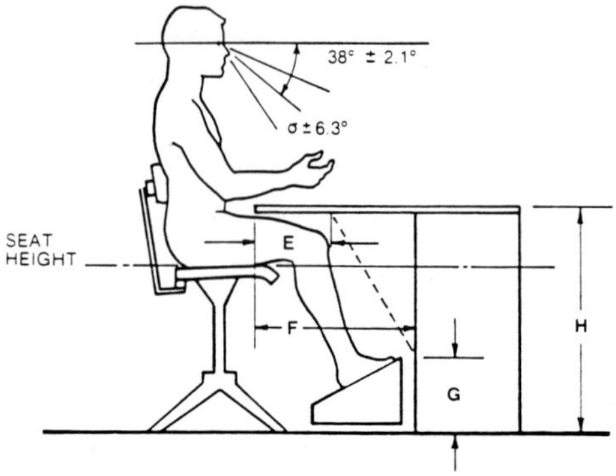

LEG ROOM:	1) DEPTH (E)	43 - 51 CM
	2) DEPTH (F)	51 - 66 CM
	2) FOOTROOM (G)	3 - 25 CM
WORK SURFACE HEIGHT:	3) (H)	64 - 81 CM

FIGURE 4. WORK SPACE DESIGN SPECIFICATIONS FOR SEATED WORKER (Adapted from Ayoub, 1971)

There is more than one approach to designing the optimum worksurface height. Konz (1979) lists three different techniques to achieve the proper worksurface height:

1) by making more than one different worksurface height available to workers;

2) by the use of platforms or adjustable chairs to elevate the worker's elbow height to the desired level; and

3) by making the workbench itself adjustable. The second option, adjusting the worker by the use of platforms or chairs, is probably the most prevalent technique due to economic factors.

Work surface height, whether it is at workbench level or higher due to the materials being used, should be four to six centimeters above the elbow height (Bendix, 1986). Ayoub (1973) gives recommendations for seated worksurface heights broken down for four types of tasks. (See Table 2).

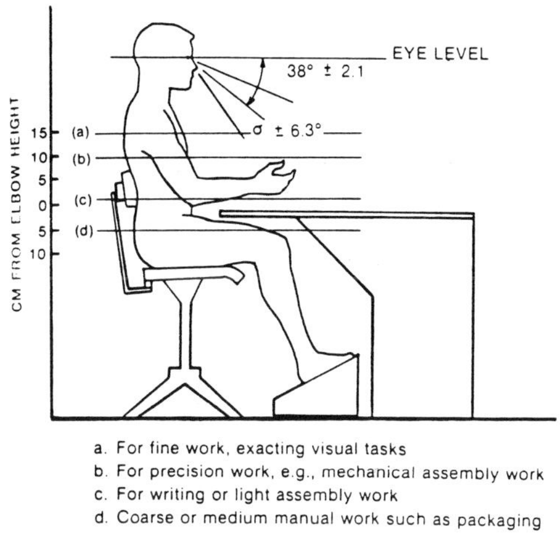

FIGURE 5. DESIGN SPECIFICATIONS FOR CORRECTED HEIGHT OF WORK SURFACE FOR SEATED OPERATOR (Adapted from Ayoub, 1971)

TABLE 2

Recommendations for seated work surface heights (Ayoub, 1973)

Type of task	Males	Females
Fine Work	39.0 - 41.5	35.0 - 37.5
Precision Work	35.0 - 37.0	32.5 - 34.5
Writing or Light Assembly	29.0 - 31.0	27.5 - 29.5
Course or Medium Work	27.0 - 28.5	26.0 - 27.5

heights in inches

The weight of the piece being worked with is an important factor in determining the worksurface height. "The more precise the work, the higher the work surface. The heavier the work, the lower the work surface" (Anderson, 1988, p. 93). The reason is fairly obvious. Fine work almost always requires the work be close in the visual field. On the other hand, it is preferable to lift heavy work as small a vertical distance as possible.

If the work surface height is above the workbench it may become necessary to either use a taller chair or put the chair on some type of risers to get the correct relationship to the work surface height. An alternative may be to lower the workbench. However, it is important that knee clearance for the seated worker is maintained.

Work surfaces do not always have to be level either. Work surfaces can be horizontal or tilted. Bendix (1986) suggests that the highest priority should be given to the slanting work surface. The incline should be 35 to 45 degrees and adjustable. The slanting work surface is desireable for situations where large reaches are required or fine work is done. The slant will reduce the length of the reach with the higher end bringing closer the further end of the materials being used. The advantage for the worker doing fine work is the increased ability to see.

Finally, the hard and/or sharp nature of most workbench edges must be taken into account. These characteristics of workbenches will cause difficulties for workers because such edges often restrict blood flow and apply pressure on nerves and tendons. For this reason Anderson (1988, p. 96) suggests, "all edges of the work surface should be well rounded and padded where an elbow or forearm might be rested."

Some of the relationship between the chair and the worksurface has been mentioned. For the worker to be at the proper work surface height, the seat may have to be raised in some manner. If there is a foot rest or foot control attached to the work bench, the relationship must be such that the worker can comfortably reach the footrest or foot controls while still being at the proper work surface height. Adjustability among the different components of the seated workplace is essential.

<u>The standing workplace</u>

There are conditions which make the use of standing workplaces desireable. Standing workplaces are best:

1) if the workplace or work station does not have knee clearance for a seated operation;

2) objects weighing more than 4.5 kg (10 lbs) are handled;

3) high, low, or extended reaches are frequently required;

4) operations are physically separated and require frequent movement between work stations; and

5) downward forces must be exerted (Cushman, et al., 1983).

Ayoub (1973) mentions another condition that can be added to this list:

6) machine controls are such that the operator must stand to able to reach them. For safety reasons, standing workplaces should be avoided if the worker is to operate foot controls.

The important feature in the standing workplace is the worksurface height. Again, the worksurface height and the workbench height may not be synonymous. In fact, with the downward forces and larger materials being used, the two are less likely to be the same. The workbench is likely to be lower so the work surface height is still at the optimum elbow height and body weight can be utilized in the application of downward force. Care should be taken to avoid placing the work bench too high. A workbench that is too high leads to an increased energy expenditure on the part of the worker (Chaffin, 1987). (See Figure 6; Tichauer, 1973).

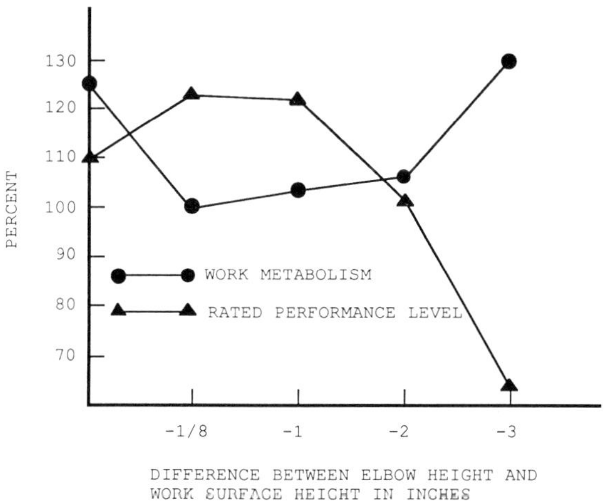

FIGURE 6. EFFECT OF CHANGE IN WORK SURFACE HEIGHT ON PERFORMANCE (TICHAUER, 1973)

The work height that is desireable varies with the task the standing worker is performing and can be broken down into three categories (Cushman, et al., 1983). For tasks involving light assembly, writing, packing, or use of small materials, the preferred working height for the hands is 107 cm (42 in). For slightly heavier tasks requiring large downward or sideways forces, the optimum work height for the hands is 91 cm (36 in). When very heavy downwards forces are required, it may be desireable to lower the work height even further to 76 cm (30 in). The best work height for the hands is 81 cm (32 in) for tasks requiring large upwards forces. Ayoub (1973) alternately lists recommendations for worksurface height for the standing worker broken down into three types of tasks. (See Table 3; also see Figure 7).

Alternately, the worker may be raised on platforms to reach the proper working height. Such a situation has the disadvantage of creating a condition where the worker might fall from the platform and is better to be avoided if possible. Flexibility in the workplace is key to a good design.

TABLE 3

Recommendations for standing work surface heights (from Ayoub, 1973)

Type of task	Males	Females
Precision Work	43.0 - 47.0	40.5 - 44.5
Light Assembly Work	39.0 - 43.0	34.5 - 38.5
Heavy Work	33.5 - 39.5	31.0 - 37.0

heights in inches

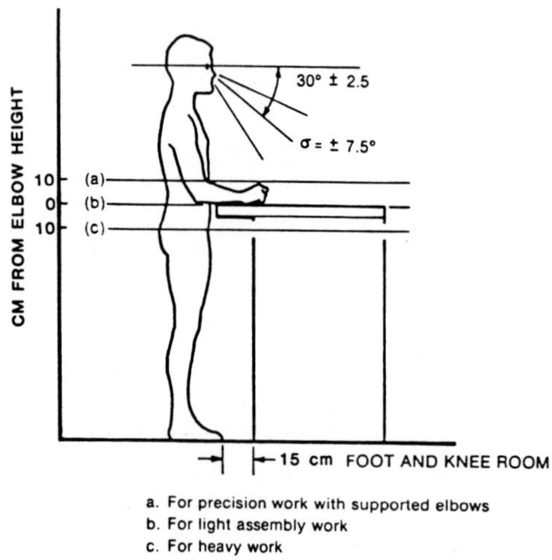

FIGURE 7. DESIGN SPECIFICATIONS FOR CORRECTED HIGHT OF WORK SURFACE FOR STANDING OPERATOR (Adapted from Ayoub, 1971)

Lastly it should be noted standing is more physically taxing than sitting. Additional supports should be given the standing worker. Special soled shoes and/or floor mats should be in the workplace of the standing worker. Carefully designed rest periods should be instituted as well depending on the type of work performed.

<u>The sit/stand workplace</u>

The sit/stand workplace is a unique combination of the sitting and standing workplace as the name suggests. Most often the sit/stand workplace takes the higher workbench from the standing workplace and a higher adjustable chair that is specially designed to keep the worker in an almost standing position but at the same time provide some of the supports a chair gives the worker. (See Figure 8; Grandjean, 1988). The sit/stand workplace is often more economical since provisions for both workplaces separately may not be feasible owing to the space constraints.

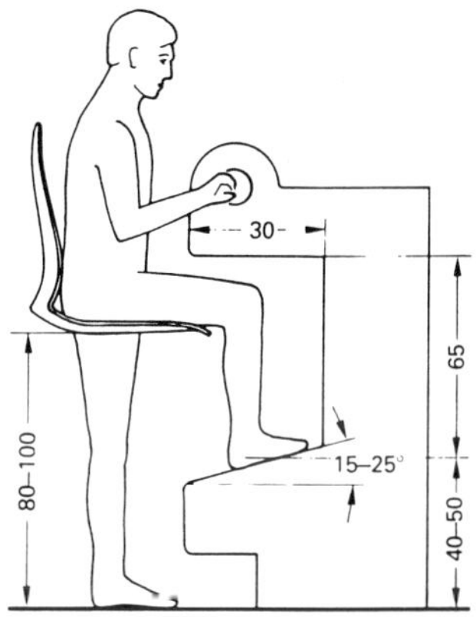

a. Knee room	[illegible]
b. Height of working field above seat	30-60cm
c. Height of working field above floor	100-120cm
d. Range of adjustment of seat	80-100cm

FIGURE 8. MACHINE THAT PERMITS ALTERNATE SITTING AND STANDING (FROM GRANDJEAN, 1988)

Cushman, et al. (1983) list the conditions in which the flexibility of the sit/stand workplace is the most desireable. These situations include:

1) when repetitive operations are done with frequent reaches more than 41 cm (16 in) forward and/or more than 15 cm (6 in) above the work surface. Operations that would be done in the sitting workplace if it were not for the reach requirements; and

2) where multiple tasks performed, some best done sitting and others best done standing.

However, it is important to note the sit/stand workplace should not be used if the task cycle being done requires the worker to stand up and sit down frequently. Such a situation will lead to increased energy expenditure on the part of the worker and should be abandoned in favor of a standing workplace.

For the sit/stand workplace careful attention should be paid the different characteristics of the tasks likely to be performed in the different positions (See Table 4).

One last detail, not to be overlooked in the sit/stand workplace, is the appropriate amount of knee clearance. In the sit/stand workplace there should be 51 cm (20 in) of forward leg clearance (Cushman, et al., 1983). Again, a good design hinges on adjustability of the different components of the workplace.

<u>Adjustability of the workplace</u>

The key to workplace design is adjustability of the design. Important components need to be available in a range of values to fit the majority workers most comfortably. The two characteristics of adjustability which should concern us the most are: 1) the range of adjustability and 2) the ease of adjustability.

TABLE 4

Conditions of the sit/stand workplace
(from Cushman, et al., 1983, p. 26)

A standing work surface for tasks that can be done while seated should be 102 cm above the floor. A seated task at a height of less than 102 cm will cause stooping and excessive static loading on the back and shoulder muscles for taller operators if the task is done from a standing posture.

Operations that consist of several different tasks requiring different working heights (sitting and standing) can be done at a workplace 91 cm above the floor with an adjustable footrest. Workplace height for standing tasks can be raised by using 10-30 cm platforms on the top of the work surface. At least 51 cm of forward leg clearance is required at a sit/stand workplace.

Work surfaces using large-size products or drawings are often 112 cm high. The work done on these tables lends itself to a sit/stand operation; provision of adequate leg clearance considerable reduces static loading on the legs and back of people as they lean over large work.

The range of adjustability should be such that a certain percentage of the population, usually 80 or 90 percent, will be accommodated within a certain workplace design. For any given dimension, such as popliteal height, the range of adjustment should be from the smallest worker of the population to the largest worker of the population being designed for. As it has been put by Konz (1983), "Let the smallest woman reach. Let the largest man fit."

The ease of adjustability is equally important as the range of adjustability. A workplace that is not easily adjustable is not adjustable. If the worker cannot easily change a seat height, then he or she will not change the seat height and the advantage of adjustability is lost.

The advantages of adjustability are fairly straight forward. Adjustability allows for individual variation and preference in majority of the worker population. It avoids a situation where only workers of certain size must be hired for a task or having to build individualized workplaces.

WORK ENVELOPES

Work envelopes are important when designing a workplace. The work envelope is the area which the worker is expected to work. It is the reach limits of the worker within this envelope that should dictate the ultimate limits of the workplace layout. A workplace larger than the area in which a worker can reach is a waste of space. Therefore, workplace reaches should be designed based on the smaller worker's reach limits.

The size of the envelope will be determined by several characteristics of the worker and the type of workplace. Recommendations usually suggest that the clearance dimensions of the work envelope be based on the largest of the workerpopulation (the 95th percentile person) and the reach limits be determined by the smallest of the worker population (the 5th percentile person) (Woodson & Conover, 1964; Konz, 1983; Sanders & McCormick, 1988).

The dimensions of the work envelope will also change with the type of workplace. For example, Van Cott and Kinkade (1972) recommend the minimum work envelope for a seated worker be 76 cm long by 76 cm width by 146 cm height.

<u>The Forward Reach</u>:

The forward reach of the worker is often the most important limiting factor in work envelopes. Usually, the worksurface with its materials and tools in front of the worker is the location where most of the worker's tasks are done. If the worker cannot reach necessary items it will increase the difficulty of the worker's task. The forward reach for most people has been assessed as being 46 cm (18 in) as long as the object being reached is not more than 46 cm (18 in) off to the side of the body's midline (Cushman, et al., 1983).

It is important to note that clothing and safety gear worn by the worker will effect the reach limits, occasionally severely restricting the worker's reach (Sanders & McCormick, 1987). Roth, Ayoub, and Halcomb (1977) showed the effects of the direction of reach and presence of restraints on the work space envelopes in detail. A good design will consider such restraining factors into the workplace design. All clothing and gear should be worn when the measures are taken that will be used in the design of the work envelopes. (See Figure 9).

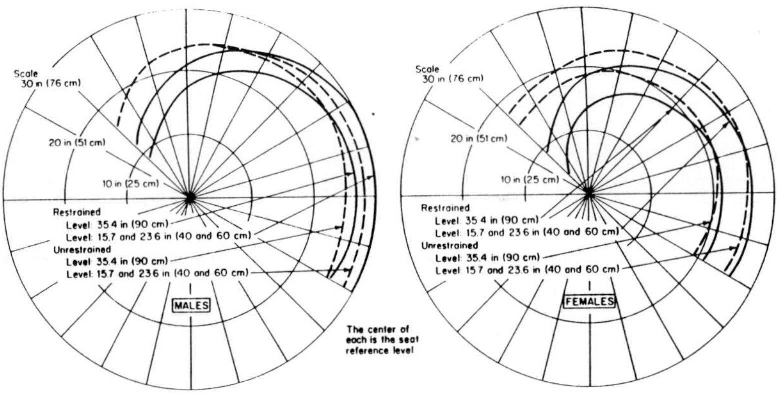

FIGURE 9. FUNCTIONAL ARM REACH FOR THE 5TH PERCENTILE OF MALES AND FEMALES AT SPECIFIED LEVELS ABOVE A SEAT REFERENCE POINT. THE THREE-DIMENSIONAL SPACE REPRESENTED BY THESE DATA CAN BE CONSIDERED AS FORMING A WORK-SPACE ENVELOPE THAT WOULD ACCOMMODATE 95 PERCENT OF THE POPULATION. (Adapted from Roth, Ayoub, & Halcomb, 1977)

Additionally, reach limits change as a result of the type of workplace as well. Reach limits have been estimated for sitting and standing workers. Distances a worker is expected to regularly reach should not exceed reach limits since the extra effort to make these reaches will lead to increased fatigue and increased performance times.

The reach to the side in a horizontal plane by either hand also has reach limits. (See Figure 10; Barnes, 1963; Squires, 1956; Sanders & McCormick, 1988). Care should be taken so the worker does not have to operate controls or reach items on one side with the opposite hand. Under these conditions, the torso will have to be twisted to make the reach. As mentioned before, twisting is undesirable as part of the work cycle due to the stresses imparted on the spine.

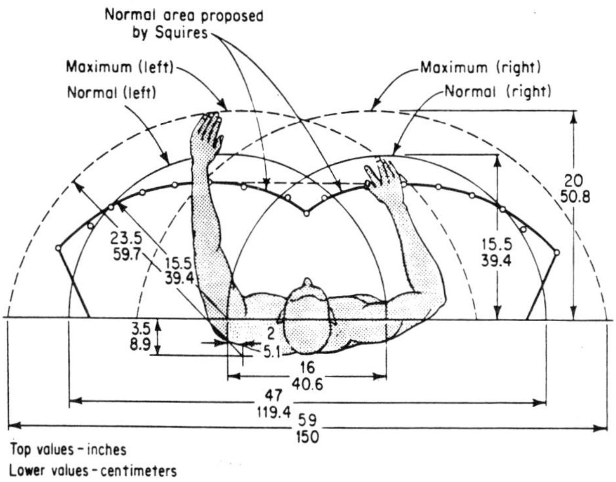

FIGURE 10. DIMENSIONS (IN INCHES & CENTIMETERS) OF NORMAL AND MAXIMUM WORKING AREAS IN HORIZONTAL PLANE PROPOSED BY BARNES, WITH NORMAL WORK AREA PROPOSED BY SQUIRES SUPERIMPOSED TO SHOW DIFFERENCES. (Barnes, 1963; Squires, 1956; From Sanders & McCormick, 1987).

The Overhead Reach:

Overhead reach is another reach workers are often required to make. The overhead reach can be very tiring due to the fact that there is little support for this kind of reach and the worker is required to support the weight of the arm in addition to perform whatever activity that necessitated the reach. Anderson (1988) advises against reaching above shoulder level or behind the body, again emphasizing the need to avoid twisting motions in the performance of workers duties.

Functional Reach:

There are effects on manual activity as a result of extended reaches. A worker can reach further than he or she can successfully perform some manual activities since the fingertips can reach what the hand cannot grasp. This fact must be taken into account when designing the workplaces (Sanders, & McCormick, 1987). The forces the workers can generate when making extended reaches is often reduced as well because of the poor mechanical advantages these reaches tend to create. Controls, especially important ones, should not be beyond the functional reach of the worker.

While workers can reach fairly far, it is farther than is desirable for performing most activities. Joints should ideally be kept at the midpoint of their range for most work activities (Anderson, 1988). The optimal workarea for a seated worker is a square approximately 25 cm by 25 cm (Ayoub, 1973). (See Figure 11). If the majority of work done is confined to this area, reaches and energy expenditures are reduced. Consequently, much of the stress and fatigue associated with work will be avoided.

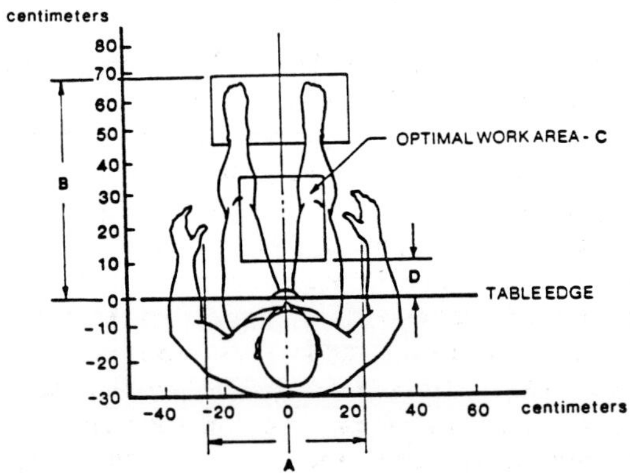

LEG ROOM: 1) WIDTH (A) MINIMUM OF 50 CM
2) DEPTH (B) 65 CM, IF LIMITED
OPTIMUM WORK AREA: 3) (C) APPROXIMATELY 25 X 25 CM
4) (D) 10 CM FROM EDGE OF WORK SURFACE

FIGURE 11. WORKSPACE DESIGN SPECIFICATIONS FOR THE SEATED WORKER
(Adapted from Ayoub, 1973)

POSTURE AND RELATED CONCERNS

Ultimately all the designing done for the workplace layout is to create an environment with good working posture for workers. Workers with good posture are more comfortable, suffer less adverse effects from carrying out the tasks of the job, and are more productive. The two major areas of concern are the posture of the back/torso and the posture of the head and neck. First, let's explore the posture of the back/torso. From previous discussions it should become obvious that the posture of the back is very important. Too much stooping, usually the result of a work surface that is too low, can increase fatigue and lead to serious injury. Chaffin (1987) has estimated that a 20 degree stooped posture can be maintained the majority of the day by most men and between 85 and 90 percent of women. Steps should be taken in the design of the workplace to avoid exceeding 20 degrees.

The posture of the head and neck is important to the worker's welfare and performance too. Several conditions can result in the neck being flexed (Chaffin, 1987). The combination of the seat being too high or too far away from the work area, the work surface being too low, and/or the visual demands of the task can lead to the neck being flexed to a greater degree than desired. Chaffin (1973; 1987) recommends that the neck inclination angle should not exceed 20-30 degrees for extended periods of time.

There are techniques available which are applicable when designing or evaluating a workplace to aid in the analysis of posture. Among the different postural analysis techniques applicable to occupational activities (Keyserling, 1986) include: techniques developed by Keyserling; Posturegram developed by Priel; OWAS developed by Karhu, et al.; Posture Targeting developed by Corlett et al.; techniques for looking at the posture of upper extremities developed by Armstrong et al.; techniques for looking at the posture of the truck developed by Nordin, et al.; and VIRA developed by Persson, et al. All of these techniques provide methods for observing and recording different working postures for analysis. The various methods employ such devices as video tape and computers to aid in postural analysis. Each of the various techniques have different strengths and areas where they are most applicable. For a complete review of these techniques see Keyserling (1986).

EVALUATION OF WORKPLACE DESIGN

The evaluation of an existing design or plans to build a new design must not be overlooked. No design is complete without a careful assessment of the design and its effects on the worker. There are some helpful tips and techniques for the assessment of a workplace design.

Anderson (1988) outlines what he calls a "cautious way to proceed" in the implementation of design developments or modifications. Following this method, an evaluator should: 1) Perform a thorough examination first (job analysis) to determine the specific problem. 2) Evaluate and select the most appropriate intervention(s). This step may require the assistance of an expert. 3) Start conservative treatment, implementing the intervention on a limited scale if possible. 4) Monitor the progress of the intervention. 5) Continue to adjust or refine the scope of the intervention as becomes necessary. For the development of a new design, "the design should start from the working point, where the hands spend most of their time" (Anderson, 1988, p. 91). An important point that can be easily overlooked when starting the large project of workplace design development.

Whether looking at an existing design or a design in the development phase, there are some general principles that should be looked for and followed in workplace design. These principles are key things an evaluator looks for while following Anderson's outline. The general principles for workplace design (from Naderi & Ayoub, 1989) are: 1) Avoid static work. 2) Avoid extreme position of the joints. 3) Avoid overloading of the muscular system. 4) Aim at best mechanical advantage. 5) Avoid unnatural postures. 6) Maintain a proper sitting posture: a) Elbow at a proper height in relation to work surface height. b) Foot rests are adjusted to prevent excessive pressure at back of thighs. c) Backrest clears the pelvis and the ribs to give freedom of movement for the operator. d) Arm rests are available for support to decrease static work and provide stability when dictated by the task. e) Clearance is provided for the thighs. 7) Permit change of posture. 8) Allow the small operator to reach and the large operator to fit. 9) Train the operator to use the physical facility. 10) Match the job demands and operator capacity.

Some additional principles are outlined by Corlett (1982) are: 11) The worker should be able to maintain an upright and forward-facing posture during work. 12) Where vision is a requirement of the task, the necessary work points must be adequately visible with the head and trunk upright or with just the head inclined slightly forward. 13) All work activities should permit the worker to adopt several different, but equally healthy and safe, postures without reducing the capacity to do the work. 14) Work should not be performed consistently at or above the level of the heart; even the occasional performance where force is exerted above heart level should be avoided. 15) Rest pauses should allow for all loads experienced at work, including environmental and information loads and the time interval between successive rest periods. 16) Work activities should be performed with the joints at about the midpoint of their range of movement. This applies particularly to the head, trunk, and the upper limbs. 17) Where muscular force has to be exerted, it should be by the largest appropriate muscle groups available and in a direction colinear with the limbs concerned. 18) Where a force has to be exerted repeatedly, it should be possible to exert it with either of the arms or either of the legs, without adjustment to the equipment.

The process of designing a workplace is a complex one. There are many things to consider and many components of the workplace, the type of workplace, reach limitations, and working postures, that must be carefully interrelated. For example, chair dimensions are important but are meaningless without taking into consideration the chair's relation to the workbench. The success of the finished workplace will be related to the attention given to design principles and the details of the different components during the development phase. Finally, to ensure the adequacy of the design over time, periodic re-evaluations must be carried out regularly and updates to the design made. This is especially true when either task characteristics or worker characteristics are changed.

REFERENCES

Anderson, V.P. 1988. Cumulative trauma disorders: A manual for the musculoskeletal diseases of the upper limbs. Taylor and Francis, Ltd, London.

Ayoub, M.M. 1971. Posture in industry. Proceedings of the Human Factors Society. Human Factors Society, Santa Monica, CA.

Ayoub, M.M. 1973. Work place design and posture. Human Factors, 15(3): 265-268.

Ayoub, M.M., Mackenzie, H.J., Deivanayagam, S., Fernandez, J.E., Smith, J.L., Selan, J., and Masten, Y. 1987. Design of Workplace. Training manual in occupational ergonomics. Institute for Ergonomics Research, Texas Tech University, Lubbock, TX.

Barnes, R.M. 1963. Motion and Time Study (5th edition). John Wiley and Sons, New York.

Barnes, R.M. 1980. Motion and time study design and measurement of work, (7th edition). John Wiley and Sons, New York.

Bendix, T. 1986. Chair and table adjustments for seated work. In N. Corlett, J. Wilson, and I. Manenica (eds.), The ergonomics of working postures. Taylor and Francis, London, pp. 355-362.

Bridger, R.S., Wilkinson, D., and van Houweninge, T. 1989. Hip joint mobility and spinal angles in standing and different sitting postures. Human Factors, 31(3): 229-242.

Corlett, E.N. 1982. Design of handtools, machines, workplace, office, and buildings. In G. Salvendy (ed)., Handbook of Industrial Engineering. John Wiley & Sons, New York.

Corlett, E.N., and Bishop, R.P. 1976. A technique for assessing postural discomfort. Ergonomics, 19(2): 175-182.

Chaffin, D.B. 1973. Localized muscle fatigue -- definition and measurement. Journal of Occupational Medicine, 15(4): 346-354.

Chaffin, D.B. 1987. Biomechanical aspects of workplace design. In G. Salvendy (ed.), Handbook of Human Factors. John Wiley and Sons, New York, pp. 602-619.

Cushman, W.H., Nielsen, W.J., and Pugsley, R.E. 1983. Workplace design. In Eastman Kodak (ed.), Ergonomic design for people at work, vol. 1. Lifetime Learning Publications, Belmont, CA, pp. 12-34.

Das, B., and Grandy, R.M. 1983. Industrial workplace layout and engineering anthropology. In T.O. Kvaleth (ed.), Ergonomics of workstation design. Butterworths, London, pp. 103-128.

Grandjean, E. 1988. Fitting the task to the man, (4th edition). Taylor and Francis, Ltd, London.

Hertzberg, H.T.E. 1972. Engineering anthropology. In H.P. Van Cott, & R.G. Kinkade (eds.), Human engineering guide to equipment design, (revised edition). McGraw-Hill Book Company, New York.

Keyserling, W.M. 1986. Postural analysis of the trunk and shoulders in simulated real time. Ergonomics, 29: 569-583.

Konz, S. 1979. Work design. Grid Publishing, Inc., Columbus, OH.

Konz, S. 1983. Work design: Industrial ergonomics, (2nd edition). Grid Publishing, Inc., Columbus, OH.

Mandal, A.C. 1976. Work-chair with tilting seat. Ergonomics, 19(2): 157-164.
Mandal, A.C. 1982. The correct height of school furniture. Human Factors, 24(3): 257-269.
Maynard, H.B. 1934. Workplace layouts that save time, effort and money. Iron Age, 134(23): 28-30.
Naderi, B., & Ayoub, M.M. 1989. Cumulative Hand Trauma Disorders. Unpublished Report, Institute for Ergonomics Research, Texas Tech University, Lubbock, TX.
Niebel, B.W. 1988. Motion and time study, (8th edition). Richard D. Irwin, Inc., Homewood, IL.
Roth, Ayoub, M.M., and Halcomb, C.G. 1977. Seating, console and workplace design. Proceedings of the Human Factors Society 21st Annual Meeting, Human Factors Society, Santa Monica, CA, pp. 83-87.
Sanders, M.S., and McCormick, E.J. 1987. Human factors in engineering and design, (6th edition). McGraw-Hill Book Company, New York.
Squires, P.C. 1956. The shape of the normal work area, Rept. 275. Navy Dept., Bureau of Medicine and Surgery, Medical Research Laboratory, New London, CT.
Tichauer, E.R. 1973. The industrial environment -- Its evaluation and control. National Institute for Occupational Safety and Health, Department of Health, Education, and Welfare, Washington, D.C., pp. 138-139.
Tichauer, E.R. 1978. The biomechanical basis of ergonomics. John Wiley and Sons, New York.
Van Cott & Kinkade. 1972. Human engineering guide to equipment design, (revised edition). McGraw-Hill Book Company, New York.
Woodson & Conover. 1964. Human engineering guide for equipment designers, (2nd edition). University of California Press, Berkeley, CA.

Workspace, Equipment and Tool Design
edited by A. Mital & W. Karwowski

SITTING AT WORK: RECORDING AND ASSESSING BODY POSTURES, DESIGNING FURNITURE FOR THE COMPUTER WORKSTATION

Karl H. E. Kroemer, Dr. Ing., Professor and Director, Industrial Ergonomics Laboratory, Human Factors Engineering Center, IEOR Dept., Virginia Tech (Virginia Polytechnic Institute and State University) Blacksburg, Virginia 24061-0118, USA

ABSTRACT

Posture-related musculo-skeletal disorders are common occupational problems. They are particularly prevalent in work situations which require extensive sitting, such as in computer workstations. While techniques for recording postures are well developed, their analysis and evaluation with regard to "healthiness" is difficult: criteria for judging the dependent variables (e.g. heart rate, fluid volume, EMG, disk pressure) are mostly questionable or missing. Still, workplaces including computer furniture need to be ergonomically designed. The "straight-and-right" sitting posture, propagated until a few years ago, can no longer be considered a health or design standard. A new set of design goals is at hand: they emphasize adjustibility, variety, and user comfort.

A "NEW OLD" PROBLEM (Grieco, 1986)

Physiologists have tried for decades to assess the appropriateness of certain body postures, yet the problem of "the one healthy body posture at work" is still unresolved. The problem is partly conceptual: Is there "one" suitable posture; and how to define what is "healthy" or otherwise appropriate for various types of work? The other part of the problem is technical: It is difficult to record and to evaluate body postures at work.

The advent of the computer in the office and shop has made many jobs more interesting and more efficient. It also generated a cluster of posture-related health complaints. As recently discussed (Grieco, 1986; Kroemer, 1988a; Millar, 1989), work-related diseases involving the human locomotor apparatus, particularly the spinal column, are by far the most frequent disorders among the working population. Low back pain has been diagnosed to have existed 5000 years ago and neck pain was discussed in the literature more than 250 years ago (Grieco, 1986). Since so many of these complaints seem to be related to job conditions, these have been labeled occupational disorders (Millar, 1989). "Fixed postures" particularly when sitting, are increasing in frequency, not only at the workplace, but as a general feature of modern life style. Grieco (1986, page 347) interprets this as *homo erectus* turning into *homo sedens*. He believes that the human spinal column, suitable for a body in four-legged motion, has not had enough time to adjust to the more horizontal position, used in bipedal locomotion; and that the current rapid transformation to *homo sedens* will lead to serious mal-adaptation of the spine. Indeed, musculo-skeletal pain and discomfort (together with eye strain, and "fatigue") constitute at least half, in some surveys up to 80%, of all subjective complaints and objective symptoms of computer operators, both in Europe and North America (Kroemer, 1988a).

Some of the underlying problems are, apparently, related to each other: difficulties in viewing characters on the computer screen (focusing distance, angle of gaze, etc.) together with straining curvatures of the spinal column (particularly in the neck and lumbar region), joined by fatiguing postures of shoulders and arms generate a stress-strain combination where causes and effects intermix, alternate and build upon each other.

Figure 1 shows some of the hypothesized interactions between person and task, work activities and work postures, and the outcomes of the activities in terms of performance and well-being. Of course, the characteristics of the work activities (visual, motoric, auditory or vocal) affect the body postures assumed at work (head position, curvatures of the spinal column, positions of upper and lower extremities). The traditional approach was to provide workstations designed so that a certain ("upright") body posture would be enforced by the design features of the furniture and equipment. However, the knowledge base of what constitutes an appropriate healthy posture is rather dubious. To gain such information, one wants to observe, record and assess work postures, obviously in relation to the required activities; and then establish guidelines for the ergonomic design of the workstation.

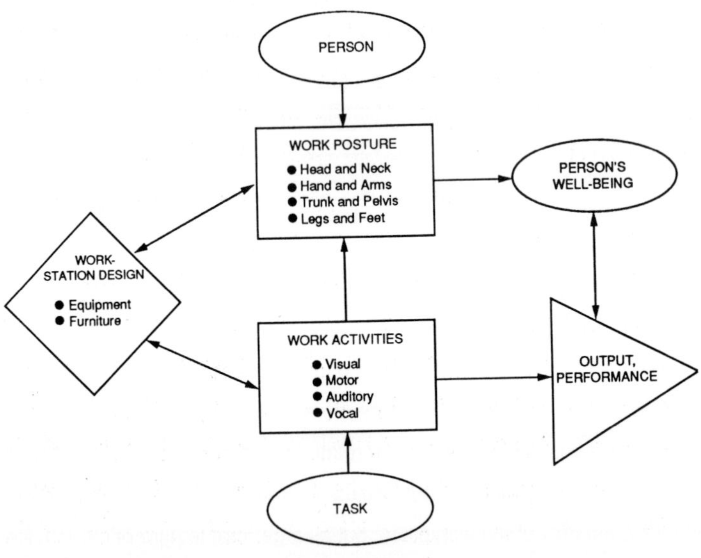

Figure 1. Hypothesized Interactions at Computer Workstations (Adapted from Kroemer, 1988a)

TECHNIQUES FOR OBSERVATION AND RECORDING OF BODY POSTURES

In 1974, the "posturegram" was developed. It is a technique to record the heights at which limbs and joints are located, and the direction and magnitude of segment movements on these height levels (Priel, 1974). The posturegram recording form relies on so-called basic postures; deviations from which are recorded. A somewhat different procedure is the 1976 "posture targetting" (Corlett and Bishop, 1976; Corlett, Madeley, and Manenica, 1979). The technique uses prepared sketches of the human body, with "targets" (like in shooting contests) located next to each major body joint. Direction and magnitude of deviations from a standard posture are recorded for each body segment on the targets.

The 1977 OWAS system also relies on observed deviations from reference body segment positions (Karhu, Kansi, and Kuorinka, 1977; Karhu, Karkonen, Sorvali, and Vepsalainen, 1981). The 1980 TRAM system provides recording forms on which the actually observed postures are entered (Berns and Milner, 1980). The 1982 ARBAN technique is primarily designed to provide an ergonomic analysis of work conditions (Holzmann, 1981). For this, work postures are videotaped, coded, and recorded. The 1985 VIRA technique assigns body positions (also read from a videotape taken on the job) to predetermined posture categories, with particular emphasis on neck and shoulders (Kilbom, Persson, and Jonsson, 1985). A 1986 computer-aided procedure (Keyserling, 1986a, b) also uses the method of providing a menu of standard postures to which actual positions recorded on videotape are compared. In 1989, Gil and Tunes (1989) describe a card-based technique for recording sitting posture.

The basic idea of these techniques is the reference to a "basic" posture of the body segment, deviations from which are recorded on paper or computer. Each of the techniques has advantages in certain situations, but also significant disadvantages: for example, the body postures may be very gross and stylized, the technique may be restricted to only certain body parts, the technique may show positions only in one plane, or the technique may require a mental conversion of the observed posture to fit a recording scheme.

All techniques described either rely on the existence of static postures that can be observed, or on the sampling of dynamic motions via videotape, which are then "frozen in time" for recording. (A number of techniques are available by which movements can be recorded, e.g., SELSPOT®, MOTION ANALYSIS®, WATSMART®, or ELITE®. However, these are not discussed here in detail since this report is concerned only with static or semi-static postures, as opposed to motions.) Several comparisons of various techniques have been published, e.g. by Christensen, Casali, and Kroemer (1984), Pedotti, Boccardi, Frigo and Menoni (1985), and Eklund (1986). In a critical review of their own technique and other methods to record and analyze postures, Punnett and Keyserling (1987) recently concluded that many methods are extremely time-consuming and labor-intensive, and that more efficient work analysis techniques should be developed.

To overcome these difficulties as far as possible, we developed in 1988 a new system that allows recording on paper or computer. It was named "Penguin" according to its appearance, see Figure 2. It encompasses all major body segments, and can be enhanced to include even such appendages as fingers: also, one may decide to concentrate, for example, on head and neck postures. For each body segment of interest, a number of preselected possible positions is shown (either on paper or on the computer screen). The observer simply marks that pictogram which most closely resembles the actually existing posture. A conversion to "numbers" (angles, curvatures, directions, etc.) is done automatically by a computer system.

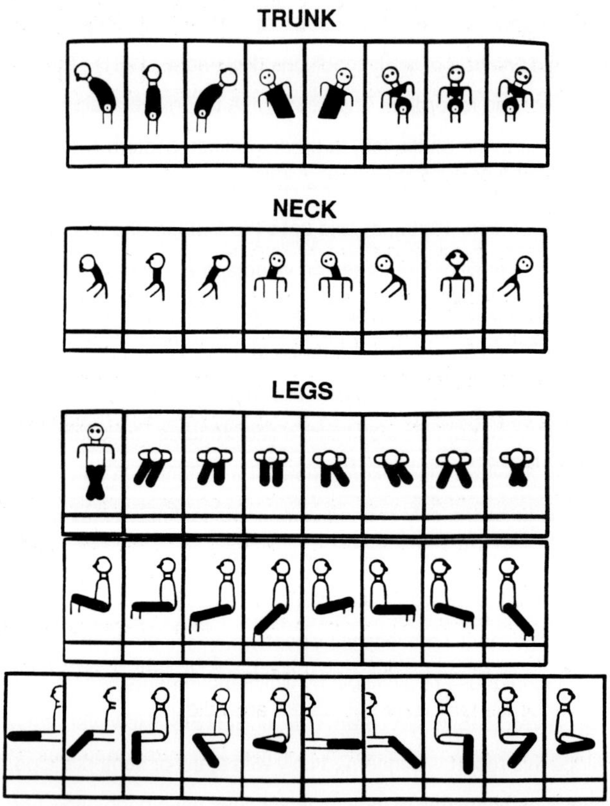

Figure 2(a). "Penguin" Pictograms of Body Postures: Trunk, Neck, Legs

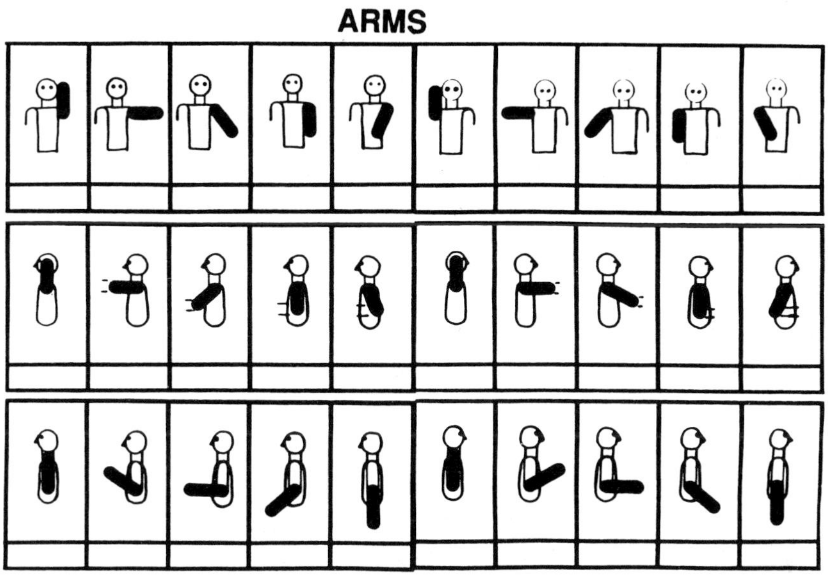

Figure 2(b). "Penguin" Pictograms of Body Postures: Arms

ASSESSING THE "HEALTHINESS" OF A POSTURE AT WORK

Some of the previously mentioned techniques to record body postures also include, implicity or explicitly, an evaluation of the suitability of the observed postures, such as in the OWAS, TRAM, ARBAN, and Keyserling techniques. Often, the methods are only briefly discussed, and the criteria are not clearly stated by which the appropriateness, acceptability, healthfulness, comfort, etc., of these postures are assessed.

Applying the rules of "experimental design", the various body postures can be considered "independent variables". Their variations result in changes in the observed "dependent variables". All other existent conditions and variables should be "controlled", because, if uncontrolled, they may "confound" the relationships between independent and dependent variables. The conditions under which independent and dependent variables are observed establish the boundary conditions within which the technique and results are valid.

Examples of dependent variables, by which the effects of postures are measured, stem from the following disciplinary areas:

<u>Physiology</u>: assessment of oxygen consumption, heart rate, blood pressure, electromyograms, fluid in (volume of) lower extremities.

Medicine: acute or chronic disorders, including cumulative trauma injuries.

Anatomy/Biomechanics: x-rays, CAT scans, stature changes, measurements of disc and intra-abdominal pressure, model calculations.

Psychophysics: interviews, structured or unstructured, subjective ratings by either the subject or the experimenter.

Some of these techniques have become "standard procedures" which have been described and used in many experiments (Bendix, Winkel, and Jessen, 1985; Christensen, Casali, and Kroemer, 1984; Drury and Francher, 1985; Corlett and Eklund, 1986; Eklund, 1986; Grieco, 1986; Tougas and Nordin, 1987; Kroemer, 1988a, b). For example:

Posture-dependent events in the lower extremity, particularly foot swelling and skin temperature, have been reported by Winkel and Jorgensen (1986a, b). They found increases of 2% in volume when the feet could be moved freely while sitting, and nearly 5% if the leg had to be kept in position. These conditions were accompanied by increases in skin temperature up to 5 degrees Celcius, with the highest value associated with inactive sitting. However, the temperature in the flexor hallucis longus muscle decreased with inactive sitting. Heart rate was also found to be higher in inactive sitting.

Recently, the electromyographic technique to assess muscular activities has been used extensively e.g., by Haegg, Suurkuela, and Liew (1987); Marras (1988); Marras and Reilly (1987) Marras, Rangarajulu, and Wongsam, (1987); Nag, Chintharia, Saiyed, and Nag (1986); Reilly and Marras (1989); Winkel and Bendix (1986). The research of the Ohio State University Group under Marras is particularly informative because EMG activities of various trunk muscles were recorded in a variety of static and of well controlled dynamic conditions. It was found that, depending on movement velocity, the eight observed muscle groups in the trunk started their activities at different times, in different sequences and with different intensities. In many cases, measurements taken under static (isometric) conditions did not conform at all with those measured while the body was in movement. This, of course, makes the application of the static measurements to dynamic (i.e., non-static) conditions highly suspicious while, in the past, however, static muscle strength data were often believed to be transferable to dynamic muscle activities.

Several authors have recently addressed the question of spinal strain, particularly of disk compression, and its relation to intra-abdominal pressure. The results of the famous experiments conducted several decades ago in Scandinavia, in which pressure transducers were inserted into the spinal disks of sitting and standing persons (summarized e.g., by Chaffin and Andersson, 1984, and by Grieco, 1986), had been looked upon as definitive. However, recent considerations and measurement results do not supplement older findings. For example, Jaeger (1987) questions the validity of some disk compression values. Adams and Hutton (1985) discuss the contributions of the facet joints (apophyseal joints) of the spinal column to load bearing, in addition to the effects of the vertebral disks. Their recommendation, quite in contrast to traditional ideas, is to flatten the lumbar spine for biomechanical reasons. Eklund and Corlett (1986) propose to use changes in stature (standing height) as an indicator of spine loading. McGill and Norman (1987) question the traditional assumptions about the relationship between intra-abdominal pressure and spinal compression. They found (as Marras did), that a simple relationship between abdominal pressure and disk pressure does not exist. It is of some interest to note that the recommendations for "safe" lifting, derived either from spinal compression norms in the USA or from intra-abdominal pressure values in the UK/Europe, lead to differing results (Freivalds, 1989).

In many studies subjective ratings were collected, often in combination with "objective" measurements. The procedures vary widely among researchers, however, many rely on the outlines and recommendations developed by Shackel, Chidsey, and Shipley (1969) and by Corlett and Bishop (1976). Hallberg (1976) developed a system for semantic description and classification of movement behavior. Information on musculo-skeletal symptoms can be collected in standardized questionnaires (Occhipinti, Colombini, Frigo, Pedotti, and Grieco, 1985; Kuorinka, Jonsson, Kilbom, Vinterberg, Biering-Sorensen, Andersson, and Jorgensen, 1987).

When reviewing the many related publications, it becomes obvious that many techniques exist in different disciplines. An attempt to list these is presented in Table 1.

Apparently, several techniques stem from classical physiology, such as measurement of oxygen consumption, heart rate, and blood pressure. The measurement procedures are well established, as are the assessment criteria: the medical experiences with suitable energy consumption, cardiac rate, and blood pressure are well accepted. However, whether indeed a given oxygen consumption or cardiovascular response or blood pressure is suitable for any given person, assuming specific postures, is highly variable and depends on many circumstances. Thus, interpretation of observed values of the "dependent variables" oxygen uptake, heart rate, or blood pressure is highly variable, and may in fact not be suffice to assess the work posture.

Observation and Recording Techniques	Measurement Procedures	Assessment Criteria	Threshold Values	Interpretation for Assessment of Posture
Oxygen consumption	E	E	V	V
Heart Rate	E	E	V	V
Blood pressure	E	E	V	V
Blood flow	E-D	E-D	V	V
Innervation	D-V	D-V	U	?
Leg/foot volume	E	V	U	U
Temperature, skin or internal	E-D	U	?	?
Electromyography	D	D	V	V
Joint diseases	E-D	D	V	V
Musculo-skeletal disorders	D	D	V	V
Cumulative trauma disorders	D	D	V	V
Spinal phenomena:				
-- disk pressure	E-V	D	V	V
-- disk disorders	D-V	D	V	V
-- facet disorders	D-V	V	U	U
-- alignment of vertebrae	D-V	V	V	?
-- spine curvature	D-V	V	U	?
-- mechanical stresses, including model calculations	D-V	V-U	V	?
Intra-abdominal pressure	E-D	V	U	?
Surface pressure				
-- buttocks	E-D	V	U	?
-- back	E-D	V	U	?
-- thighs	E-D	V	U	?
Body posture				
-- of head/neck, arms, trunk, legs	E-D	V	?	?
-- changes in posture	E	?	?	?
-- change in stature	E	E	V	?
Sensations (ratings) of				
-- ailments	E-D	V	D	D
-- pain	E-D	V	D	D
-- discomfort	E-D	V	D	V
-- comfort, pleasure	E-D	D	D	V

Legend: well established E
being developed D
variable or unknown V, U
questionable ?

Table 1. Techniques for Posture Assessment

Measurement of leg or foot volume, as another example, is a well established procedure. However, what increases in volume are acceptable or not is a judgment that depends on many circumstances. Hence, no given threshold values are known by which the suitability of the body posture can be determined on the basis of leg or foot volume.

A similar problem exists in the use of electromyography. The basic procedures are well established. However, how much muscular activity observed in the EMG is acceptable or even desirable is open to interpretation. For example, one concept holds that activities, particularly if varying over time, are desirable because they represent "good and healthy exercises" (as in "dynamic sitting") thought to promote "muscle tone and training" and to facilitate fluid supply to the intervebral disks and to improve blood circulation (Grieco, 1986; Kilbom, 1986; Winkel and

Bendix, 1986). On the other hand, EMG activities have also been interpreted as being a sign of excessive muscular activity. Typically, muscular activities in the trunk while sitting are at or below 10 percent of their maximal capacities. In spite of this low level of activity, postures that showed even lower activity levels than others have been judged to be "better", an argument often used in the literature (see, e.g., Andersson, Schultz, and Ortengren, 1986; Chaffin and Andersson, 1984; Occhipinti, Colombini, Frigo, et al., 1985)

This example of the use and interpretation of electromyographic signals is typical for many other assessment procedures, such as discussions of spinal phenomena, intra-abdominal pressure, or the assessment or measurement of body curvatures: while the measurement procedures are fairly well established, the criteria for assessment are highly variable, as are the threshold values that separate "healthy" from "unhealthy" postures. How much increase in heart rate or blood pressure, or in foot volume, is acceptable, tolerable, or potentially dangerous? Which curvature of the spinal column is suitable, or should be avoided? Over what periods of time can certain head angles, or neck contortions, become medically suspect? Can a given health complaint be traced to specific aspects of a sitting or standing posture? What are the permissible doses or threshold values? Which postures are recommended? In other words: which are the criteria by which one may judge the suitability of certain body postures at work?

Thus, a new challenge is upon us. We need to judge the independent "posture" variables by their dependent outcomes. And we need to do so in a systematic, reliable, demonstrable manner.

With the exception of acute medical or pathological conditions which can be traced to unsuitable postures, the assessment of dependent threshold values has been and is subjective, a matter of "experience and best estimate". This has been amply evident from the successions of "authoritative statements" regarding sitting posture. Since the middle of the last century it had been generally accepted that one should stand or sit with a "straight" (erect, upright) trunk, because this was believed to be natural, healthy, and aesthetically pleasing. While strong deviations (such as severe scoliosis, sway back, or reversed lordosis) were demonstrated as unhealthy, no clear evidence was ever provided for the suitability of an erect spinal posture. Furthermore, the demand for "erectness" or "straightness" contradicts the "natural" curvature of the spinal column which in the lateral view is S-shaped with lordoses in the cervical and lumbar regions and a kyphosis in the thoracic part. Thus, there exists a contradiction in terms in such that the "straight" spinal column is in fact bent.

The measurement of the spinal curvature is still a problem on a person in a fixed posture, and nearly impossible on a moving person. Interpretation of the measured curvatures is a complex task even with refined and specific classification criteria applied (Occhipinti, Colombini, Molteni, Menoni, Boccardi, and Grieco, 1988).

Associated with the command to sit upright was the idea that there should also be right angles at knee, hip, and elbow joints. Design templates of "straight-and-right" seated humans were used for many workplaces, in particular for office furniture. In fact, such a model is still shown in the current ANSI standard (Human Factors Society, 1988). While this was an easy approach for the simple-minded, it had never been supported by sufficient experimental evidence, while existing evidence that would have suggested other approaches was misinterpreted or neglected. For example, recommendations to "open up the hip angle" to values larger than 90° were quoted in the literature for decades, but not realized in furniture design. A forward-declining seat pan has been propagated since 1884. Provision of a suitable back rest has been a matter of concern in the literature for longer than a hundred years. (For recent discussions, see, e.g., Grandjean, 1988; Grieco, 1986; Eklund, 1986; Kroemer, 1988a, b).

In summary: truly "unhealthy" work postures can be identified from physiological, medical, biomechanical, etc., data in a fairly straightforward manner, particularly if they indicate musculo-skeletal or other disorders associated with the posture. However, for other than those extreme conditions, the information is inconclusive, speculative, clearly individual-related, thus depending on the personal judgment of the expert. This is due to the fact that no objective threshold data have been defined which separate the healthy from the unhealthy postures. Thus, in the majority of cases, the suitability of a work posture is assessed by personal judgment, not by objective criteria.

"What are the criteria on which posture assessments are based? ... it is not the posture per se which is assessed but its tolerability in the particular spatial, temporal and operational conditions under which it occurs... We would like to put forward the hypothesis that postures complying with the following requirements may be considered tolerable: (i) those which do not involve feelings of short-term discomfort; (ii) those which do not cause long-term morpho-functional complaints." (Colombini, Occhipinti, Molteni et al., 1985, p. 278.) Regarding techniques to analyze working postures, Colombini, Occhipinti, Molteni et al. say (1985, p 283): *"techniques ... have not yet been fully developed and evaluated: the main problem awaiting*

solution being the criteria for interpreting the data. ... To conclude, it appears ill-advisable, in view of the present knowledge concerning posture analysis, to adopt a set of rules covering criteria and standard reference values in the immediate future."

ERGONOMIC DESIGN FOR SUITABLE WORKING POSTURES

The analysis and interpretation of objective data provides very limited information regarding the long-term healthiness or tolerability of working postures (Colombini, Occhipiniti, Molteni et al., 1985), but at least it indicates which "unhealthy" conditions must be avoided. The remaining large range of various and variable body postures is "acceptable" to the medical or ergonomic expert, and, more importantly, to the operator. Thus, for the majority of cases, the criteria of comfort, acceptance, and self-selection of postures establish ergonomic design guidance: this has been called "free posturing" (Kroemer 1987).

Many work tasks in the shop, or outside, require the operator to move about, to exert considerable forces and power, often to handle and transport bulky or heavy materials. This work is usually performed while standing or walking, such as in quarry work, construction work, material handling, etc. Several new publications address the large variety of work demands and conditions which therefore will not be further discussed here. (See, for example, Colombini, Occhipinti, Cristofolini, and Grieco, 1987; Grandjean, 1988; Kroemer, McGlothlin, and Bobick, 1989).

An operator may be seated at work if the task requirements and the working conditions are fairly restricted. In most cases, the movement and force requirements are fairly low. Often (and particularly in the office) the operator interacts with the equipment in terms of information exchange and control inputs.

For the computer workstation, most of the output from the human to the machine is currently provided by hand operation of controls, occasionally also by foot controls (Kroemer 1988a). The requirement to place hands and feet at given locations fairly strictly defines the location of the trunk with respect to the workstation. Visual requirements to see the computer screen, a source document, or even the keyboard also enforce a fairly close geometric association of the head (and hence of the trunk) with the equipment. (Acoustic coupling of the operator either by providing voice inputs to the machine or by providing acoustic feedback to the human ears, would very much loosen up the otherwise fairly strict geometric placement requirements.) But even within the boundaries established by task and given conditions (which change with technical development), ergonomic design leeway exists.

Current design principles for computer workstations can be summarized as follows:

1. Allow the operator to freely select various sitting postures, workstation adjustments, and even to get up and move about as part of the job. Abandon the idea of an operator sitting erect, with right angles at major body joints, maintaining this fixed position over long periods of time (Kroemer, 1988a, b).
2. Design for a variety of body dimensions of the users. Use anthropometric information appropriately (Kroemer, 1988a, 1989; Pheasant, 1986). Do not design for the "average person" phantom.
3. Consider that new technology develops quickly and should be usable at the workstation. For example, radically new keyboards or input devices may be available soon (Kroemer, Fathallah, and Langley, 1988), displays and their locations may change quickly (Hill and Kroemer, 1986), voice communication with the computer is imminent.

ERGONOMIC DESIGN OF THE OFFICE (COMPUTER) WORKSTATION

"Egyptian tomb reliefs illustrate clearly, that even at this time, dignified man had to sit with back straight, eyes horizontal and lower legs vertical. How was it, in spite of all the evidence against it being either comfortable or natural, that this stilted posture came to be accepted as standard, whether for sitting on a throne, for dining, for contemplation in the privacy of the boudoir or for working in an office?" (Part of the editorial introduction of the March 1986 issue of *Ergonomics.*) *"Most of the operators do not maintain an upright trunk posture...In fact, the great majority of the operators lean backwards even if the chairs are not suitable for such a posture"* (Grandjean, Huenting, and Nishiyama, 1984, p. 101).

AVC Design Strategy

No longer restricted by the false design principle of forcing or assuming a person to sit upright, new design recommendations for office workstations have been developed (Eklund, 1986; Kroemer, 1988a, b; Tougas and Nordin, 1987). The "AVC" design strategy allows "free posturing" by emphasizing

* ADJUSTIBILITY to fit a wide range of user body sizes and task requirements;

* VARIETY of user-selected postures and movements while sitting, standing, and walking;

* COMFORT and user acceptance instead of imposed fixed body postures.

The AVC strategy includes that indeed the whole workstation must be designed so that seat, table, desk, computer, and all other pieces of furniture and equipment fit each other. It is convenient, however, to discuss the major workstation components separately - see Figures 3, 4, and 5.

Seat Pan

The surface of the seat pan must support the body weight comfortably and securely. Pressure points would result from using hard surfaces, but can be avoided by using upholstery that elastically or plastically adjusts to body contours.

The only limitation to the size of the seat pan is that it should be so short that the front edge does not press into the sensitive tissues behind the knee. The height of the seat pan must be adjustable widely to fit various body sizes and preferences. Hence, the lowest position should be about 38 cm and the highest up to 58 cm. One must be able to easily adjust the seat height while sitting on the chair.

The seat pan is usually essentially flat, between 38 and 42 cm deep and at least 45 cm wide. A well rounded front edge (often called a waterfall design) is mandatory to avoid pressure to the underside of the thighs. Often, the rear and side borders of the seat pan are slightly higher than its central section, which is usually achieved by allowing the center area to become more compressed by the weight of the seated person. Figures 3 and 4 illustrate major dimensions of seat pan and back rest.

Seen from the side, the seat pan is often kept fairly horizontal. However, it can be individually comfortable and desirable to be able to tilt the seat pan slightly backward, or even strongly forward. Seat pans that tilt backward facilitate leaning on the backrest. Seat pans that are tilted forward allow the hip angle to be opened, even to "semi-sit" on a strongly downward-tilted seat. To counteract slippage, shin pads may be incorporated in the structure, or the seat surface may be shaped to fit the human underside.

Individual preferences vary widely and should be accommodated. Shifting, for example, from a strongly forward to a backward tilted seat pan, and alternately leaning on a backrest, facilitates subjective comfort and well-feeling although it may disrupt the uniform appearance of a standard office posture.

Back Rests

The back rest serves to carry some weight of the upper trunk, arms, and head, thus allowing muscle relaxation. The purpose can be fulfilled only when the trunk truly reclines on the backrest, not if one sits upright.

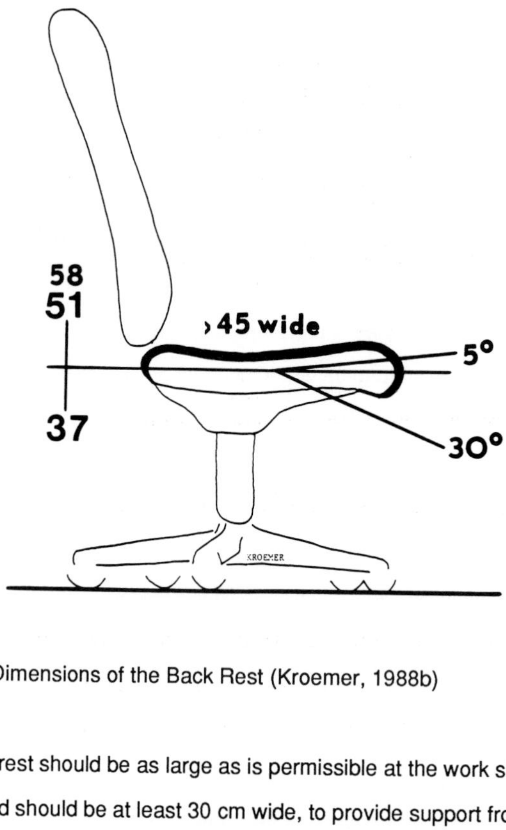

Figure 3. Major Dimensions of the Back Rest (Kroemer, 1988b)

The back rest should be as large as is permissible at the work site. Hence it can be up to 85 cm high, and should be at least 30 cm wide, to provide support from head and neck down to the lumbar region. For this, its shape follows back contours, specifically in the neck and lumbar regions. An adjustable pad for lumbar lodosis, such as an inflatable cushion, is appreciated by many users. The lumbar pad should be adjustable from 15 to 23 cm above the seat pan, while a cervical pad should be adjustable from 50 to 70 cm above the seat pan.

One must be able to adjust the angulation of the back rest easily while seated. The adjustment range should be from upright to 30 degrees backward. Further declination for rest and relaxation is desirable, but requires that stability of the seat be assured so that it cannot tip over.

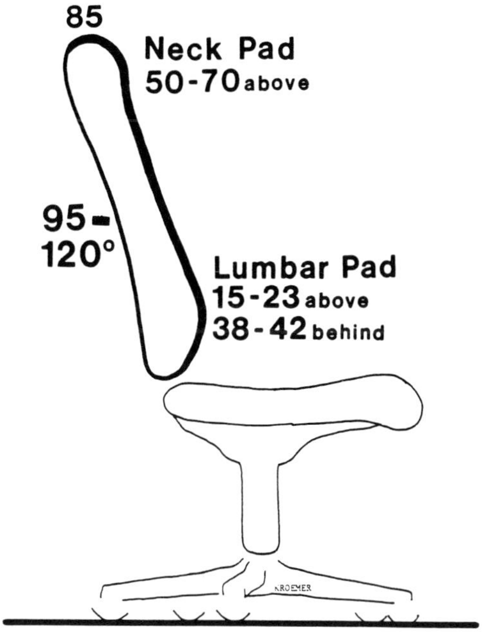

Figure 4. Major Dimensions of the Back Rest (Kroemer, 1988b)

Arm Rests

Arm rests allow the support of the weight of hands, arms, and even of portions of the upper trunk and head. Thus, arm rests are often used for short periods of time. On the other hand, on some workstations, arm rests "get in the way" and are not useful.

Foot Rests

Need for a foot rest usually indicates that the height of the workstation, particularly of the seat pan is too high to accommodate the seated person. Hence, presence of a foot rest usually indicates a deficient workstation design.

If foot rests are unavoidable, they should be so high that the seated individual can have the thighs nearly horizontal. This often requires adjustability of foot rest height, or provision of foot rests of several heights so that the person may select the appropriate one. Foot rests shall provide a large surface so that one can shift the location of the seat, but should not consist of a single bar or a small surface.

Design for Motion and Change

In workstation design, it is necessary to consider that persons want to change body positions, even to stand up and move about after having sat for a while. Maintaining a given posture, even if it appears comfortable in the beginning, becomes stressful as time goes on. Changes in posture are necessary, best by brief periods of physical activity.

Some people even prefer to stand at work, though often only for a limited period of time. This requires a workstation where the working height is at or slightly below elbow height of the standing individual, i.e., between 90 and 100 cm. A foot rest at about two-thirds knee height, approximately 30 cm, is often well liked because one can prop one foot up on it temporarily.

The Computer Workstation

Because of its expected prevalence and its specific work requirements, the computer workstation requires special design care (Kroemer, 1988a, b). The seat is only part of the workstation, albeit particularly important. The heights of the support for the input device, of the support for the computer display, and of the support for source documents all need to be adjustable with respect to each other. Furthermore, the horizontal locations with respect to each other need to be adjustable. Much of this can be done by moving the seat forth and back, but the distance and height arrangement of the computer display is often a difficult task and needs to be considered a specific design feature. Figure 5 depicts the most important design features of an ergonomic computer workstation.

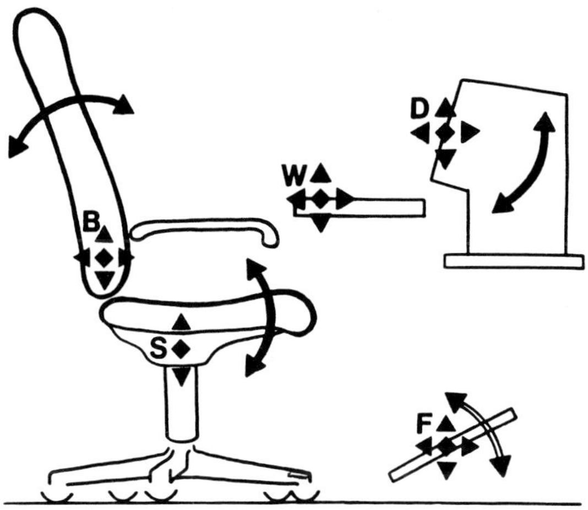

Figure 5. Adjustment Features of a Computer Workstation

SUMMARY AND CONCLUSIONS

Many techniques are available to record body postures at work. However, the assessment and evaluation for "healthfulness" is not well developed, except for extreme negative cases. The straight-and-right sitting posture is neither healthy nor comfortable. A new design strategy emphasizes user Adjustibility, Variety, and Comfort: AVC is of special importance for the design of computer workstations.

ACKNOWLEDGEMENT

An earlier version of this treatise was prepared for and presented at the Italian National Seminar "Lavaro e Patologia del Rachide", Milan 29-30 May 1989. Permission for publication was granted by Prof. Dr. A. Grieco on 30 June 1989.

REFERENCES

Adams M. A. and Hutton W. C. (1985). The Effect of Posture on the Lumbar Spine. *Journal of Bone and Joint Surgery,* Vol. 67-B, No. 4, 625-629.

Andersson, G. B. J., Schultz, A. B., and Ortengren, R. (1986). Trunk Muscle Forces During Desk Work. *Ergonomics*, Vol. 29, No. 9, 1113-1127.

Berns, T. A. R. and Milner, N. P. 1980). *TRAM: A Technique for the Recording and Analysis of Moving Work Posture.* (Report 80:23, pages 22-26). Stockholm, Sweden: Ergolab.

Bendix, T., Winkel, J. and Jessen, F. (1985). Comparison of Office Chairs with Fixed Forwards or Backwards Inclining, or Tiltable Seats. *European Journal of Applied Physiology and Occupational Physiology*, 54, 78-385.

Chaffin D. B. and Andersson, G. B. J. (1984). *Occupational Biomechanics.* New York, NY: Wiley.

Christensen, C., Casali, J. G., and Kroemer, K. H. E. (1984). A Methodology for Work Chair Assessment. *Proceedings of the Human Factors Society 28th Annual Meeting* (pp. 832-835). Santa Monica, CA: Human Factors Society.

Colombini, D., Occhipinti, E., Molteni, G., Grieco, A., Pedotti, A., Boccardi, S., Frigo, C., and Menoni, O. (1985). Posture Analysis, *Ergonomics* Vol. 28, No. 1, 275-284.

Corlett, E. N. and Bishop, R. P. (1976). A Technique for Accessing Postural Discomfort. *Ergonomics,* Vol. 19, No. 2, 175-182.

Corlett, E. N. and Eklund, J. A. E. (1986). Change of Stature as an Indicator of Loads on the Spine. Chapter 20 in N. Corlett, J. Wilson, and I. Manenica (Eds.), *The Ergonomics of Working Postures: Models, Methods and Cases* (pp. 232-242). London: Taylor & Francis.

Corlett, E. N., Madeley, S. J., and Manenica, I. (1979). Postural Targetting: A Technique for Recording Working Postures, *Ergonomics,* Vol. 22, No. 3, 357-366.

Drury, C. G. and Francher, M. (1985). Evaluation of a Forward Sloping Chair. *Applied Ergonomics,* Vol. 16, No. 1, 41-47.

Eklund, J. (1986). Industrial Seating and Spinal Loading. Doctoral Dissertation, University of Nottingham. Linkoeping: University of Technology (ISBN 91-7870-14409).

Eklund, J. A. E. and Corlett, E. N. (1986). Experimental and Biomechanical Analysis of Seating. Chapter 28 in N. Corlett, J. Wilson, and I. Manenica (Eds.), *The Ergonomics of Working Postures: Models, Methods and Cases* (pp. 319-330). London: Taylor and Francis.

Freivalds, A. (1989). Understanding and Preventing Back Trauma: Comparison of U.S. and European Approaches. In K. H. E. Kroemer, J. D. McGlothlin, and T. G. Bobick (Eds.), *Manual Material Handling: Understanding and Preventing Back Trauma* (pp. 55-63). Akron, OH: American Industrial Hygiene Association.

Gil, H. J. C., and Tunes, E. (1989). Posture Recording: A Model for Sitting Posture. *Applied Ergonomics*, Vol. 20, No. 1, 53-57.

Grandjean, E. (1988). *Fitting the Task to the Man. (4th Edition).* London: Taylor and Francis.

Grandjean, E., Huenting, W., and Nishiyama, K. (1984). Preferred VDT Workstation Settings, Body Postures and Physical Impairments. *Applied Ergonomics*, Vol. 15, No. 2, 99-104.

Grieco, A. (1986). Sitting Posture: An Old Problem and a New One. *Ergonomics,* Vol. 29, No. 3, 345-362.

Haegg, G. M., Suurkuela, J., and Liew, M. (1987). A Worksite Method for Shoulder Muscle Fatigue Measurements Using EMG, Test Contractions and Zero Crossing Technique. *Ergonomics*, Vol. 30, No. 11, 1541-1551.

Hallberg, G. (1976). A System for the Description and Classification of Movement Behavior. *Ergonomics* Vol. 19, No. 6, 727-739.

Hill, S. G. and Kroemer, K. H. E. (1986). Preferred Declination of the Line of Sight. *Human Factors*, Vol. 28, No. 2, 127-134.

Holzmann, P. (1981). ARBAN-A New Method of Analysis of Ergonomic Effort. *Applied Ergonomics* Vol. 13, No. 2, 82-86.

Human Factors Society (Ed.) (1988). *ANSI Standard 100, American National Standard for Human Factors Engineering of Visual Display Terminal Workstations.* Santa Monica, CA: The Human Factors Society.

Jaeger, M. (1987). *Biomechanisches Modell des Menschen zur Analyse und Beurteilung der Belastung der Wirbelsaeule bei der Handhabung von Lasten.* Biotechnik Series 17, No. 33. Duesseldorf: VDI Verlag.

Karhu, O., Kansi, P., and Kuorinka, I. (1977). Correcting Working Postures in Industry: Practical Method for Analysis. *Applied Ergonomics,* Vol. 18, 199-201.

Karhu, O., Karkonen, R., Sorvali, P., and Vepsalainen, P. (1981). Observing Working Postures in Industry: Examples of OWAS Application. *Applied Ergonomics,* Vol. 12, No. 1, 13-17.

Keegan, J. J. (1952). Alterations to the Lumbar Curve Related to Posture and Sitting. *Journal of Bone and Joint Surgery 35:A3*, 589-603.

Keyserling, W. M. (1986a). Postural Analysis of the Trunk and Shoulder in Simulated Real Time. *Ergonomics* Vol. 29, No. 4, 569-583.

Keyserling, W. M. (1986b). A Computer-Aided System to Evaluate Postural Stress in the Workplace. *Am. Industrial Hygiene Assoc. Journal,* Vol. 47, No. 10, 641-649.

Kilbom, A. (1986). Physiological effects of extreme physical inactivity. In *Proceedings of Work with Display Units.* Stockholm, May 12-15, 1986, (pp. 486-489). Stockholm: Swedish National Board of Occupational Safety and Health.

Kilbom, A., Persson, J., and Jonsson, B. (1985). Risk Factors for Work-Related Disorders or the Neck and Shoulder with Special Emphasis on Working Postures and Movement. Zadar, Yugoslavia: International Symposium on the Ergonomics of Working Posture.

Kroemer, K. H. E. (1987). "Ergonomic" Furniture for Computer Workstations. Workshop Background Paper, Annual Meeting of the Human Factors Society, October 19, 1987. Blacksburg, VA: Virginia Tech, IEOR Department

Kroemer, K. H. E. (1988a). VDT Workstation Design. Chapter 23 in M. Helander (Ed.), in *Handbook of Human Computer Interaction* (pp. 521-539). Amsterdam and New York: Elsevier.

Kroemer, K. H. E. (1988b). Ergonomic Seats for Computer Workstations. In F. Aghazadeh (Ed.), *Trends in Ergonomics/Human Factors V.* (pp. 313-320). Amsterdam: Elsevier.

Kroemer, K. H. E. (1989). Engineering Anthropometry. *Ergonomics*, Vol. 32, No. 7, 767-784.

Kroemer, K. H. E., Fathallah, F. A., and Langley, L. W. (1988). A New Keyboard with Chorded Ternary Keys. In *Proceedings of the Human Factors Society 32nd Annual Meeting* (pp. 1005-1008). Santa Monica, CA: Human Factors Society.

Kroemer, K. H. E., McGlothlin, J. D., and Bobick, T. J. (Eds.) (1989). *Manual Material Handling: Understanding and Preventing Back Trauma.* Akron, OH: American Industrial Hygiene Association.

Kuorinka, I., Jonsson, B., Kilbom, A., Vinterberg, H., Biering-Sorensen, F., Andersson, G.,and Jorgensen, K. (1987). Standardised Nordic Questionnaires for the Analysis of Musculoskeletal Symptoms. *Applied Ergonomics,* Vol. 18, No. 3, 233-237.

Lueder, R. K. (1983). Seat Comfort. A Review of the Construct in the Office Environment. *Human Factors,* Vol. 25, No. 6, 701-711.

Marras, W. S. (1988). Predictions of Forces Acting Upon the Lumbar Spine Under Isometric and Isokinetic Conditions: A Model-Experiment Comparison. *International Journal of Industrial Ergonomics, 3,* 19-27, Amsterdam: Elsevier Science Publishers.

Marras, W. S. and Reilly, C. H. (1988). Networks of Internal Trunk-Loading Activities Under Controlled Trunk-Motion Conditions, *Spine*, Vol. 13, No. 6, 661-667.

Marras, W. S., Rangarajulu, S. L., and Wongsam, P. E. (1987). Trunk Force Development During Static and Dynamic Lifts. *Human Factors,* Vol. 29, No. 1, 19-29.

McGill, S. M. and Norman, R. W. (1987). Reassessment of the Role of Intra-abdominal Pressure in Spinal Compression. *Ergonomics*, Vol. 30, No. 11, 1565-1588.

Millar, J. D. (1989). Occupational Safety and Health Objectives for the Year 2000. Centers for Disease Control, Atlanta, GA: Letter of 23 February 1989.

Nag, P. K., Chintharia, S., Saiyed, S., and Nag, A. (1986). EMG Analysis of Sitting Work Postures in Women. *Applied Ergonomics,* Vol. 17, No. 3, 195-197.

Occhipinti, E., Colombini, D., Frigo, C., Pedotti, A., and Grieco, A. (1985). Sitting Posture: Analysis of Lumbar Stresses with Uupper Limbs Supported, *Ergonomics*, Vol. 28, No. 9, 1333-1346.

Occhipinti, E., Colombini, D., Molteni, G., Menoni, O., Boccardi, S., and Grieco, A. (1988). Clinical and Functional Examination of the Spine in Working Communities: Occurrence of Alterations in the Male Control Group. *Clinical Biomechanics 4.*

Pheasant, S. (1986). *Bodyspace.* London: Taylor and Francis.

Priel, V. Z. (1974). A Numerical Definition of Posture. *Human Factors,*Vol. 16, No. 6, 576-584.

Punnett, L. and Keyserling, W. M. (1987). Exposure to Ergonomic Stressors in the Garment Industry: Application and Critique of Job-Site Work Analysis Methods. *Ergonomics*, Vol. 30, No. 7, 1099-1116.

Reilly, C. H. and Marras, W. S. (1989). Simulift: A Simulation Model of Human Trunk Motion. *Spine*, Vol. 14, No. 1, 5-11.

Shackel, B., Chidsey, K. D., and Shipley, P. (1969). Assessment of Chair Comfort. *Ergonomics*, Vol. 12, No. 2, 169-306.

Tougas, G. and Nordin, M. C. (1987). Seat Features Recommendations for Workstations. *Applied Ergonomics,* Vol. 18, No. 3, 207-210.

Winkel, J. and Bendix, T. (1986). Muscular Performance During Seated Work Evaluated by Two Different EMG Methods. *European Journal of Applied Physiology,* Vol. 55, 167-173.

Winkel, J. and Jorgensen, K. (1986a). Evaluation of Foot Swelling and Lower-Limb temperatures in Relation to Leg Activity During Long-Term Seated Office Work. *Ergonomics*, Vol. 29, No. 2, 313-328.

Winkel, J. and Jorgensen, K. (1986b). Swelling of the Foot, its Vascular Volume and Systemic Hemoconcentration During Long-Term Constrained Sitting. *European Journal of Applied Physiology and Occupational Physiology*, Vol. 55, 162-166. Springer-Verlag.

Workspace, Equipment and Tool Design
edited by A. Mital & W. Karwowski

DESIGNING FOR WORKSPACE SAFETY

L.L. SMITH[1] and S. DEIVANAYAGAM[2]

[1]TASC (The Analytic Sciences Corporation), Resource Management Division, 1101 Wilson Blvd., Suite 1500, Arlington, Virginia 22209

[2]Industrial Engineering Department, Tennessee Technological University, P.O. Box 5011, Cookeville, Tennessee 38505-5011

ABSTRACT

This Chapter presents descriptions, discussions, and guidelines on workplace safety. The fundamentals of safety analysis and for designing for safety are introduced. The basic provisions of the Occupational Safety and Health Act of 1970 are provided. A series of design guidelines concerning major attributes of a typical workplace are presented in two sections addressing ambient and physical parameters. The general types of guards and warnings are described with reference to typical applications. Useful guidelines on hazard elimination and control are provided. The basic premises of industrial hygiene and health hazard are described. Finally, a general program of safety management is outlined.

INTRODUCTION

According to the National Safety Council there were 33,300 accidental deaths in the U.S. in 1988 which were not motor vehicle related (World Almanac, 1989). The leading types of accident, in descending order, were falls, drowning, fires/burns, ingestion of food/objects, firearms, solid/liquid poison, and gas poison. Many of these accidents occurred in the victims' workplaces.

The consideration of human welfare in the workplace should be one of the most important priorities of the designer. Unfortunately, in actual practice safety in the workplace may very well be among the last things taken into account. Safety in the context of the workplace encompasses many types of considerations which impact human welfare. These types are of three general classes: injury, temporary health, and permanent health. Injury refers to those situations where a human being incurs physical injury as a result of an accident at work. The temporary health class refers to situations where a human being suffers a temporary physiological debilitation (e.g., fatigue, blurred vision, or short-term deafness) due to exposure to an unsatisfactory attribute of a workplace. The permanent health class refers to situations where a human suffers some sort of permanent physiological debilitation (e.g., deafness, blindness, or loss of the use of a limb) due to exposure to an unsafe attribute of a workplace.

Human welfare in the workplace can be affected by any aspect of a workplace -- if it is not designed, maintained, or used properly. Even though the human being in the workplace may be conscientious, experienced, and skilled, the likelihood of possible safety hazards due to poor design or practice is still significant. It is quintessential for workplace designers (including system designers, tool designers, job designers, architects, decorators, furniture designers, etc.) to make every effort to follow the basic principles of safety in their design activity.

The basic principles of safety are:

- Eliminate the hazard
- Guard against it
- Warn about it
- Reinforce in the user's mind the presence and the nature of the hazard and of the safety measures and practices provided.

If a possible risk is anticipated, the first principle admonishes the designers to allow for it to become an inherent part of the product. In many systems (such as aircraft, weapons, and ships) some of the dangers and hazards cannot be avoided, or eliminated. In the simple (relatively speaking) context of a workplace, however, virtually most known hazards can be avoided through careful, diligent design practice. There usually remains a large variety of workplace aspects or components which pose little threat to life and livelihood but which in their own inimitable ways pose some risk to the humans who will be present nonetheless. For example, Table 1 lists a variety of incidents which are common occurrences in a typical workplace.

TABLE 1

INCIDENT	**CAUSING AGENT**
PINCH	DOOR, DRAWER, WINDOW, CHAIR, TOOL
PUNCTURE	SCISSORS, STAPLER, LETTER OPENER
CUT	PAPER, SCISSORS, GLASS
SCRAPE	FURNITURE
FRACTURE	ESCALATOR

In general, these types of incidents can be minimized in their frequency and also in their severity by prudent proactive design practice. Salvendy (1987) describes 11 basic safeguarding measures which, when implemented, contribute to ensuring that a workplace has limited opportunity for injury-related incidents. These measures are:

- Openings or gaps should be less than one-fourth inch wide
- Barriers and enclosures should be positively fixed
- Barriers and enclosures should be movable or adjustable prior to fixing
- Barrier components should be interlocking
- Critical controls should have presence-sensing capability
- Hazardous conditions can be avoided through the use of hand contact push buttons
- Use pullouts and restraints whenever possible
- Use awareness devices whenever possible
- Ensure floormats lay flat and are securely fastened
- Use lockout devices whenever possible
- Use emergency stop devices whenever possible.

Hammer (1980) provides a concise set of key terms and definitions which are pertinent in this context. This set contains five terms: hazard, danger, damage, safety, and risk. Hammer's definitions, slightly modified, are:

- Hazard -- A condition with the potential of causing injury to personnel, damage to equipment or structures, loss of material, or lessening of the ability to perform a prescribed function.
- Danger -- An expression of relative exposure to a hazard with respect to the quality and thoroughness of precautions taken.
- Damage -- The severity of injury or of the physical, functional, or monetary loss which might result if control of a hazard is lost.
- Safety -- An expression of the degree to which a hazard condition is being controlled.

- Risk -- An expression of the possibility of a loss due to the loss of control over a hazard.

The types of injury shown in Table 1 can usually be avoided through effective safety oriented workplace design. Proper safety design practice also significantly affects the frequency and severity of accidents and incidents in the workplace. The last three principles of safety design are particularly applicable and effective in nearly all workplace situations.

This Chapter presents workplace design guidelines which address the objectives of worker and user safety. These guidelines are based on human factors theory and principles and on empirical results. This Chapter contains design guidelines which address the safety design principles of guarding and warning. The fourth principle, reinforcement, refers to a joint responsibility between employer and employee for continuously reminding each other of the possible and potential safety hazards present in the workplace. This philosophy is inherent in the OSH Act described in the first section of this Chapter and represents the important notion that guards and warnings have little value and effect in a complaisant environment. Finally, the discussion presented in this Chapter will be oriented toward how each of the design guidelines described address the three classes of impact on human welfare which were identified previously.

The human being is very perceptive and sensate. The five basic senses of vision, audition, smell, touch, and taste serve as the chief harbingers of impending risk and danger. Hammer (1981) provides a comprehensive list of examples of how our senses can keep us safe. It is important to note that Hammer's list demonstrates that safety precautions and provisions are a necessary component to a safe workplace. Hammer's list is presented in Table 2.

There are seven major sections in this Chapter. First, OSH Act Compliance, presents a brief overview of the Occupational Safety and Health Act of 1970. Second, Guidelines, provides a series of design guidelines concerning the major attributes of a typical workplace. Third, Workplace Hazards, outlines several of the more common hazards which may be found in typical workplaces. Fourth, Guards and Warnings, highlights how special design features such as

TABLE 2

SENSE	MEANS	DESCRIPTION	EXAMPLE
Vision	Illumination	A Hazardous area is more brightly illuminated than non-hazardous surrounding areas.	Having well-lit highway intersections, obstacles, stairs, and transformer substations.
	Discrimination	Paint a physical hazard in a bright color or in alternating light and dark colors.	A structure (such as a pole), piece of equipment, or fixed object which could be hit by a moving vehicle.
	Notes in instructions	Warning and caution notes are inserted in operations and maintenance instructions and manuals to alert personnel to hazards.	A warning in a car owner's manual to block the wheels before jacking the car to change a tire.
	Labeling	Warnings are painted on or attached to equipment.	"NO STEP" markings on hydraulic or pneumatic lines; high voltage; jacking points.
Audition	Alarms	A siren, whistle, or similar sound device provides warning of existing or impending danger.	A siren indicates that there is a fire in a plant; a siren or whistle warns personnel to clear an area where blasting is to take place.
	Buzzer	Alerts a person that a specified time has passed or that time has arrived to take the next step in a sequence of actions.	Some compressed air packs contain buzzers that sound when the pressure in the tank has decreased to a predetermined level, or after a preset time has passed.
	Shout	Voice action to warn of danger.	One person warns another of an obstruction.

TABLE 2, Continued

SENSE	MEANS	DESCRIPTION	EXAMPLE
Smell	Odor detection	Presence of an odorous gas can indicate the presence of a hazard.	An odorant is added to refined natural gas (which has no odor) so that leaks can be readily detected.
Feel	Vibration	Rough running of equipment can indicate the presence of a problem and impending failure.	Vibration of a rotating shaft can signal a loss of lubrication, wear, and damage.
	Temperature	Excessively high temperature can warn of a problem.	A maintenance man may be able to detect by its temperature a bearing that is acting abnormally; a temperature increase in an air conditioned space may warn of equipment problems.
Taste	Ingestion	May indicate that material taken in the mouth is dangerous.	Acid, bitter, or excessively salty taste may indicate that material is not proper for consumption.

guards and warnings can have a favorable effect on the safety of a workplace. Fifth, Hazard Elimination and Control, provides useful guidelines on how to minimize the deleterious effect of common workplace hazards. Sixth, Hygiene and Health Hazard, introduces the principles of industrial hygiene and long-term health hazards. The discussion on hygiene demonstrates the various ways the day-to-day health of workers can be affected by conditions in the workplace. The discussion on health hazards illustrates how long-term health problems can occur due to exposure to certain workplace conditions. Finally, a brief outline of a safety management program is provided and discussed.

OSH ACT COMPLIANCE

The Occupational Safety and Health (OSH) Act of 1970 and the Occupational Safety and Health Administration (OSHA) of the U.S. Federal Government exist to ensure that the basic health and livelihood of the American worker are protected. The Act itself is a massive document which contains many detailed sections. Its level of detail concentrates on specifications for the proper manufacture of many types of tools, equipment, and materials which are both common to workplaces and also were, prior to the involvement of the federal government in this manner, the top contributors to severe injury in the workplace.

In addition to establishing a minimum design standard for many typical components of the workplace the OSH Act also established a requirement for all employers to have a safety responsibility and awareness program. The Act provides broad guidelines on how employers may conduct safety awareness programs. The Act encourages each employer to adopt a proactive safety attitude and provides many suggestions on how employers can get their employees involved in safety awareness and improvement programs and efforts.

In spite of the depth and scope of the OSH Act, there are still innumerable opportunities for workers and users to become injured not only from unsafe workplaces but also from unsafe attitudes and practices. It is important to emphasize that a workplace which is specifically designed to be in compliance only with OSHA standards is by no means an optimally safe workplace. Optimal safety in a workplace can only be achieved via the consideration of human factors guidelines which are specifically oriented to the capabilities and limitations of human beings.

The OSH Act has 18 major subparts, 15 of which address the design and manufacture of typical workplace tools, equipment, and materials. These subparts are:

- D: Walking-Working Surfaces
- E: Means of Egress
- F: Powered Platforms, Manlifts, and Vehicle-Mounted Work Platforms
- G: Occupational Health and Environmental Control
- H: Hazardous Materials

- I: Personnel Protective Equipment
- J: General Environmental Controls
- K: Medical and First Aid
- L: Fire Protection
- M: Compressed Gas and Compressed Air Equipment
- N: Materials Handling and Storage
- O: Machinery and Machine Guarding
- P: Hand and Portable Power Tools and Other Hand-Held Equipment
- Q: Welding, Cutting, and Brazing
- R: Special Industries
- S: Electrical.

GUIDELINES

This section presents a series of design guidelines which support optimal workplace safety. The guidelines are grouped into four categories of workplace attributes: ambient, physical, tools, and clothing.

The Ambient Workplace

The ambient workplace comprises those parts of the workplace with which the worker is in contact but with which the worker usually interacts passively and over which the worker usually has little or no control. Specifically, ambient workplace attributes include the air, humidity, temperature, light, sounds and noise, vibration, and the particulate matter in the area of a workplace.

Guidelines for Workplace Air. Poor air quality in a workplace can have an insidious effect on worker safety. Worker alertness can be seriously decreased. The level of worker conscientiousness can also be degraded significantly by poor air quality. Low oxygen levels or air with unusually large proportions of carbon dioxide or other gases can cause dizziness, disorientation, poor balance, confusion, and deterioration of visual and auditory acuity. Salvendy (1987) states that "In the United States and many other countries, the major source of information on acceptable concentrations of substances in work room air is the American Conference of Governmental Industrial Hygienists (ACGIH)."

The air in a workplace should be well circulated and mixed with the best quality air source possible. This source is usually the open air outside the building as long as it is fresh. Fresh air typically implies that it is not contaminated by outside factors, such as pollution, rotting garbage, stagnant water, decomposing materials, pollen, smoke, etc. Further, fresh air should be as close to natural proportions of oxygen, carbon dioxide, nitrogen, and hydrogen as possible. In most normal circumstances a workplace can be infused with acceptable fresh air simply by ensuring that the ambient air is well circulated. The ACGIH publishes guidelines and recommendations in terms of Threshhold Limit Values (TLV) for substances in work room air. These TLVs are the eight hour time-weighted average concentrations which will protect nearly all people exposed (Salvendy, 1987). If exposure to concentrations of alien or toxic substances is unavoidable in the workplace, the safety specialist should obtain the latest ACGIH publication and adhere to appropriate guidelines provided in the OSH Act and in other documents such as Salvendy (1987) and VanCott and Kinkade (1972).

Guidelines for Workplace Humidity. Unusually high humidity in the workplace can significantly increase the rate of worker fatigue and its magnitude. Fatigued workers are much more prone to poor performance and accidents than otherwise. Unusually low humidity in the workplace can affect (albeit to a lesser degree) a worker's physical well-being which results in reduced alertness, attention span, and/or conscientiousness.

Workplace humidity need not be mechanically controlled as long as the source of ambient air is relatively unhumid. A typical range of comfortable relative humidity is 40 to 60 percent. If the source of ambient air has a relative humidity outside this range or if the conditions of the work activity significantly alter the ambient air's relative humidity outside this normal range, then some form of artificial humidification or dehumidification, as appropriate, of the workplace air is recommended.

It is important to recognize that relative humidity is dependent upon air temperature and the guidelines provided in this subsection should be considered in conjunction with those for workplace temperature in the subsequent subsection. Consultation with a certified heating and air conditioning engineer is recommended before any workplace is constructed or modified. It is important to recognize that heating and air conditioning professionals are qualified to build air handling systems which not only work safely and properly but also comply with local building ordinances and codes. These professionals are not, however, well versed in optimal human factors considerations or in workplace safety principles in general. Hence, optimal design is usually a synergism and often a symbiosis of several contributors.

Guidelines for Workplace Temperature. As observed in the previous subsection, consideration of workplace temperature should be made in conjunction with the consideration of workplace relative humidity. Workplace temperature, if not maintained at a normal level, can have a deleterious effect on worker performance. If the workplace temperature is allowed to drop or increase to extreme levels, a significant degradation in worker performance, attitude, and in some cases in worker health may result. These effects on a worker's mental and physical condition have a direct correlation to the worker's ability to execute work tasks in a safe and diligent manner. Therefore, each employer has a responsibility to provide a workplace which is as close to being optimally heated/cooled as practicable.

For most types of sedentary or light work activity a workplace temperature should be maintained in the range of 70 to 75 degrees Fahrenheit (21.5 to 23.5 degrees Celsius). This range is appropriate as long as the workers typically work in normal indoor type clothing, e.g., trousers and shirts (or dresses). If, however, there exist certain inherent work conditions which either generate heat (such as a metal foundry) or require specially low temperatures (such as food processing), certain exceptions to this recommendation for workplace temperature apply. If the normal workplace is hot, and the workers cannot "dress down" (usually for safety reasons; see the guideline subsection on clothing) special provisions must be made to mediate the high workplace temperature as much as possible. In some cases, the maximum allowable (beyond which the efficacy of the overall work system is obviated) effect of such provisions has been insufficient to appropriately alleviate the detrimental effect on the workers. In these cases, the employers have sought alternate means of executing the human tasks, i.e., these activities have been automated in some manner. If the normal workplace is unusually cold, and additional clothing and protection are insufficient compensation, then similar alternate means of accomplishing the human task should be seriously considered.

Guidelines for Workplace Light. Improper lighting in a workplace can negatively affect both work quality and performance and can cause long range physiological effects on the workers. There are generally accepted ranges of ambient light which are recommended for various types of work dependent upon the rigors that the work effort places upon the eyesight of the workers. For the most part, it is far more detrimental to have too little light than too much. Conditions with insufficient light present contribute significantly to lower productivity and product quality. It also has been found that human beings generally have a more positive attitude and cheerful disposition in workplaces which have an abundance of light.

Light is radiant energy which is in a frequency range to which the human eye is sensitive. There are many physical dimensions of light and many ways in which each may be measured. For the purposes of this discussion light will be regarded as the illuminance type and will be expressed in units of foot-candles. VanCott and Kinkade (1972) and Sanders and McCormick (1982) present definitive explanations about light and radiant energy to which the reader may refer for further information. Workplace design considerations should include the type of light, the source of the light, the placement of the light source, the adjustability of the light intensity, the reflectance of the workplace components, and the likely position of the workers with respect to the light source and the components. These considerations should be made in order to provide the worker with adequately illuminated worksurfaces, materials, and tools with minimal instances of glare, each instance having as little glare as possible.

The recommended levels of illuminance are categorized according to the size of the smallest aspect of the work activity. Because the size of an object is relative (from the viewer's point of view) to the distance from the eye, the categorization of sizes is in terms of subtended angle. Subtended angle is the number of seconds, minutes, or degrees of arc which an object "subtends," i.e., the measure of that angle of a triangle which may be constructed between the worker's eye and the object being viewed. Thus, a 12-inch ruler viewed at a distance of one foot subtends the same angle to a viewer as would a yardstick at a distance of three feet. The following table (modified from VanCott and Kinkade, 1972) presents the recommended light illuminance levels for several categories of viewed object subtended angles.

TABLE 3

SUBTENDED ANGLE*	TASK CONDITIONS	TYPICAL EXAMPLE	GENERAL ILLUMINANCE (FT-C)
<23" (<.002)	Small detail, low contrast, prolonged periods, high speed, extreme accuracy	OBJECT: Very fine copper wire TASK: PC board wiring	100
23-45" (.002-.004)	Small detail, medium contrast, medium periods, moderate speed and accuracy	OBJECT: Human hair TASK: Threading a needle	100
45"-1'30" (.004-.008)	Detailed inspection, watch repair, machining	OBJECT: Lines on a micrometer TASK: Reading fine print	75-100
1'30"-3' (.008-.015)	Sewing, studying, proofreading, inspecting for imperfections	OBJECT: Button thread TASK: Reading text, QC Checks	50-100
3'-6' (.015-.03)	Reading, parts assembly, general office and laboratory work	OBJECT: Boldface type TASK: Reading signs, using tools	25-75
6'<(.03<)	Hallways, restrooms, areas of assembly	OBJECT: Light switch, door knob TASK: Playing checkers	10-20
	Recreational facilities, homes	OBJECT: Glassware TASK: Bowling, vacuuming	5-10
	Restaurants, bulk supply warehouses	OBJECT: Chair, table TASK: Operating a forklift	2-5

* Expressed in seconds (") and minutes (') of arc at a viewing distance of approximately 25 inches with the height or breadth of the item in inches in parentheses.

<u>Guidelines for Workplace Sound</u>. Worker performance may be seriously affected by unnecessary and excessive ambient sound. The

term "noise" is usually used in human factors (or ergonomic) contexts to refer to unwanted or unnecessary sound. Human beings are constantly vigilant to sounds. In other words, humans have to consciously disregard unwanted sounds. It is noteworthy that humans usually have no difficulty ignoring visual stimuli in which they have no interest. This sensitivity to sound serves humans both well and unsatisfactorily depending upon the circumstances. In the circumstance of emergency, auditory channel alarm systems (such as a smoke alarm) are most effective since humans are more likely to hear the alarm signal than see a visual signal. In the circumstance of working in an environment with a lot of unnecessary sounds (or noise), on the other hand, the concentration and resulting quality of the worker can be negatively affected.

There are two types of noise: basic and excessive. Examples of basic noise are:

- Conversations of other employees within earshot
- Sounds of small office equipment and machines, e.g., typewriters, computer printers, copy machines, telephones, etc.
- Sounds of activities outside of the building, e.g., automobile traffic, aircraft overflights, birds, water fountains, building construction, etc.

Excessive noise implies that the "hearer" is in close proximity to the source of the noise. Examples of excessive noise are:

- Noise made by high speed machinery
- Noise of aircraft engines and other unmuffled engines
- Explosions and reports
- Excessively amplified sounds.

Both types of noise have a detrimental effect on worker performance. Excessive noise can have a serious long-range effect on a human's auditory acuity and other physiological factors. Excess or conflicting ambient noise conditions can degrade the effectiveness of both the communication between workers and also the interaction between workers and the auditory signals of the workplace. Some of these signals and certain communications (such as "Watch out!") can be extremely critical to worker safety. In order to minimize the negative impact on safety that can be caused by workplace noise the following guidelines are presented.

First, minimize the noise at the source to the maximum extent possible. This activity involves such actions as sound proofing, insulating, shrouding, etc. Second, minimize the intensity of the noise at the receiving end to the maximum extent possible. This activity involves the use of protective devices such as earplugs, ear protectors, etc. Both of these recommendations when implemented will, in addition, reduce the likelihood of long-term hearing damage or loss. Third, all emergency or danger signals (e.g., fire alarms, and overload, over pressure, or overheat signals, etc.) should be designed to have a sound intensity level which is set at approximately halfway between the ambient noise intensity level and the maximum deciBel level which would cause pain, i.e., approximately 135dB at 300Hz, according to VanCott and Kinkade (1972).

Noise is usually measured in terms of frequency and intensity. Frequency is expressed in units of cycles per second or Hertz (Hz). Intensity is expressed in units of deciBels (dB). Human sensitivity to sound waves in the air is limited to the range of frequency of 150 to 8000 Hz. Airwaves of frequencies outside of this range are imperceptible by humans. There is little possibility of causing discomfort or injury to humans by exposing them to sounds with frequencies outside of their sensitivity range. This passivity is not true of the intensity dimension of sound, however. Even though sounds of low intensity, regardless of frequency, have little physiological impact on humans, sounds of high intensity can usually cause temporary loss of hearing and often can cause ruptured eardrums and/or long-term (and even permanent) loss of hearing or sensitivity in certain frequency ranges. It is recommended by most empiricists that ambient sound conditions be maintained in frequencies less than 3500 Hz and at intensity levels below 70 dB.

It is interesting to note that human perception of sound is phenomenologically different than the scientific dimensionality and measurement of sound waves. In other words, physiologists measure the human perception of the sound wave parameter of frequency in terms of a parameter called pitch. In addition, human perception of sound intensity is measured in terms of loudness. Neither parameter of pitch or loudness is measured in a metrical ratio scale. In other words, human perception of sound is neither regular nor linear.

Guidelines for Workplace Vibration. Vibration is a phenomenon which is caused by an object which is oscillating in a narrow band at a high frequency. Most of us have experienced vibration in many different ways. For example, the perception of the hum of an electric motor such as a grinder is as much due to vibration as it is to sound. The perception of "noise" during the takeoff of an aircraft as perceived by its passengers is primarily due to vibration. Humans perceive vibration as an integration of both the senses of hearing and of touch.

For the most part vibration is harmless, unless the human being is in direct contact with the source of the vibration (such as when holding a tool or being very close to a large vibrating piece of equipment or machinery). Direct contact with the vibration can cause physiological harm to that part of a human's body touching or holding the vibrating object even though there is not any associated harmful potential to his or her hearing. Vibration is insidious in that harm or damage can be occurring even though the human, since there is no discomfort perceived by the auditory system, feels perfectly "safe."

Devices such as tools which vibrate should be dampened, if possible, or the interface with the human operator should be insulated as much as possible. Further, the operators should not be required to be in direct contact with the vibrating device for extended periods of time, regardless of the dampening or insulation. This objective may easily be accomplished by providing the operators with alternative duties with other devices in a work process flow or by imposing mandatory rest periods of sufficient length to allow for the human body members to recover from the vibration's effects.

Guidelines for Workplace Particulate Matter. Particulate matter is usually dust. However, there are many work conditions which generate such products as smoke, sawdust, machine filings, gases, water vapor, gasoline vapor, chemical powder, grain powder (flour), and so on. The human nose is equipped with hair and mucous membranes specifically designed to protect the sensitive and delicate tissues of the lungs from contamination due to alien particulate matter in the air which is breathed. This protection system cannot effectively guard against conditions which are either excessive or sustained for long periods of time or when the workers tend to breathe through their mouths. Humans may be affected in the short-term by reacting to particulate matter as it may deteriorate the efficiency of the lungs or as it may alter the delicate chemical balance of the blood. In addition, humans can develop significant long-term physiological damage from particulate matter, e.g., black lung disease, emphysema, chronic bronchitis, and lung cancer.

Every workplace situation which has particulate matter inherently present should be critically reviewed by competent specialists to minimize the magnitude and frequency of the presence of the particulate matter. [Refer to the discussion on ACGIH and TLVs in the subsection on Workplace Air.] If the results of this type of review do not reduce the particulate matter to acceptable levels (which are different for each type of matter), then special provisions must be made for the workers to be protected from direct contact with the matter. Typical protection devices are eye protectors, masks, and breathing filters. In some cases the workers need to be provided with a means of being completely separated from the source of the matter. These cases are, understandably, extreme cases. For example, the diving suit, the space suit, and laboratory cases with arm-length rubber glove inserts are examples of protection devices designed to allow humans to work in environments which either had particulate matter so dense (as underwater), so sparce (as in outer space), or so poisonous (as with plague specimens in vitro) that would have entirely prevented human involvement otherwise.

The Physical Workplace

The physical workplace consists of the specific things with which the worker interacts in the course of doing the work. For the purposes of this Chapter, this subsection will address all aspects of the physical workplace according to this definition except tools, which will be addressed in a separate subsection. The components of the physical workplace can be classed into two categories: fixed and movable.

Guidelines for Fixed Physical Workplace Components. Fixed physical workplace components consist of walls, doorways, doors, floors, carpets, ceilings, windows, stairs, elevators, escalators, machinery, poles, ramps, rails, beams, pipes, cranes, winches, racks, shelves, fixtures, etc.

For the purposes of safety, for the most part, fixed components are most conducive to contributing to injuries and accidents as opposed to long-term health problems. It is important to note, however, that many accidents and injuries may be severe enough to disable a worker for a significantly long period of time.

Generally, it is fundamental to understand that the phenomenon of injury to the worker due to an accident with respect to a fixed physical component is one where the component is passive and the worker is active. In other words, the worker is not injured because the wall hit him/her; rather, it is because the worker hit the wall. Thus, for maximum safety it is important for all fixed components with which a worker may be in contact to be properly protected so as to eliminate the likelihood of injury and minimize the severity thereof as much as physically possible. For example, there are many actions which an employer (or workplace designer) can take to make the workplace as safe as possible for the employees:

- Wrap insulating material around pipes which are transmitting hot materials.
- Make doorways high enough for most people to pass through without ducking (taking into account the height effects of headgear and vehicles upon which the workers may be riding).
- Eliminate all sharp points and edges from walls, doors, doorways, beams, rails, etc.
- Provide cushions and bumpers when and where contact is likely.
- Minimize the frequency and the magnitude of the presence of material on the floor which would cause slipperiness or unsure footing for the worker.
- Eliminate materials which break easily into sharp objects or pieces such as glass in doors or walls which are likely to be abruptly contacted during the course of normal work.
- Thoroughly insulate and ground electrical components.
- Effectively illuminate stairs and other potentially treacherous paths and passages.

- Effectively mark with standard nomenclature hot, cold, poisonous, and dangerous items (such as shafts, pits, and drop-offs), and otherwise hazardous work components and/or conditions.

Guidelines for Movable Physical Workplace Components. Movable components are comprised of tables, desks, chairs, stools, benches, carts, cases, boxes, bins, equipments (such as computers, meters, sensors, etc.), and materials (including products, assemblies, pieces, parts, wire, wood, metal, fluids, adhesives, lubricants, insulations, cleaners, etc.). These components are the things with which the workers are in direct contact continuously throughout the work process.

As with the fixed physical components of the workplace, the movable components are more conducive to causing injury to the worker than any other type of component. For the most part, the best mechanism to protect the worker from typical injuries is to provide some sort of tool or clothing (such as gloves, goggles, earplugs, etc.) which is designed to protect the worker from risk or injury. Guidelines for these types of devices are provided in the next two subsections. This subsection specifically addresses those movable components of the workplace which are not considered to be tools or clothing.

Generally the most common movable workplace components which are designed specifically with respect to safety considerations are the tables, desks, chairs, and stools, i.e., the places on which the workers sit and at which they work. Most of these components are designed and built today in full consideration of human factors principles, one of which is safety. Basically, from a safety point of view such factors as are listed below are either eliminated or protected in some effective manner:

- Pinch points
- Sharp edges or points
- Heaviness (if to be lifted)
- Resistance (if to be pulled or pushed as a drawer)
- Breakability into sharp pieces
- Instability (as chairs and stools which might tip over)
- Instability (as a high place from which someone might fall)

• Rough edges and abrasiveness.

Tools

Tools are the devices which workers use to help them accomplish the required work. For the most part workers use tools with their hands. Thus, it is important for tool designers to utilize basic human factors principles in the designing and the making of tools. Poor tool design can lead to both poor work performance and quality as well as to long-term physiological problems such as carpel tunnel syndrome, tendonitis, and synovitis. There are several safety considerations which are pertinent to tool design and use. This subsection provides a number of safety related guidelines as pertain to tools.

Tools are either mechanical or power-assisted (i.e., are electrically or motor driven). Mechanical tools are mostly traditional tools such as hammers, chisels, pliers, screwdrivers, and so on. Power-assisted tools are relatively common these days such as electric drills, electric saws, chain saws, and shop tools such as air hammers, pneumatic drills, table saws, drill presses, grinders, and lathes.

The first rule of safety for all kinds and types of tools is for every worker to understand the application and the limitations of each tool. In other words, there is a strong possibility of jeopardizing the workers' (and any co-workers') safety if any tool is used for purposes (or outside the limits) other than for what it was intended. In addition, each tool should be designed and manufactured to minimize the user's likelihood of contact with sharp points, edges, hot surfaces, pinch points, abrasive surfaces, etc. Further, if the tool is electrical, it should be properly insulated. If the tool is driven by a gasoline (or other type of fuel) motor, special measures should be taken to eliminate hazards due to escaped fumes, hot exhausts and exhaust pipes, or explosions. If the tool is pneumatic, it should be insulated and the employee should be provided with ear and eye protection. Vibrating tools (such as chain saws, drills, air hammers, grinders, riveters, etc.) should be insulated to minimize the negative effects of vibration on the user. In addition, all power-assisted tools should have special provisions for emergency stoppages and for accidental instances of loss of control. A "dead man" switch is an example of this type of special provision.

The most important principle in ensuring and maintaining workplace safety with regard to tool design and use is that of proper training and orientation of the workers. No worker should be allowed to use any tool at work unless the supervisor is completely confident that the worker is both capable of using that tool and is also thoroughly knowledgeable of how the tool is to be used in the particular work operation at hand. Many of the accidents in the workplace which involve a tool are attributable to lack of training, understanding, or knowledge about the tool, and/or its role in the work process.

Clothing

A very important component in the workplace composition is the clothing which the workers wear. In particular, for the purposes of ensuring safety, clothing refers to those articles which the workers are wearing to either assist them in performing required work or enable them to sustain a reasonable work attitude and pace in unusual or extreme work conditions. These two types of clothing both serve to protect the worker and are necessitated by safety considerations.

Examples of clothing types which assist the workers are goggles, safety glasses, earplugs, hard hats, masks, gloves, kneepads, steel-toed boots, etc. Examples of the types of clothing which allow for the workers to work in conditions which otherwise would be too extreme are flak jackets, self-contained breathing apparatuses, thermal suits, pressure suits, welders' hoods, sound-proof helmets, and so on.

In the choice of clothing (and when necessary in the design of clothing) which comply with a safety consideration, it is important to keep in mind that the solution to one problem may cause another. For example, the selection of a particular pair of gloves which may be intended to protect a worker from burns while handling a hot piece of metal may inadvertently add a risk of the worker getting his or her hand tangled up in some of the equipment or machinery due to there being an excessive amount of loose material about the base of the glove. Therefore it is important to carefully review the merits and the drawbacks of each article of clothing considered for a particular worker in a particular situation.

Special attention and concern should be paid to those articles of clothing which a worker may wear over or near the eyes and ears. The reason for this concern is twofold: first, the eyes and ears are the most sensitive areas of a human being and special care is necessary to ensure that the incorporation of an alien substance or surface into a worker's workplace configuration is as compatible as possible; second, since the workers depend almost exclusively on interpreting stimuli in the workplace through the visual and auditory channels, it is essential that any article of clothing which is to be used around or over the eyes or ears does not interfere with this process.

Finally all employees are to be instructed on the proper care and maintenance of the various protective clothing articles provided for their safety. They should be well aware that safety is a shared responsibility between the employer and the employees.

WORKPLACE HAZARDS

This Section describes a number of common or typical workplace situations which pose hazardous threats to workers or users. Since workplaces can be of innumerable types it is impossible in the limited space available to provide a representative example of each. Rather, each example will be characterized in terms of its potential risk to the human being so that regardless of venue, the implication of safety hazard or risk can be understood.

One of the most common mishaps in the workplace and one of the most debilitating to the victim is a fall. Most falls occur in occupations where the worker is inherently at risk of a fall such as being up a ladder, in a tree, on a scaffolding, etc. The OSH Act standards provide excellent guidance to minimize the risk of falls in this type of work activity. The risk of fall is tangible in most other work situations, too. It is in these presumably innocuous situations that the risk of a fall is most insidious.

It is extremely important for all situations where a drop type fall might occur (e.g., stairways, catwalks, etc.) to be very well lit, well marked, provided with railings, and so on. In the case of stairs, the standard runner-to-riser ratio of 1.25:1 (where runner is the depth of the step and riser is the height of the step) is a minimum acceptable benchmark. It is extremely important for all stairways in a single facility to have identical ratios, once a particular ratio has been set. Human beings are quite adaptable and can easily negotiate a stairway with a particular runner-to-riser ratio (within reasonable limits). This adaptability, however, is not without its own peculiar limitations. Even though a human can usually walk up to a particular stairway which he or she has never before negotiated and use it without error, this success is based upon the human's mental expectation that the steps are all the same depth and height. It is this mental expectation which impedes our adaptability. Thus, stairways of variable runner-to-riser ratios and even varying ratios between stairways within a particular facility are strongly advised against.

It is also possible for humans to fall on level floors. Smooth floors should be kept dry or have roughened surfaces. Carpeting should be laid smoothly, tightly, and without ripples or kinks. Foreign objects and matter should be kept out of the walkways. All areas where humans work or frequent should be properly illuminated. Chairs and stools should be designed and manufactured to have low centers of gravity and wide, stable feet. Employees should be admonished from climbing unauthorized structures, from sitting or standing on inappropriate objects, and from participating in any sort of horseplay in unauthorized areas.

Many falls do not cause major injury to the victim. However, the variability of the numerous parameters of a fall situation are such that no useful generalization may be made other than that the incidence of major injury is favorably reduced if the risk of a fall is reduced. The most common major injury that is incurred in a fall is the fracture of one or more bones. Fractures can occur in incidents which do not involve a fall, too. The types of non-fall situations where fractures occur are multitudinous but may generally be categorized as situations where the human (or worker) is either working with heavy material or equipment and incurs a fracture as a result of misuse, exceeding his or her physical limit, or having some object fall onto or against him or her.

Another common accident type is the abrasion, cut, or puncture. Most of the time this type of accident occurs when the victim appropriately contacts tools, materials, or some part of an object in the physical workspace. As was pointed out in the Section on Guidelines, many of these potential hazards are easily avoidable.

In addition, many potential incidents are often prevented by the human's natural warning system of senses. [Refer to Table 2 in the Introduction Section.]

In spite of the fact that a human's senses are always turned on, humans have developed a subconscious filtering mechanism which actually blocks the brain from processing sensations that are sent from certain parts of their bodies which are not of current interest. This filtering is indiscriminate in that when a human is concentrating on, say threading a needle, his or her brain filters out all other stimuli ranging from the innocuous, such as the tightness of his or her shoes, to the otherwise alarming, such as the sting of a mosquito. The only sense which is not totally affected by the filtering is that of audition. Hence, if the mosquito had only buzzed by the human's ear before landing, his or her brain may have allowed him or her to be alerted of its presence in time.

Humans are inherently variable, too. Not only is there vast variability from human to human, each individual human also has multidimensional variability. Each human varies from moment to moment as well as between events.

The importance of this discussion for the safety specialist is to recognize the existence of human variability and to understand the human filtering function both in the context that human sensing as a safety measure can be maximized by the accentuation of the human's attitude. If the human can have and maintain a proper attitude about work performance commensurate with safety considerations, the likelihood of accidents and their associated severity will be minimized.

GUARDS AND WARNINGS

The second and third basic principles of safety listed in the Introduction Section of this Chapter refer to guarding and warning against the hazard. This Section discusses the basic types of guards and warnings which are typically used in the workplace.

Guards

The general theory behind guarding against the hazard is that if the hazard is unavoidable, then the risk of injury may be somewhat abated if the human being (or the user) is protected from the hazard as much as practicable. Obviously, to protect the user from the hazard as much as possible (such as not letting the person within ten miles of the hazard) is infeasible if, indeed, work and work products are expected. Thus, the principle of guarding is essentially a compromise from the design point of view: Add in special system features ordinarily not present, which specifically protect the user from as many possibilities of injury (and levels of severity) as can be accomplished without significantly impeding the user's requirement to work in the hazardous environment.

The principle of guarding is quite straightforward in work situations where there is an evident risk, e.g., where workers in a foundry wear protective clothing, where welders wear goggles or helmets with visors, where divers wear masks and breathing apparatuses, and where drivers wear seat belts. In situations where there is not an evident risk, the concept of guarding becomes more obscure. On the one hand, human beings are not particularly cooperative with respect to guarding devices: ear protectors are commonly not worn in noisy environments unless employers require it (and sometimes severe penalties have to be imposed to ensure proper use); protective clothing is often not worn properly; and helmets or hard hats are often left back at the shed. On the other hand, humans, as thinking and reasoning beings, often have their own opinion of the potential risk in a particular situation and eschew special provisions such as guarding which has been provided for their general safety.

Guards, as has already been discussed, may involve altering the clothing worn by and equipment used by the human and may also involve altering the equipment and machinery themselves. In other words, in addition to typical guarding devices such as goggles, earplugs, gloves, boots, and helmets, the following items are also regarded as guards: heat shields, spray shields, "dead man" switches, two-hand switches, and isolation panels. In the typical workplace which is not in an industrial setting, guards are not usually necessary. The low incidence of injuries and associated severity thereof and the generally low perception of risk both contribute to this situation.

Warnings

The warning principle of safety refers to the method of providing effective, tangible, and pertinent information to a user concerning the potential for injury with respect to a particular situation or piece of equipment. Most people encounter warnings many times a day. These warnings may be found on medicine bottles, on food containers, in television broadcasts, on electrical appliances, on rear view mirrors, on road signs, in instruction manuals, and so on.

There are three key rules for effective warnings. First, use warnings only in situations where a real hazard exists. Second, display the warning in a strategic location. Third, be as meaningful as possible with as few words and pictures as possible. Commensurate with these rules, warning designers have adopted a number of usually visual cues which add significantly to the effectiveness of a warning. These cues are of the symbol, color, or pattern type. For example, the skull and crossbones symbol has been commonly associated with poisons; the colors yellow and red have become associated with caution and danger conditions; and the hashed pattern of alternating slants of yellow and black have become recognizable as the way in which many guards are marked.

Peters (1984) has expanded upon these three basic guidelines and specified 15 principles for the warning designer. The warning message must be readable, understandable, comprehensible, practical, effective, compatible, conspicuous, visible, and durable. In addition, the warning should result in the desired modification to behavior; it should serve as a reinforcement of knowledge or training received previously; and it should be easily perceived as a danger signal. Finally, Peters provides a discussion on the types of warnings and the effectiveness of novelty.

All the best guidelines notwithstanding, Peters (1986) reminds the warning designer that "one of the most frequent complaints of those attempting to put warnings and instructions on a product is [the] lack of available space." Thus, warning messages invariably are a test of the designer's mastery of his language. Peters goes on to say that "an effective warning is one that is easily understood, comprehensible, and serves to modify what might be unsafe behavior."

HAZARD ELIMINATION AND CONTROL

The first general principle of safety design, as provided previously, was to eliminate the hazard. It is far wiser to eliminate a hazard than it is to try to deal with it. Many workplace situations with tangible risk may not be eliminated, however, without significantly affecting the mission of the workplace. Even though, in general, this is a valid point, a prudent designer may exert innovation and acumen to be able to achieve an optimal combination of "necessary evils" and safety-oriented workplace design components.

Hazard Elimination

The usual case of elimination occurs when a particular workplace is being designed and therefore does not yet exist. This situation is simpler than when the workplace already exists. The identification of safety hazards in a hypothetical (or design) situation requires talent and experience. The specialist applies his or her knowledge and experience to the design problem at hand and essentially "places himself or herself" in the situation." This exercise allows for the specialist to "visualize" the role of the humans in the planned workplace. The specialist "looks" at both the key points of interface but also at the dimensions of the workplace with respect to the dynamics that are anticipated.

The safety specialist basically conducts this exercise in a number (usually two or three) of pareto-type reviews. The first such review is one which considers major hazards, i.e., hazards which would cause a catastrophic result. Many of these hazards are identified in OSH Act guidelines. Once the prospective design is qualified to be safe enough for humans to use without major risk, the specialist then reviews the design for hazards which may cause minor injury to humans. Examples of this type of injury were presented in Table 1 and were discussed in the Workplace Hazards Section of this Chapter. The next level of review that may be conducted by the safety specialist is one which addresses the possibility of causing discomfort to humans. The sections which provide various guidelines in this Chapter address the types of situations where human factors engineers and designers have found that humans can be affected negatively if proper considerations are not made.

The safety specialist uses this review approach to identify hazards in a workplace design which should be eliminated, but has not yet executed the elimination. The process of elimination, in the design stage, is reasonably straightforward. It is important to note that the safety specialist must have design authority in order for proper hazard elimination recommendations to be effective. In every case of elimination, there is some component or aspect of the workplace which is involved. Therefore, the safety specialist has an incumbent responsibility to provide alternate design recommendations when eliminating an identified hazard. Thus, the elimination process usually requires some form of evaluation. Once the specialist has identified a hazard which he or she will advocate to be eliminated from the prospective workplace, he or she must also evaluate the situation from a human factors and safety point of view in order to find a solution which will allow the humans to still work in the workplace effectively as well as safely.

This process of hazard elimination is quite common and usually is quite effective as long as the specialist has design authority. It is quite important to observe that if the specialist does not have design authority, this process is suboptimal in that it potentially wastes the system workplace design efforts by allowing for a design to be drafted before the safety specialist begins to review it. Ideally, the human factors specialist and the safety specialist (often being the same person) should be involved in the conceptualization of the workplace design so as to prevent the most common types of hazards from being part of the hypothetical workplace in the first place.

Hazard Control

Once a workplace has been designed, built, and placed into use there is often a need to conduct a safety review. Sometimes this is routine "maintenance" sponsored by a proactive management and sometimes this type of review is perturbated by worker complaints or because there has been an accident. In either case there exists a requirement for hazard control. The process of hazard control generally consists of either a human factors evaluation or a safety assessment depending upon the background of the person performing the review.

A human factors evaluation considers the human being in the construct of the work environment. The basic capabilities and limitations of the human are considered with respect to what is expected in terms of work activity and performance and to what types of things with which the human is provided to use (e.g., tools, controls, displays, signals, etc.). The types of examples presented in Table 2 illustrate how a human factors specialist resolves problem areas identified via this type of evaluation.

A safety assessment specifically addresses the work situation with respect to a standard or common set of accident/incident factors. Many safety specialists have a checklist that they use to identify what parts of a particular work situation are in need of redesign or modification. The items on these types of checklists are workplace components or situations which have been identified in previous accident/incident investigations as either contributing or causing factors.

In either case, the result is usually a recommendation to alter the workplace in some manner. If properly effected the end result is expected to reduce (at a minimum) the frequency and the severity of subsequent accidents and incidents. This process is the foundation of hazard control.

It is important to note that hazard control most often results in workplaces which have a reduced likelihood of severe accidents or incidents. If time, resources, and expertise are available a more substantive study of the workplace and work activity may be conducted which would concentrate on the causes of poor performance or accidents. Causal analysis can often result in hazard elimination and significant improvement in overall work performance.

HYGIENE AND HEALTH HAZARD

Hygiene

Hygiene, more appropriately called industrial hygiene in this context, generally refers to the maintenance of worker health in the workplace. This general definition encompasses the full spectrum of minor illnesses such as colds, influenza, and strep throat; and long-term health problems such as cancer, heart disease, emphysema, and carpel-tunnel syndrome; and episodic work related disorders such as infection, eye irritation, and minor asphyxiation. There is a substantial responsibility on the employer to provide a workplace which is not only safe from hazards to injury but also which is safe from hazards to health and hygiene. By the same token each employee has a tangible responsibility to respect the health and livelihood of others in the workplace and therefore should be prudent in his or her conduct. Specifically, an employee who is ill with a communicable disorder should not come to work until he or she is either well or no longer contagious as a result of medication. In addition, employees should not only refrain from unsafe conduct in the workplace (e.g., failing to follow procedures or acting recklessly) but also feel responsible to report others who are violating this principle.

According to Olishifski and McElroy (1971) there are numerous dimensions of industrial hygiene. Generally these dimensions fall into the category of ambient conditions described previously. They introduce their topic by describing its breadth and scope. There are such hygiene impacting types of agents as dusts, fumes, smoke, aerosols, mists, gases, vapors, solvents, noise, radiation, lasers, temperatures, and atmospheric pressure. Such agents can be naturally occurring in the workplace environment or may be products of abnormal events. These agent generating sources can be of the types: explosions, corrosions, flammable liquids, toxic chemicals (and associated reactions), oxidation, gas generation, fires, and spills. The general means of impingement on the human are by inhalation, absorption, or ingestion.

Health Hazard

Health hazard is most commonly regarded as a subset of hygiene which specifically refers to those hygiene situations which are long-term. Long-term hygiene or health hazard situations are most often situations which not only take an extended period of time to promulgate but also result in conditions which are either permanent or take a considerable period of time from which to recover. In most cases the health condition incurred by the human (or worker) is the result of being at work (as opposed to some sort of long-term health condition by the human, independent of work, such as lung cancer due to cigarette smoking).

Jones (1971) categorizes health hazards into two general categories: chemical and physical. He says that almost every occupation involves some exposure to potentially dangerous chemical substances such as solvents, dusts, and gases. He goes on to say that "certain occupational diseases result from physical environmental factors such as temperature, noise, and radiation."

He listed six basic types of radiation and characterized the harmful effects associated with each as follows:

- Gamma rays, x-rays, and atomic particles all cause permanent damage or death to cells
- Ultraviolet rays affect skin and the eyes and can be carcinogenic
- Intense visible light can cause inflammation of conjunctiva and cornea of the eye and temporary or permanent loss of sight
- Infrared rays can burn skin and cause temporary or permanent loss of sight
- Laser beams can burn skin and cause heat injury to any part of the eye
- Microwaves can produce heat in exposed tissue and produce cataracts.

The means of contracting a long-term health condition which results from exposure to some agent or condition at work is either one of the three types provided by Olishifski and McElroy and described previously (inhalation, absorption, or ingestion) or by repeated motion in a non-ergonomic situation. This latter type of health hazard refers to situations of work activity such as tool use which calls for a constantly fixed wrist angle or finger position. The most common example of this type of activity is the use of a power tool such as a hand drill. Ergonomic principles call for the wrist angle to be as straight as possible for sustained work activity and for the minimization of fixed finger positions under strain. If a non-ergonomic work situation exists, it is extremely likely that workers will develop one or more physiologically debilitating conditions, some of which can be permanent. Some of these conditions are carpel-tunnel syndrome, tendon synovitis, and tendon sheath rupture.

Fradkin (1985) reports on exposure guidelines for indoor air pollutants. These guidelines are based on 1984 U.S. Environmental Protection Agency figures. Fradkin cites five different types of pollutants with indoor guidelines [See Table 4] and three types for which guidelines were not available: biological aerosols, polycyclic hydrocarbons, and volatile organics.

TABLE 4

POLLUTANT	GUIDELINES (Maximum Allowable)
Asbestos and other fibres	0.2 fibres/ml for 8 hours
Carbon-monoxide	0.1 parts per million
Ozone	Less than 0.12 parts per million no more than once a year
Radon	0.01 working level
Inhalable particulates	150 to 350 mg/cubic meter for 24 hours

Jones (1982) reports on the long-term health hazard potential of video display terminals. He cites a study conducted by the National Institute for Occupational Safety and Health (NIOSH) which surveyed employees of Blue Shield of California who worked with video display terminals. The study tallied the following health complaints by workers:

- Eyestrain -- 93 percent
- Pain or stiffness in the neck or shoulders -- 90 percent
- Headaches -- 89 percent
- Severe fatigue -- 83 percent
- Tearing or itching of the eyes -- 79 percent
- Burning of the eyes -- 78 percent
- Blurring of vision -- 77 percent.

It is a very important responsibility of the employer to be alert to each of these types of health hazards. Most humans are, on the one hand, quite flexible and adaptable and, on the other hand, are both desirous of doing good work and also of ensuring their job security. Therefore, it is not often likely that employees will alert management (either by complaint or by suggestion) of health hazard situations or potentials. In the specific case of long-term health deteriorations it is usually too late by the time the condition is either apparent to the employee or is affecting his or her work performance. In other words, since the somewhat instantaneous observance of something amiss as is available from accidents and incidents is not likely from the presence of a health hazard, the employer must be proactive in addressing workplace safety particularly with respect to the insidious nature of long-term health hazards.

SAFETY MANAGEMENT

According to Simonds and Grimaldi (1971) "Owners should expect their management to take whatever steps may be necessary ... to guard against major catastrophes and to keep the company's accident experience as low as possible ..." It is a primary responsibility of management to have a proactive and comprehensive safety program. This program should be oriented to accident/incident and health hazard prevention as opposed to minimization or control. Any company which waits until a number of employees have been either injured or killed before it begins to look into design modifications or other safety oriented practices is doing itself and its employees an egregious disservice.

The fourth principle of safety discussed previously is reinforcement. This principle refers to the role of management to actively remind its employees of not only the hazards which are inherent to the workplace but also of the provisions and programs developed to control (and hopefully eliminate) risks of accident and/or injury. The process of reminding may take on many forms which usually should involve orientation, training, and regular and periodic re-training. These reinforcement activities should endeavor to instill in each employee a thorough respect for the hazards extant in the workplace, a comprehensive knowledge of the safety measures in place (including the warnings and guards being used), as well as refinement of the knowledge of the practices and procedures of the workplace which management has specifically designed to minimize risk to the workers.

Hammer (1981) provides a comprehensive list of safety checks which may be used by employers to either initiate a safety management program or to conduct a safety assessment of a workplace for the purposes of hazard control as discussed previously. Hammer's list of safety checks, slightly modified and expanded, is presented below:

- Obstructed lanes for emergency vehicles

- Obstructed emergency exits and passages to the exits
- Fire doors blocked or opened
- Unlocked doors or gates to enclosed hazardous areas
- Accesses blocked to fire fighting or other emergency equipment
- Boxes, bales, cartons, or other items piled too high that they might interfere with sprinkler systems
- Boxes, crates, bales, drums, cartons, or other packages piled too high that they might fall
- Loose wires, piping metal, tools, or other material, equipment, or supplies which might cause tripping or falling
- Wet, oily, or heavily waxed floors which might cause slipping or falling
- Obstructions or loose objects on stairs or at the head of stairs
- Poorly lighted stairs and stairways
- Missing or inoperative egress or exit lights
- Loose or broken stair treads, hand rails, or guard rails
- Barriers, lights, and other protective devices missing at excavations, manholes, ditches, or other openings into which a person or vehicle might fall
- Presence of oily rags, excelsior, or other highly flammable materials outside proper containers
- Uncovered containers of solvent, fuels, or other flammable liquids
- Cigarette butts, matches, or other evidence of smoking in non-smoking areas
- Failure of welders to have fire extinguishers close at hand
- Missing fire extinguishers or evidence that they have been used, are not in their assigned locations, or due (or past due) for inspection
- Steam, oil, water, fuel, or chemical leaks
- Unusual odors or fumes
- Improperly stored or inadequately secured gas cylinders
- Broken electrical fittings, outlets, plugs, or other devices

- Broken glass or other sharp, unprotected edges, points, or surfaces
- Chairs, stools, platforms, etc. which are unsteady, unsecure, unstable, or tippable in any way
- Heaters, torches, welders, etc. which are being used improperly or without proper safety protection.

Petersen and Goodale (1980) provide a concise set of "Do's" and "Don'ts" for safety professionals. Adherence to these recommendations would provide an effective basis for a successful safety management program. The "Don'ts" are:

- Make examples of people who violate safety rules
- Impose safety standards and procedures on employees
- Check to see if people are working safely
- Develop and install new safety procedures
- Ignore safe work practices
- Stimulate safety by emotional appeals and elaborate campaigns
- Make safety exclusively the safety professional's responsibility.

And Petersen and Goodale's "Do's" are:

- Help people see the need for rules and the consequences of violations
- Involve employees in solving safety problems and setting standards
- Provide a self-monitoring system
- Make improving safety procedures part of each person's job
- Recognize safe work practices
- Stimulate safety by making safe work a part of everyone's job goals
- Share a responsibility for safety with all employees.

A very important part of a safety management program is safety training. Denton (1982) offers that "safety training prepares new employees to do their jobs better and more safely." Denton goes on to say that safety training should contain material and information such as:

- Safety rules, regulations, and responsibilities that apply to employees
- Employee's rights and responsibilities under the OSH Act
- Personal protective equipment available at the plant
- Location of medical and emergency aid
- Insurance and Workmen's Compensation benefits
- Hazards and safeguards peculiar to the employee's job
- Incentives available for safe performance
- Discipline present for unsafe performance.

In addition to ensuring that the workplace provided for the workers is as safe as possible, an effective safety management program would take into judicious consideration as many safety oriented design guidelines (such as are provided in this Chapter) as possible in the designing of new and existing workplaces.

SUMMARY

The design and management of safe workplaces should be a primary responsibility of all employers and management. In addition, each employee has an incumbent responsibility to be well informed of the safety measures and practices provided by his or her employers, to adhere to these measures and practices to the best of his or her ability, and to be alert to notice and report unsafe practices or conditions to the proper organizations.

The state-of-the-art in safety guidelines and practices is constantly evolving as more and more accident/incident data is collected and analyzed. In addition, the breadth and scope of ergonomic design guideline information is constantly increasing and improving. Therefore, it is essential for the safety specialist to have access to the literature of the professional safety and ergonomics communities as well as of the various U.S. Government, state, and municipal agencies which address safety such as OSHA, NIOSH, ACGIH, and the National Safety Council which have been referred to in this Chapter.

REFERENCES

Denton, D.K., 1982. Safety Management, Improving Performance. McGraw-Hill Book Company, New York, New York.

Fradkin, A., 1985. Something in the Air. Office Ergonomics, May/June, 1985.

Hammer, W., 1980. Product Safety and Management Engineering. Prentice-Hall, Inc., Englewood Cliffs, New Jersey.

Hammer, W., 1981. Occupational Safety Management and Engineering, Second Edition. Prentice-Hall, Inc., Englewood Cliffs, New Jersey.

Jones, E., 1982. Is Your Computer Terminal Hazardous to Your Health? Popular Computing, July, 1982.

Olishifski, J.B., and McElroy, F.E., (Eds.), 1971. Fundamentals of Industrial Hygiene. National Safety Council, Chicago, Illinois.

Peters, G.A., 1984. 15 Cardinal Principles to Ensure Effectiveness of Warning Systems. Occupational Health & Safety, May, 1984.

Peters, G.A., 1986. Warnings and Instructions for the Packaging Specialist. Hazard Prevention, January/February, 1986.

Peterson, D., and Goodale, J. (Eds.), 1980. Readings in Industrial Accident Prevention. McGraw-Hill Book Company, New York, New York.

Salvendy, G. (Ed.), 1987, Handbook of Human Factors. John Wiley and Sons, Inc., New York, New York.

Sanders, M.S., and McCormick, E.J., 1982. Human Factors Engineering and Design, Fifth Edition. McGraw-Hill Book Company, New York, New York.

Simonds, R.H., and Grimaldi, J.V., 1963. Safety Management, Accident and Cost Control, Revised Edition. Richard D. Irwin, Inc., Homewood, Illinois.

VanCott, H.P., and Kinkade, R.G., 1972. Human Engineering Guide to Equipment Design, Revised Edition. Joint Army-Navy-Air Force Steering Committee, Washington, D.C.

The World Almanac and Book of Facts 1990, 1989. World Almanac, New York, New York.

Workspace, Equipment and Tool Design
edited by A. Mital & W. Karwowski

WORKSPACE DESIGN FOR MAINTENANCE

Sheik N. Imrhan, Ph.D., P.E.

Department of Industrial Engineering, The University of Texas at Arlington, Arlington, Texas 76019

ABSTRACT

The accelerating costs of maintenance, especially in the military and in the aircraft industry, have forced attention towards maintenance techniques which are cost effective, efficient and safe. Workspace design has been identified as one of the more important areas requiring improvement. This paper describes the principles of workspace design for maintenance activities - specifically those involving human visual access, physical access, mobility, and physical strength and endurance. The importance of biomechanics and anthropometry in defining workspace in 3-dimensions has been emphasized. Guidelines are given for workspace design; and the appropriateness of 3-dimensional computerized models for future data compilation discussed.

INTRODUCTION

No one today can realistically expect to design a machine which will work without stoppage, forever. Engineers traditionally have viewed this ability to work from the concept of reliability; that is, the probability of successive performance of the system. Efforts to enhance reliability of a system have traditionally concentrated on developing and using components with high reliabilities, and networking components (redundant arrangements) to achieve high reliabilities. Machines and their components, however, fail and must be replaced or repaired. Reliability theory, using sophisticated statistical techniques, has helped to predict failure times. Components of a machine are, therefore, replaced (serviced) before they are expected to fail. For relatively simple machines this may be done efficiently and with a fair degree of cost effectiveness. With very complex machines, not only has prediction of failures become more difficult, but also the very act of maintenance (servicing and repairing) have become mind-boggling and costly. The increasing complexity of machines over the last 40 years has progressively reduced the percentage of persons in the population who can even comprehend their operation, much less maintain them. This has been exacerbated by their complex relationships with the environment and with humans. At one time it was hoped that the rapid improvement in the reliability of machines and components would practically obviate the need for maintenance activities (Morgan et al. 1963). Unfortunately this has never been realized and, perhaps, it never will be realized. The growth of machine complexity (Crawford and Altman, 1972; Morgan et al., 1963;and Oborne, 1981) has far outpaced the improvements in reliability techniques. New problems in downtime have been proliferating, and the concept of maintenance as a means to reduce downtime has now assumed elevated importance compared to the concept of reliability. A look at a practical definition of maintainability will illustrate why.

According to Morgan et al. (1963) maintainability can be defined as "the degree of facility with which an equipment or system is capable of being retained in, or restored to, serviceable operation. It is a function of parts accessibility; interval configuration; use and repair environment; and the time, tools, and training required to effect maintenance." This definition is still appropriate today; and, by implication, maintenance is predominantly a human activity.

Maintainability Versus Operability

Historically, maintenance has been given much less attention than operational aspects of human-machine systems by ergonomists. Design for ease of maintenance has been more of an art than a science; it has evolved to a greater extent by common sense than by scientific investigation (Oborne, 1981). The increasing need to perform maintenance activities, combined with the high cost of human labor, has made maintenance one of the most costly aspects of human-machine systems. For example, aircraft maintenance costs in the U.S. military have been estimated at about 35% of the life cycle costs of military systems (McDaniel and Askren, 1985); and as early as 1970 it has been reported that the U.S. Department of Defense allocated 1/4 of its budget to maintenance costs (Smith et al., 1970). The technical complexity of modern aircrafts, compounded by the vast array of hardware choices made possible by computer aided design systems, has accentuated this problem (Adams, 1988b). As a result, the U.S. Air Force has embarked on the development of a computerized model, CREW CHIEF, to define, describe and predict human strength and anthropometric requirements for many basic maintenance tasks, including the use of common hand tools, and especially in unnatural body postures (Adams, 1988b; and Deivanayagam, 1988).

Designing for maintenance is much more difficult than designing for operability. The maintenance workplace is much more variable and unpredictable; and technicians must often work within limited workspaces created by what are usually considered good designs for operability and marketability. Greater and rapidly changing variability in equipment and system design render standardization more difficult. Often designing for operability and cost effectiveness may require criteria which are in direct opposition with those required for maintainability. One of these is space: within a machine, between components; between different machines in the workplace; and between a machine and nearby objects in the environment. The present trend is towards miniaturization and compactness (Tichauer, 1978), driven by the need to lower the cost of production; to enhance the ease of manipulation of machines during operation and during transportation; and to satisfy the demands of a gradually shrinking workplace size. This trend runs counter to the fact that performance in maintenance tasks are degraded when workspace is limited. Workspace requirements must therefore be given importance at the planning and design stages, to avoid prohibitive costs of maintenance and delayed downtime, later on.

MAINTENANCE REQUIREMENTS FOR WORKERS

Like any other skilled worker, a maintenance technician must be equipped with a variety of skills and knowledge to perform his task efficiently - psychomotor, cognitive, mental, physiological and physical (Oborne, 1981). There are many aspects of the work environment which are also necessary for his successful performance, such as appropriate tools, good environmental conditions (absence of extremes of heat, cold, noise, illumination, vibration, etc.), suitable co-workers, and adequate workspace.

Inadequate and poorly designed workspaces can affect many of the attributes of the worker -- his speed, accuracy, precision, coordination, control, strength, endurance, and motivation. These will certainly translate into increased cost of maintenance, as downtime is prolonged (from extended maintenance periods) and quality of work deteriorates (requiring shorter intervals between successive maintenance sessions). The system may fail more often, necessitating a greater number of workers, or more highly skilled ones.

The purpose of this chapter is to examine the importance of workspace design for maintenance and provide data and guidelines for such designs. The discussion of the other conditions is beyond the scope and purpose of this chapter, and can be found in texts such as Morgan et al. (1963), VanCott and Kinkade (1972) etc. This has been a much neglected topic in human factors/ ergonomics textbooks, which tend to concentrate almost entirely on the operational aspects of ergonomic designs of machines and human-machine-environment systems (Hutchinson, 1981; Gandjean, 1987; and Sanders and McCormick, 1987). It is hoped that this discussion will highlight the areas where data is badly needed and the implications of having such data on hand, in addition to explaining the ergonomic principles of workspace design for maintenance. Since the concept of efficiency in human performance (in terms of speed, accuracy and precision) is intricately intertwined with safety, some emphasis in the discussion will be placed on the biomechanics of maintenance work to illustrate how inadequate workspace design can contribute to work related musculoskeletal disorders.

TYPES OF TASKS

In order to fully appreciate the effects of workspace design for maintenance one must recognize the types of tasks involved. These include, but are not limited to (McKendry et al., 1960):

1. Trouble shooting to identify faults to be corrected.
2. Inspecting hardware to detect and locate faults.
3. Manipulation of parts.
4. Adjusting and repairing.
5. Removing.
6. Replacing.
7. Servicing.

How well a maintenance job is performed depends, to a large extent, on the worker's inherent capacities, and his ability to utilize those capacities within the constraints of the maintenance environment. The types of tasks which can be affected by workspace design include (Crawford and Altman, 1972; and Deivanayagam, 1986):

1. Visual access -- to inspect and observe faults.
2. Physical access -- into spaces with the whole body or part of the body, such as the hand
3. Muscular strength -- in applying forces and torques directly onto machine parts or via handtools to loosen, remove or attach parts.
4. Mobility -- the ability to move the body or part of the body to maintain equilibrium, gain leverage for force (or torque) application, and for moving equipment and other maintenance items.

Apart from visual access, these tasks are mainly manual, involving a wide variety of muscular actions, using the upper body mainly for generating mechanical power, and the lower body for stabilization. The impact of the volume and arrangement of space available will, therefore, have significant effects on performance. This concept will be developed in greater detail later, in the section on body posture.

ACCESSIBILITY

One of the earliest conditions that must be satisfied for a maintenance technician to perform his tasks effectively is establishing visual and physical contact with the hardware being repaired or serviced. Physical accessibility may involve the whole body, as in removing/replacing heavy machinery which must be lifted and carried manually; or the hands, as in using a hand tool for tightening or loosening nuts and bolts. This has been one of the most perplexing ergonomic design problems, because of the rapidly changing relationship between person and machine, as the latter steadily tends towards compactness and miniaturization (Tichauer, 1973), whereas the human body cannot change size and shape, at will.

Adequate workspace requirements are necessary for (Morgan et al. 1963; VanCott and Kinkade, 1972; Alexander, 1985; and Galer, 1987):

1. General and local illumination of machines and components.
2. Establishing visual access to hardware, especially in local areas where small components are to be serviced or repaired.
3. Face to face communication with other members of the maintenance team; it is often the most effective means of communication where instructions are to be conveyed by gestures, as in the direction and supervision of a team of technicians by a supervisor.
4. Placement of maintenance equipment where they can be conveniently retrieved while working; this becomes critical when workers must work in poorly illuminated environments.
5. Uninhibited movement of one or more workers for retrieving tools and establishing communication links, in the flow or sharing of materials and tools.
6. Changing of body posture in order to gain effective biomechanical leverage and relieve fatigued muscles and tissues.
7. Clearance for whole body when entering or exiting workspaces; or for part of the body, such as the hand, when establishing contact with machine components.
8. Clearance for tools that are larger than the hand or arm, as the case may be.
9. Clearance for easy removal and replacement of machines components.

Visual access

Visual access is required for

1. Seeing other workers with whom communication is necessary to coordinate various aspects of a job; very often verbal communication, combined with gestures of the hand, is the most effective means of communication, especially when working in noisy environment.
2. Seeing parts of equipment being worked on or simply locating problem sites for maintenance, during inspection stage, such as cracks on hardware, leaking fluids, worn belts and displaced and misaligned parts.

Inability to see co-workers may occur when huge immovable equipment like cabinets and opaque physical barriers like walls are present in an area where more than one worker is required to work on a single task. During the design stages such difficulties may be anticipated and eliminated by using cabinet tracks which allow cabinets to be removed fairly easily; and apertures in walls which can be sealed by hinged doors when not needed. The necessity to see parts being worked on is more crucial as the parts get smaller, and when hand tools are to be used. Smaller parts require greater accuracy in positioning the hand; and tool and nearby parts can easily be mistaken when the worker is unable to place his head in a position to establish a direct line of sight with them. During design stages, such direct line of sight and similarity in shapes and sizes of nuts, bolts, and other parts must therefore be considered. When using a handtool, such as a wrench or screwdriver, the maintenance technician does not have the full use of sensory feedback that occurs as when the fingers are in direct contact with parts. Clear visual contact with such parts must be established to reduce errors of identification of parts, and time taken to perform the task, in addition to eliminating (or minimizing) the danger of inadvertently coming into contact with sharp edges, exposed wires or other forms of hazards.

When general and local illumination are below that required for maintenance tasks, accuracy and speed of performance may also deteriorate. Local illumination is often a problem when spaces between parts are limited and do not present a linear path for flashlight beams and other sources of light. The worker may then have to move the flashlight, say, around corners within the machine -- a near impossible task for compact machinery with limited space for such objects.

Problems with visual access occur not only with the amount of space available but also with the orientation of the machinery. Technicians may have to resort to unnatural and fatiguing postures, like stooping or kneeling while twisting the trunk, to see parts that are being maintained, and also to exert forces and torques with the arm(s). The nature and effects of these constrained postures are discussed in detail elsewhere in this chapter. A solution to this in the design phases of the machine is to provide for a mechanical system which can reorient the machine in all planes of space (Van Cott) and at various heights above the floor, so that the worker can assume advantageous body postures and local skeletal configurations, while in good visual contact with the part(s) being maintained.

Vision and age

Workspace design should also take into account the visual acuity of the potential workforce. The near point of the eye of a given worker determines the limit to which the eye can brought to the object under scrutiny so that a sharp focus can be achieved; and the closer the eye is brought to the object, the greater is the area of distinct vision, within the vertical angle of 1 deg (Grandjean, 1987). Physical obstructions and inadequate clearances for the head, and part or all of the upper body, can severely limit the proximity of the eye to the object. This is especially important when the potential workforce is of a wide age range. Grandjean (1987) lists average near point distances at various ages, as follows:

at 16 years: 8 cm
at 32 years: 12.5 cm
at 44 years: 25 cm
at 50 years: 50 cm
at 60 years: 100 cm

The drastic effect of age beyond 44 years is evident, and must be considered because of the increasing importance of older workers in the workplace. Presbyopia (near point beyond 25 cm) can be corrected by spectacles or contact lenses. For those who prefer to wear spectacles, additional clearance must be added for the effect of the size of frame of the spectacles.

Serious musculoskeletal problems may occur if the worker's neck must be flexed and the head thrust forwards (as in attempting to bring the workpiece within focus of the eye) for prolonged periods of time. The moment due to the mass of the head about the atlanto-occipital joint can produce high tensions in the cervical extensor muscles and, consequently, fatigue and pain (Kumar and Scaife, 1979; and Chaffin and Anderson, 1984). A knowledge of the preferred line of sight in humans can help in the arrangement of work stations for maintenance. Recent research (Hill and Kroemer, 1986) has shown that the preferred line of sight associated with subjective feelings of comfort depended on body posture and the object distance from the eye, averaging 34 deg below the Frankfurt plane. The need to extend the neck to look above the Frankfurt plane is common, where workspace is poorly arranged; and fatigue of the ciliary muscles of the eye is common where the eye must be rotated to the extremes within the eye sockets. As a rough guideline, the 30 deg cone about the line of sight (U.S. Department of Defense, 1981) may suffice for optimal placement of workpiece.

Continuous close work, also, may also lead to fatigue of the ciliary muscles due to frequent and rapid convergence and accommodation, especially in older workers in whom these two functions deteriorate gradually (Singleton, 1982; Oborne, 1981; and Grandjean, 1987).

Design guidelines for accessibility

From the previous discussion design guidelines can be summarized as follows:

1. Enough access space must be provided (i) for the maintenance worker to see what he/she is doing; (ii) for worker to see what other team members are doing; and (iii) to allow free communication by gestures, as

in relaying non-verbal instructions, between workers. Design features should be incorporated to prevent physical obstructions to visual access; for example, use of cabinet tracks, roll out racks, slides or hinges.

2. Where local illumination is necessary to see things, enough space must be provided for the largest lamps to be manipulated. Consideration must be given to the method of use of the lamps; for example, additional space must be provided for large hands grasping flashlights.
3. Older workers with visual defects may require greater eye-object separation.
4. Physical access into workspaces must be based on realistic anthropometric dimensions; task analyses should be performed to determine the body segment(s) that will be used for specific tasks, and the type of motions involved.

BODY POSTURE AND BIOMECHANICS

The success and efficiency of many maintenance tasks involve the ability of the worker to generate forces or torques for loosening and tightening mechanical parts, such as connectors, bolts, nuts, and screws; for holding and stabilizing one section of a machine while working on another section; and for moving equipment around. Some of the most important of these tasks involve:

1. Holding - with the fingers or with the whole hand.
2. Gripping - with the hand to generate and transmit forces and torques.
3. Pinching - with the fingers.
4. Twisting - with the wrist, for generating rotary torques.
5. Pushing.
6. Pulling.
7. Lifting.
8. Pressing Down.
9. Carrying - often combined with lifting. Tasks 4-9 are hardly accomplishable without gripping.

In all these situations the human body acts as a complex series of lever systems (Tichauer, 1973; Chaffin and Anderson, 1984) whereby the forces of muscular contraction are transmitted to overcome external resistances. When the musculoskeletal system is stressed, either through powerful instantaneous muscular contractions (as in attempting to lift a very heavy object) or through repetitive stressful contractions, the worker is at risk to musculoskeletal injuries (Chaffin and Anderson, 1984). As will be argued later in this section, poor body posture and/or mechanically disadvantageous local skeletal configurations contribute greatly to injuries and strains. It is therefore important to know how the design of workspaces can influence postures assumed by the maintenance technician during work.

The typical body postures assumed by technicians during maintenance tasks are:

1. Standing
2. Sitting
3. Kneeling
4. Squatting
5. Stooping
6. Lying on side
7. Lying supine
8. Lying pronated

Body postures during work are determined by the relationships between the worker's anthropometric dimensions and the nature of his physical and visual connections with the hardware (Pheasant, 1986). These, in turn, are determined not only by the amount of space available, but also by its arrangement with respect to the machines being operated on, and the surrounding workplace. For example, in many cases the whole body must not only enter an area, but the parts being worked on must be in the positions where the worker may work comfortably for fairly long periods, and be able to use his body as an efficient mechanical system -- an optimal posture. General body postures like kneeling, squatting, stooping and, perhaps, lying down are not postures which a worker would assume voluntarily; they are imposed by the nature and amount of workspace available; the greater the deviation of the workspace arrangement from optimal, the greater the deviation of the human body posture from an optimal one. If, for example, a machine under inspection cannot be raised above waist height, then the worker is forced to squat or stoop to lower his eye level. It is therefore appropriate, at this point, to explain the biomechanics of an optimal posture so that one can gain a better understanding of the relationship between workspace, working postures and work output.

<u>Muscular work and body posture</u>

The importance of posture in maintenance work relates to the manner in which it affects the strength (the ability to generate forces and torques to be transmitted to the machine), and endurance (the ability to maintain a certain level of work output without fatigue) of the worker. These, in turn, determine the worker's accuracy, control, speed and, hence, productivity and efficiency for tasks in which physical and physiological attributes are important.

The main principles which determine how strength and endurance are dependent on posture and, hence, on workspace design are:

1. Length - tension properties of muscles.

2. Leverage effects (or mechanical advantages) of the skeletal lever systems of the body, especially the system actively engaged in muscular power output, like the arm while using screwdrivers.
3. Speed of dynamic muscular contractions.
4. Strength-endurance relationships.

The length-tension properties state that muscles have their greatest tension-producing capabilities when they are fully stretched (Astrand and Rodahl, 1977). On the other hand the mechanical advantage (or leverage) of the lever system is usually greatest at some intermediate position in the range of motion (Kroemer, 1986). The interaction between these two mechanisms of muscular force (or torque) generation in the human body indicate that the leverage property may exert greater influence than the length - tension property (Imrhan and Ayoub, 1988). The implication is that, for a single joint contraction, the greatest strength producing capability occurs when the joint is at about the mid-range of motion. For multiple joint contractions, and especially if the motion is dynamic, the situation is more complicated. There is very little data available to make strong conclusions, but the scant available evidence (Ayoub et al., 1981 and 1982; Imrhan and Ayoub, 1988) indicates that the greatest strength generation also occurs at intermediate positions; the greater the speed of contraction, the further this point is along the range of motion.

These biomechanical principles imply that, where great muscular forces or torques must be applied for maintenance tasks, whether they be twisting, rotating, pulling, pushing, lifting, pressing down or even holding, adequate and properly designed space must be provided for the technician to assume postures which optimize the biomechanical properties of the musculoskeletal system. If the space is limited the technician may be forced to bring segments of his limbs closer together, like holding the elbow(s) inwards, thereby reducing muscle stretch as well as reducing the leverage. If the machine or component being worked on cannot be brought closer than the outer limits of the technician's reach, then he would have to maintain extended limbs, thereby reducing leverage effects. Examples of such cases in maintenance can be found in use of hand tools for tightening/loosening parts (nuts, bolts, etc.) where the arm must exert rotation, pull or push on the lever arm of the tool, which by necessity must be either close to the body or at arms length; when the arm must be elevated, as when the workpiece is above shoulder height level, musculoskeletal stresses at the shoulder are intensified (Tichauer, 1973; Ayoub et al., 1976). At other times the body may be required to maintain static tension in order stabilize a mechanical system, or part of it. Asymmetric or twisted body positions are also very common in maintenance. They may occur when the worker tries to reach parts which may be positioned deep within machines, where direct visual access and physical access are obstructed by physical barriers, such as other parts of machines. In these cases, great muscular tension, and passive tensions in the soft tissues about the muscles, must be generated to maintain a desired work output. These restricted workspaces do not allow variations in posture, and attempts to use muscle groups other than the ones specialized for the task, to relieve the latter of fatigue, restrict the ability to perform (Corlett, 1981). Work may therefore be limited to a restricted number of muscle groups repeatedly. Adequate workspace allows variation in posture (holding the limbs in or out, sitting slightly to one side or askew, kneeling or squatting, etc.) and comfortable access to the work area. Workspace design should, therefore, always consider the necessity for changes in posture to relieve fatigue.

Endurance relationships

It is wrong to even assume that, in constrained postures, the desired work output can be maintained above maximal voluntary effort, for the long periods (Corlett, 1982) characteristic of maintenance tasks. The worker may unconsciously respond to his perception of musculoskeletal strain; he may try to maintain a level which is just bearable and, in so doing, work far below his optimum level for comfortable postures. When such unusual postures have to be maintained for long periods, local muscular fatigue and general physiological fatigue exert their effects, work output is reduced, errors are made more often, and accidents may increase (Grieve and Pheasant, 1982; Grandjean, 1987).

Handtool use and body posture

Among the few studies on the effects of common maintenance tools and typical body postures in maintenance tasks are those of Mital and Sanghavi (1986), Mital (1986), Mital and Channaveeraiah (1988), and Adams (1988a). Those studies indicated that the type of tool had significant effects on MVC torque exertion capabilities -- wrenches generating greater torque (9 times) than screwdrivers -- but body postures, unless they were very different, such as standing versus squatting, had little effect.

The high rate of handtool injuries in the U.S. has been described by Aghazadeh and Mital (1987). Though not stated explicitly, it is likely that many of those injuries occurred when workers were fatigued, especially by poor working postures imposed by workspace design inadequacies.

Hand-arm configuration

The adverse effects of restricted workspace are not limited to whole body postures, but are equally important for local skeletal configurations, at the hand/machine interface. For torque or force generation whole body postures allow the body weight to exert its influence as, for example, in leaning towards a screw driver handle in tightening screws, or onto the lever arm of wrenches in the direction of its motion. The force due to body weight is transmitted to the hand at the hand/machine interface, thereby augmenting the muscular force generated by arm muscles alone. There are many situations, however, where workspace is too limited to permit the whole body in close proximity to a machine part, or to assume a comfortable position to exert the composite effect of its lever systems. A typical example is in bending over a motor vehicle engine, stretching the upper body and reaching downward and forward for the hand (often with a handtool) to access a machine part, through limited arm clearance(s). In these situations work is performed by a limited number of muscle groups and, consequently, there is early onset of fatigue. In many cases the arm may be placed in unnatural configurations which magnify the strain on the particular musculoskeletal system. The connection between cumulative trauma disorders and repetitive forceful actions have been well researched (Tichauer, 1966 and 1973; and Armstrong and Chaffing, 1979). Cumulative trauma disorders of the upper extremity, like carpal tunnel syndrome, can easily result from maintenance work which frequently involves working with a deviated wrist, imposed by insufficient clearances and poor postures.

Even where the general body posture is good (as in standing with some support) the torque production is significantly reduced when the arm must assume an unnatural configuration. The data of Adams (1988a), for example, indicated that the orientation of connectors to the right of the body's mid-saggital plane was better than directly in front or facing away from the subject (reverse frontal plane), for hand torque generation using finger grips.

Hand clearance

The size of hand clearance also affects other parameters of performance. Kama (1963 & 1965) found that the time taken to remove and replace a component decreased sharply as the aperture available for hand clearance increased; and time increased as depth of aperture increased. More recently, Adams (1988b) showed that the clearance between the hand and any physical obstruction affects the torque capabilities of the hand, in addition to the type of hand grip used. Specific recommendations (Table 1) were given for hand clearances for torque application for electrical connectors of the following diameters: 2.0 in (large); 0.9 in (medium) and 1.5 in (small).

Hand clearance space restrictions in shelf openings have also been found for maximal acceptable weights lifted and placed onto the shelf. Mital and

Wang (1989) found a decrease of up to 9% in the load lifted when hand clearance changed from unrestricted space to 15 mm shelf separation, and an additional 4% when the clearance narrowed to 3 mm.

TABLE 1
Hand clearance dimensions for torque application on circular electrical connectors.

Type of grip	Connector diameter 5.1 cm(2 in)	3.81 cm(1.5 in)	2.29 cm (0.9 in)
Full wrap around	≥ 25	30 - 35	30 - 35
Fingertip	≥ 15	20 - 25	≥ 25

Forearm orientation

The orientation of the forearm while exerting torque is imposed by the body posture relative to the tool/workpiece orientation. Data is extremely rare on the effect of forearm orientation on torque exertion. Recent studies (Mital and Channaveeraiah, 1988; and Imrhan, unpublished data) have shown the effect is significant but depends on the task, which in turn depends on the muscular actions and type of hand/handle interface involved. It seems, however, that torque exertion with the forearm supinated is weaker than in the neutral or pronated orientations. Prudent design of the workspace should aim at a body posture which avoids the supinated arm orientation. For example, a chassis which can be pulled out of its cabinet and rotated (Woodson, 1981), rather than a fixed one, can achieve this.

Wrist position

The use of handtools in restricted handspaces is often accompanied by deviated wrist positions. Simply gripping the tool firmly, in these cases, can present problems; MVC grip forces are significantly less in deviated wrist positions compared to the neutral (straight) position. Terrell and Purswell (1976) found decrements of 30% for flexion, 22% for extension, 18% for radial deviation and 15% for ulnar deviation. Torque generated by handtools, such as screw drivers, and by the bare hand are strongly dependent on handgrip force (Pheasant and O'Neill, 1981; Imrhan and Loo, 1988); hence, design of access spaces must aim at preserving neutral wrist positions. Where deviated wrist positions cannot be avoided, greater clearance spaces must also be provided for the hand, compared to neutral positions.

TABLE 2

Clearance dimensions for selected tool use in various wrist positions (adapted from Eastman Kodak Co., 1955)

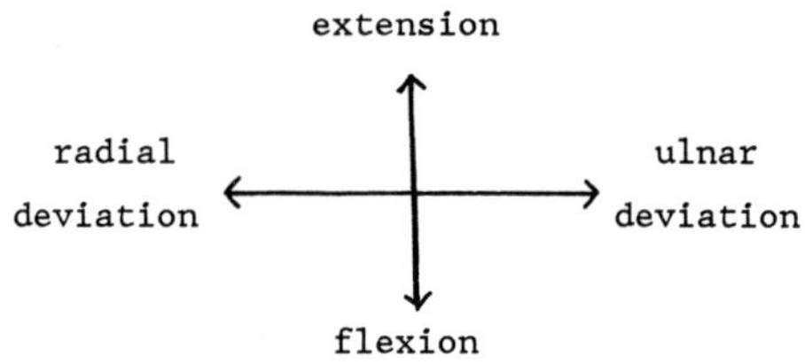

Tool and Task	Wrist Position							
	Radial Deviation		Ulnar Deviation		Extension		Flexion	
	cm	in	cm	in	cm	in	cm	in
Turning Screwdriver [20-cm (8-in.) Length] or Spinate Wrench [15-cm (6-in.) Length]	3.8	1.5	6.2	2.4	6.2	2.4	4.2	1.7
Grasping, Turning, and Cutting with Needle-Nosed Pliers[14-cm (5-in.) Length] or Wire Cutters [13-cm (5-in.) Length]	5.3	2.1	7.3	2.9	4.6	1.8	6.8	2.7
Turning Socket Wrench [10-mm (3/8 in.) Base, 7-cm (3.2 in.) Shaft]	5.4	2.1	8.3	3.2	8.3	3.2	7.3	2.9
Turning Allen Wrench [5-cm (2-in.) Length]	3.6	1.4	8.6	3.4	9.4	3.7	6.4	2.5

Very small clearances for accessing machine parts to be removed or replaced may sometimes require great pinch forces, in conjunction with a deviated wrist, thus aggravating the conditions leading to tendinitis in the fingers and carpal tunnel syndrome. A flexed wrist creates great compressive forces between the extrinsic finger flexor tendons and the trochlea, and compresses the median nerve. When pinching with the fingers the compression increases if the whole hand is also involved in a grasp (Armstrong and Chaffin, 1979). A limited clearance for the hand may also influence the type of pinch used -- more space may be required for the lateral and chuck pinches than the pulp pinches. Swanson, Matev and deGroot (1976) and Imrhan and Loo (1989) have shown that the pulp pinch force with the index finger is only about 2/3 to 3/4 the forces from the chuck and lateral pinches, for adults and the elderly. A little more clearance space, therefore, can translate into less muscular and tendon strain, and reduced incidences of cumulative trauma disorders, in maintenance tasks, by allowing for the more powerful pinches. When physical obstructions to access with the arm imposes a bent wrist while pinch forces are being applied, muscular and tendon strain may be aggravated. Imrhan (unpublished data) has shown that pinch forces declined by 15-52% depending on the combination of wrist position and type of pinch.

HANDTOOL USE AND HAND CLEARANCE

Intimately connected with hand orientation and performance is the subject of handtool use in restricted access spaces. This section will deal only with those aspects of handtool use, which can be optimized by good workspace design. A more detailed description of handtool design has been dealt with in another chapter by A. Mital.

The effective torque generated with a hand tool increases as the moment arm of the tool increases. The use of shorter-arm tools has been imposed by the restricted arc of motion in restricted workspaces, and by physical obstructions from machine parts. Hence, in the design stages of machines, more consideration may be given to the facility for separating obstructing parts to create greater space for tool movement and manipulation. Table 2 illustrates clearance differences for a few selected hand tools.

Two other points should be noted in contemplating the use of tools with longer lever arm: (1) the ability to manipulate the tool may decrease as the lever arm gets longer, as for wrenches; additional clearance may be required to avoid contacts with opposing surfaces such as switches, controls, protruding objects, or breakable objects like glass panels; and (2) the increase in torque is not proportional to the increase in the lever arm of the tool due, probably, to change in the body's leverage. The latter effect is still under investigation (Deivanayagam and Weaver, 1988). It is important, however, to determine the best types of tools and their characteristics for anticipated maintenance tasks, and to design the workspace to accommodate their use. Designing for the use of wrenches should be preferred over the use of screw drivers, if great torque exertion is important; the data of Mital (1986) indicate that, in general, one can exert about 9 times as much torque with commonly used wrenches than with commonly used screw drivers. Using wrenches would also encourage the use of the larger stronger muscles of the arm instead of the smaller more injury prone hand and finger muscles associated with screwdriver use. A practical design problem here, however, is the greater work-space required for wrench usage relative to screwdriver usage.

Design guidelines for attaining optimal body postures and biomechanical advantages

The size and arrangement of workspace must be designed to conform to the human body's biomechanical and physiological characteristics. The design guidelines here are, therefore, similar to those used in other types of work requiring biomechanical strength (Ayoub et al. 1976) like manufacturing, for example.

1. Enough space must be provided to prevent cramped body postures where leverage is poor and working muscle are shortened towards their limits. For whole body postures, machine design features should incorporate mechanisms for raising, lowering, rotating and removing parts so that biomechanical strength can be optimized. The human body should hot have to bend and twist for long periods.
2. Where manual forces must be generated, the hand should work at about chest-to-shoulder height to prevent lifting of the shoulder(s) and upper arms(s). The exact position would depend on the degree of visual effort required. The more precise the work the higher (proximity to the eyes) the workpiece should be moved.
3. Enough space must be provided to allow variations in posture to relieve fatigued muscle groups without hindering performance.
4. Arm/hand access spaces must not be based merely on minimum clearance dimensions, but must allow for arm/hand movements in all planes of space.
5. Clearance space dimension must also incorporate space for hand tools (the largest handtools likely to be used).

ANTHROPOMETRIC CONSIDERATIONS

From the discussions in the previous sections it should be clear that designing workspace for maintenance requires accurate information on certain characteristics of the machine, maintenance tasks, the environment and the people who are expected to perform those tasks. The various types of maintenance tasks should be anticipated in order to determine tools and other equipment requirements; the amount of space needed for their flow, from storage areas to work areas and within the work area; and the amount of workers needed for particular tasks. The amount of workspace needed for every task to be performed by a human depends not only on equipment but also on the body size, range of motion and strength capacities of the worker. Hence a detailed task analysis may be necessary.

Cataloguing the various types of tasks can be facilitated by studying the maintenance records for the particular system, and by interviewing technicians who have actually worked on those or similar tasks. If the machine or system being designed is a new one, this information may be derived from studying tasks resembling the one to be designed (Crawford et al., 1963). Tools and other maintenance support equipment should be studied carefully for possible improvements in selection or design, since they may influence the volume and arrangement of access spaces. Limited access areas from which machine parts must be removed or replaced may, in turn, be influenced by the type of tools required (Van Cott, 1973), as was discussed earlier.

The size and arrangement of workspace needed for a particular task depend largely on the overall body size of the worker, and on the dimensions of specific parts of the body, like the hand, for accessing machine parts to be removed. Usually a designer is faced with the task of finding relevant anthropometric data quickly, and the selection of appropriate data bases becomes of paramount importance. Finding the correct data has never been easy. Very little of the available anthropometric data bases, such as those mentioned in Sanders and McCormick (1987), are suitable for the design of workspaces. Their measurements are mostly 1-dimensional; some of the engineering sources offer some 2-dimensional measurements. On the other hand, workspace is 3-dimensional. Available 3-dimensional anthropometric data (Dempster, 1955; Kennedy, 1964; and Roth et al., 1976) may not be suitable since they refer to the person in postures (seated upright) which are hardly ever used by maintenance workers; in fact, those data were compiled for the operational aspects of the workplace. They do not incorporate the effects of joint mobility, external loadings, and the body's geometrical configuration in defining workspace envelopes (kinetospheres). For example, standing workspace envelopes are limited by balance which are, in turn affected by foot placement, upper body position and weights in the hand, or forces exerted by the hand (Grieve and Pheasant, 1982), and not necessarily by the length and mobility of of limbs alone.

It is very appealing when one can accurately predict (estimate) data which are either unavailable as direct measurements, or difficult to measure. NASA (1978) gives numerous regression equations for predicting static and 1-dimensional anthropometry. Unfortunately, dynamic (functional) measures cannot be predicted accurately from static anthropometry (Grieve and Pheasant, 1982). Predicting kinetospheres from any type of data seems to be even more remote. Workspace measures, like functional anthropometry, must therefore be obtained from special purpose studies, even though this could severely limit their application outside the original areas of concern (Pheasant, 1986).

Until appropriate anthropometric data for workspace design are developed, designers must resort to careful selection from the present sources. Eastman Kodak Co. (1985) has published some useful 1- and 2-dimensional data for maintenance--type working postures and access clearances. They have been adapted from a few original sources, and include: lying down supine for inspection work and for using handtools; lying down prone; standing erect; standing with foot braced; kneeling; entering an open top vessel; arm entry;

hand entry; and finger entry. The dimensions are given in Table 3. Pheasant (1986) also gives tables of clearance dimensions, and Garrett (1971) contains useful hand anthropometry data.

The exact percentiles of the population accommodated were not given, but it was stated that the 'largest person should be accommodated'. The use of qualifiers such as 'accommodate most workers', 'accommodate most people', and 'accommodate a very tall person' emphasizes the inadequacies of available data. Kodak and Eastman Co. (1985) further states that 'more space is needed if the arms have to exert force or use hand tools'.

The chapter, in this book, on "Workspace clearance and access dimensions and design guidelines" provides a summary of other useful 1- and 2- dimensional data.

Factors affecting reach and mobility

When maintenance work must be done in environments in which special apparel must be worn -- in extreme cold and heat; at high altitudes; in space (Stramler, 1986); underwater, etc. -- reach envelopes may be significantly modified. For example, wearing gloves may increase hand clearance dimensions by 2cm (Eastman Kodak Co., 1985); winter jackets can limit arm reach by 5cm (Sanders, 1977); and restraints like harnesses used in vehicles can reduce arm reach by as much as 24% (Garg et al., 1982). Seeman et al. (1966) have reported that the reduction in mobility imposed by the use of pressurized suits can increase performance times, even in simple tasks, by a factor as great as 2 1/4.

Special apparel may also affect workspace requirements in a more indirect and unpredictable way, which can be determined only by direct measurements. For example, the reduction of grip force when using gloves (Tichauer, 1978; and Sudhakar et al., 1988) may force maintenance workers to modify their body posture to gain better leverage to compensate for the loss. Additional space may, therefore, be needed depending on the range of postures adopted. This example merely emphasizes the great need for data on dynamic anthropometry specific to maintenance tasks.

ANTHROPOMETRIC DATA BASE SELECTION

In addition to the inadequacies mentioned earlier, available anthropometric data bases possess certain limitations characteristic of the sampling techniques used. The designer of work spaces for maintenance must be aware of them in order to avoid using incorrect design parameters (Bradtmiller, 1985).

Data from combined surveys (Diffrient, 1974) may contain dimensions which have been measured differently.

TABLE 3

Selected clearance dimensions in typical maintenance postures (adapted from Eastman Kodak Co., 1985)

Activity		Dimensions (cm)	(in)
<u>A: Whole Body Postures</u>			
Width for walking	1 person	61	24
	2 persons abreast	137	54
	3 persons abreast	183	72
Height for walking	accommodates very tall person wearing thick-soled shoes and hard hat	203	80
Lying supine			
- elbow completely flexed	vertical (fore-aft)	46	18
	body length clearance	193	76
- elbow in mid-flexion	vertical	61	24
- elbow completely extended	vertical	81	32
Standing naturally	vertical	203	80
	horizontal (reach)	76	30
Standing with legs braced	vertical	203	80
	horizontal	102	40
Kneeling	vertical	122	48
	horizontal	117	46
Prone arm reach	lying down	243	96
<u>B. Hand and Arm Access Ports</u>			
Hand/arm entry/exit			
Both arms	75% depth of reach + 15cm (6in)		
Arm to elbow	11cm (415 in) diameter		
Arm to shoulder	13cm (5 in) diameter (add 8cm (3 in) for winter clothing)		
Hand with extended fingers	5cm (2 in) X 10cm (4 in) rectangular		
Hand with clenched fist	10cm (4 in) X 13cm (5 in) rectangular		
Hand holding small object	10cm (4 in) X 10cm (4 in) square		
Hand holding larger object	4.5cm (1.75 in) clearance around object; add 2cm (0.75in) for gloved hand		
Push button access	4cm (1.5 in) diameter		
Two finger twist	9cm (3.5 in) diameter		
<u>C. For Entering Open-Top Vessels</u>			
Circular hatch	76cm (30 in) dia		
Rectangular hatch (horizontal entry)	61cm (24 in) X 51cm (20 in)		
Square hatch (vertical entry)	61cm X 61cm		
Add 10-20cm (4.8 in) if heavy clothing is worn.			

N.B. see chapter on "Workspace clearance and access dimensions" for more details.

Data from surveys on one sex cannot necessarily be applied to another sex by using multiplication factors, when considering different anthropometric dimensions. For example, the female/male ratio for mean stature is not necessarily the same value for extended arm reach. The absence of arm reach data, say for females should be determined in a more direct way. This concept is also true when trying to estimate one body dimension from another across different data sets, or estimating across nationalities or age groups. (Pheasant, 1986) provides a good discussion on this topic.

Differences in dimensions of population subgroups could also cause errors. Thus maintenance workspaces for civilian population (as in commercial vehicle maintenance) might not be properly designed using data based on surveys on military subjects, on whom most of the large engineering anthropometric studies have been conducted.

The ethnic composition should also be considered, especially in the U.S. where the proportion of smaller size Asians have increased significantly over the last 15 years. Some sources (NASA 1978; and Pheasant, 1986) give anthropometric data for different nationalities and ethnic groups. The studies of Davies (1980a,b) and Courtney (1984) on hand dimensions and machine guardings emphasize the significant differences among nationalities for hand clearance dimensions.

Specifying the exact percentiles to be accommodated is necessary since all data bases do not publish the same percentile values (eg. 5th and 95th). By assuming a normal distribution for anthropometric measurements (a reasonable assumption for large samples) any percentile can be estimated from the median (or mean) and standard deviation values, or from stated percentiles (eg. 5th and 95th).

COMPUTERIZED MODELS FOR WORKSPACE DESIGN

Traditional methods of workspace design analyses have not been adequate for maintenance tasks and environments, which require specifications of volumes in space, with the human body in unnatural postures. Mock-ups, anthropometric dummies, partial manikins, stick figure manikins, 2-dimensional drawing board manikins, and manufacturers' drawings are not versatile enough to represent 3-dimensional spaces and the user population; they cannot be manipulated with the speed necessary to compare various alternatives; and they cannot respond quickly enough to rapid design developments (Rothwell, 1986). The availability of personal and mini computers, and rapid software development in computer aided designs (CAD) have now provided the means to overcome those limitations. Rothwell, (1986) and Chaffin and Evans (1986) discuss the utility of CAD models in workspace design.

Anthropometric and biodynamic CAD models hold great promise for the design of workspace for maintenance. Numerical inputs can be converted into graphical displays of the human form, and can be manipulated freely to define the physical characteristics of the human model.

CAD programs can be used to analyze partial body models when specific areas of the body are of interest; or whole body models, for general workspace analysis. The human body may be modeled as a system of interconnected links (Calvert, 1982), with each link representing a body segment, with respect to one or more of the following: length; breadth; circumference; volume; density; mass; and internal modeling of bone, muscles, ligaments, skin and subcutaneous fat. Body surfaces and volumes can be modeled by 'enfleshing' stick figures; they are better than stick figures for analyzing clearances; the former are limited to distance-based analyses such as reach and visual interference (Lane, 1982).

Visual access can be analyzed by CAD models by relocating target objects and/or the human model within the environment (Rothwell, 1986) The ability of these CAD models to alter human dimensions facilitates the evaluation of special clothing and personal equipment (thick jackets, helments, goggles, etc.) in clearance spaces.

Finally, it is possible to perform rotations and translations of the human graphical models to view the workspace form a variety of angles.

Examples of CAD Models

A few examples of CAD models which are useful for the design of maintenance workspace are SAMMIE (System for Aiding Man/Machine Interaction Evaluation), the Graphic Man model, and BUBBLEPERSON-TEMPUS. Chaffin and Evans (1986) and Galer (1987) outline the capabilities of these models. SAMMIE allows the size of the human figure to be scaled to various population subgroups using segment link rates; heuristic rules can then be used to shape the body into realistic postures, allowing the determination of visual and reach envelopes. The Graphic Man model can be scaled to represent various population subgroups. A realistic graphic model of the human form can be produced for reach analysis; and biomechanical analyses can be performed. BUBBLEPERSON allows easy change of body segment size and shape, as well as manipulation of the total body form. This form can be used from a larger interactive graphic system called TEMPUS, to describe objects being handled as well as evaluating reach and physical body interference problems within a given workplace.

CONCLUSIONS

The science of workspace design for maintenance seems to be in its infancy. Many factors have contributed to this state of affairs - the historical emphasis in cost of production; the more recent emphasis on designing for operability; trends towards miniaturization and compactness in machines; historical emphasis on reliability of systems, with little regard for maintainability; inadequate anthropometric data; technical difficulties in integrating workspace data with the speed and flexibility required for quick decisions; and lack of knowledge of the ergonomics of workspace design in engineers who design machines and other equipment. The dramatic rise in cost of maintenance has, however, caused attention to be focused on maintainability.

This chapter has described how prudent workspace design contributes to effective maintenance. The emphasis was placed on the worker-environment interaction, and how the science of ergonomics can contribute to design for maintainability. The recent development of computer aided design models of the human in relation to his workplace is a boon to workspace design. When this is supplemented with 3-dimensional anthropometric analyses, many of the problems which were ignored by designers may then be resolved. The task is an extremely difficult and tedious one. Some of the other raw materials for accomplishing this include more detailed 3-dimensional task analyses and anthropometry. Data are very rare in these areas; but it must be now be acknowledged that the traditional 1- and 2-dimensional view of work are inadequate for the conceptualizing 3-dimensional workspace.

The design of hand tools and other maintenance equipment may influence workspace requirements, and vice-versa. This is an area of workspace design which is virtually untapped.

The increasing participation of the elderly in the workforce (Fozard and Fisk, 1988) presents special problems in the design of maintenance workspace. Decrements in muscular performance and range of motion of body segments have hardly been investigated in this age group. Extrapolating from adult data is unlikely to be realistic. Hence research in this area must be initiated now, in anticipation of the age composition of the workforce of the future.

RECOMMENDATIONS FOR FUTURE RESEARCH NEEDS

It has been argued, in this chapter, that the available data that can be used for specifying workspace design parameters for maintenance tasks are very inadequate, vague and sometimes irrelevant. The following research areas can provide the kind of data that can help fill the gap:

1. Specific maintenance task analyses, to determine manpower requirements; equipment requirements; human motion characteristics; and human strength requirements.
2. Human motion characteristics data for specific body postures and specific tools, especially where muscular strength is important.
3. Clearance data, especially for specific tools being grasped by the

hand, in 2- and 3- dimensions.

4. Three-dimensional workspace envelopes for specific tasks, and for the use of specific tools.
5. Strength capacities for simulated maintenance tasks, in both optimal postures and space constrained postures.
6. Three-dimensional anthropometric data related to specific motions encountered in maintenance work; anthropometric (and strength) data must be measured on samples representative of the industrial population, with respect to sex, age range, ethnic origin (for local use), and nationality (for international use).
7. Anthropometric and strength data should be in the form which can be easily used in 3-dimensional computer models.

REFERENCES

Adams, S.K., 1988a. Hand grip torque for circular electrical connectors:the effect of obstructions. Proceedings of the Human Factors Society, 32nd Annual Meeting, pp 642-645, Santa Monica, California.

Adams, S.K. and Peterson, P.J., 1988b. Maximum voluntary handgrip torque for circular electrical connectors. Human Factors, 30(6):733-745.

Aghazadeh, F. and Mital, A., 1987. Injuries due to handtools. Applied Ergonomics, 18(4), 273-279.

Alexander, D.C., 1986. The practice and management of industrial ergonomics. Prentice-Hall, Inc., Engewood-Cliffs, New Jersey.

Armstrong, T.J. and Chaffin, D.B., 1979. Some biomechanical aspects of the carpal tunnel. Journal of Biomechanics, 12:567-570.

Astrand, P. and Rodahl, K., 1977. Textbook of work physiology. McGraw-Hill Book Co., New York.

Ayoub, M.M. McKenzie, H.J. and Deivanayagam, S., 1976. Training manual in occupational ergonomics. Institute for Biotechnology, Texas Tech University, Lubbock, Texas.

Ayoub, M.M., Gidcumb, C.F., Reeder, M.J., Beshir, M.Y., Hafeez, H.A., Aghazadeh, F.A. and Bethea, N.S., 1981. Development of an atlas of strengths and establishment of an appropriate model structure--males. (Final Report, McDonnell-Douglas Aircraft Co., Contract No. AS-20501-C). Institute for Biotechnology, Texas Tech University.

Ayoub, M.M., Gidcumb, C.F., Reeder, M.J., Beshir, M.Y., Hafeez, H.A., Aghazadeh, F.A. and Bethea, N.S., 1982. Development of an atlas of strength and establishment of an appropriate model structure--females. (Final Report, McDonnell-Douglas Aircraft Co., Contract No. AS-20501-C). Institute for Biotechnology, Texas Tech University.

Bradtmiller, B., 1985. Data base selection in applied anthropometry. In R.E. Eberts and C.E. Eberts (eds.). Trends in Ergonomics/Human Factors II, Elsevier Science Publishers B.V., North Holland, pp 41-64.

Calvert, T.W., Chapman, J. and Palta, A., 1982. Aspects of the kinematic simulation of human movement. IEEE Computer Graphics and Applications, 2(9), 41-50.

Chaffin, D.B. and Anderson, G.B.J., 1984. Occupational biomechanics. John Wiley, New York.

Chaffin, D.B. and Evans, S., 1986. Computerized biomechanical models in manual work design. In Proceedings of the Human Factors Society 30th Annual Meeting, pp 96-100, Santa Monica, California.

Corlett, E.N., 1981. Pain, posture and performance. In E.N. Corlett and J. Richardson (Eds.) Stress, work, design and productivity. John Wiley & Sons, New York.

Corlett, E.N., 1982. Postural considerations in workspace design. In Easterby, R., Kroemer, K.H.E., and Chaffin, D.B. (Eds.). Anthropometry and Biomechanics: Theory and Application. Plenum Press, New York.

Courtney, A.J. and Ng, M.K., 1984. Hong Kong female hand dimensions and machine guarding. Ergonomics, 27(2):187-193.

Crawford, B.M. and Altman, J.W., 1972. Design for maintainability. In H.P. Vancott and R.G. Kinkade, Human Engineering guide to equipment design, pp 585-631. U.S. Department of Defense, Washington, D.C.

Davies, B.T., Abada, A., Benson, K., Courtney, A. and Minto, I., 1980. Female hand dimensions and guarding of machines. Ergonomics, 23(1):79-84.

Deivanayagam, S., 1986. Maintainability ergonomics: what and why. IE Ergonomics News, Vol XX (3), 1-2.

Deivanayagam, S. and Weaver, T., 1988. Effects of handle length and bolt orientation on torque strength applied during simulated maintenance tasks. In F. Aghazadeh (Ed.) Trends in Ergonomics/Human Factors V, Elsevier Science Publishers, North Holland, pp 827-833.

Dempster, W.T., 1955. Space requirements of the seated operator. Technical Report 55-159, U.S. Air Force, WADC.

Diffrient, N., Tilley, A.R. and Bardagjy, J.C., 1974. Human scale 1/2/3. The MIT Press, Massachusetts Institute of Technology, Cambridge.

Eastman Kodak Co., 1985. Ergonomic design for people at work: volume I, Van Nostrand Reinhold Co., New York.

Fozard, J.L. and Fisk, A.D., 1988. Human factors in the aging population. Human Factors Society Bulletin, 31(11), 1-2.

Galer, I.A.R., (ed.), 1987. Applied ergonomics handbook. Butterworths, London.

Garg, A., Bakken, G.M., and Taxena, U., 1982. Effect of seat belts and harnesses on functional arm reach. Human Factors, 24(3), 367-372.

Garrett, J.W., 1971. The adult human hand: some anthropometric and biomechanical considerations. Human Factors, 13(2), 117-131.

Grandjean, E., 1987. Fitting the task to the man: an ergonomic approach. Taylor and Francis, Philadelphia.

Grieve, D. and Pheasant, S., 1982. Biomechanics. In W.T. Singleton (ed.), The body at work, pp 71-200, Cambridge University Press, Cambridge.

Hill, S.G. and Kroemer, K.H.E., 1986. Preferred declination of the line of sight. Human Factors, 28(2), 127-134.

Hutchinson, R.D., 1981. New horizons for human factors in design. McGraw-Hill Book Co., New York.

Imrhan, S.N. and Ayoub, M.M., 1988. Predictive models of upper extremity rotary and linear pull strength. Human Factors, 30(1):1139-1153.

Imrhan, S.N. and Loo, C.H., 1988) Modelling wrist-twisting strength of the elderly. Ergonomics, 31(12), 1807-1819.

Imrhan, S.N. and Loo, C.H., 1989. Trends in finger pinch strength in children, adults and the elderly. Submitted for publication in Human Factors.

Imrhan, S.N., 1989a. In preparation.

Imrhan, S.N., 1989b. In preparation.

Kama, W.N., 1963. Volumetric workspace study: I. optimum workspace configuration for using various screw drivers. Report No. AMRL-TDR-63-68(I), Aerospace Medical Research Laboratories, WPAFB, Ohio.

Kama, W.N., 1965. Volumetric workspace study: II. optimum workspace configuration for use of wrenches. Report No. AMRL-TRD-63-68(II), Aerospace Medical Research Laboratories, WPAFB, Ohio.

Kennedy, K.W., 1964. Reach capability of the USAF population. Phase I, the outer boundaries of grasping reach envelopes for the shirt sleeved, seated operator. Report AMRL-TDR-64-59, WPAFB, Ohio.

Kroemer, K.H.E., 1986. Human muscle strength: definition, generation and measurement. In Proceeding of the Human Factors Society 30th Annual Meeting, pp 977-981, Santa Monica, California.

Kumar, S. and Scaife, W.G.S., 1979. A precision task, posture and strain. Journal of Safety Research, 11(11), 28-36.

Lane, N.E., 1982. Issues in the statistical modeling of anthropometric data for workplace design. In R. Easterby, K.H.E. Kroemer.

McDaniel, J.W., and Askrein, W.B., 1985. Report submitted to U.S. Airforce Aerospace Medical Research Laboratory, Wright Patterson AFB, Ohio.

McKendry, J.M., Cooso, J.F. and Grant, G., 1960. The design and evolution of maintainable packaging methods for electronic equipment. Ergonomics, 255-272.

Mital, A., 1986. Effect of body posture and common hand tools on peak torque exertion capabilities. Applied Ergonomics, 17(2):87-96.

Mital, A. and Sanghavi, N., 1986. Comparison of maximum volitional torque exertion capabilities of males and females using common hand tools. Human Factors, 28(3), 283-294.

Mital, A. and Channaveeraiah, C., 1988. Peak volitional torques for wrenches and srewdrivers. International Journal Industrial Ergonomics, 3, 41-64.

Mital, A. and Wang, L.W., 1989. Effects on load handling of restricted and unrestricted shelf clearances, Ergonomics, 32(1):39-49.

Morgan, C.T., Cook III, J.S., Chapanis, A. and Lund, M.W. (eds.), 1963. Human engineering guide to equipment design. McGraw-Hill Book Co., Inc., New York.

National Aeronautics and Space Administration (NASA), 1978. Anthropometric source book (volume 1) anthropometry for designers; (volume 2) a handbook for anthropometric data; (volume 3) annotated bibliography (NASA reference publication 1024).

Oborne, D.J., 1981. Ergonomics at work. John Wiley and Sons, New York.

Pheasant, S.T., 1983. Sex differences in strength - some observations on their variability. Applied ergonomics, 14:205-211.

Pheasant, S., 1986. Bodyspace: anthropometry, ergonomics and design. Taylor and Francis, Philadelphia.

Pheasant, S.T. and O'Neill, D., 1981. Performance in gripping and turning: a study in hand/handle effectiveness. Applied Ergonomics, 6, 205-208.

Roth, J.T., Ayoub, M.M., Dixon, J.J. and Halcomb, C.G., 1976. Methodology and reach profiles for restrained and unrestrained seated operator: improved seat and workplace design. (AIRTASK A 62736 N/WF 55.525.403). Pacific Missile Center, California.

Rothwell, P.L. and Hickey, D.T., 1986. Three-dimensional computer models of man. In Proceedings of the Human Factors Society 30th Annual Meeting, pp 216-220, Santa Monica, California.

Salvendy, G., 1971. Hand size and assembly performance. American Institute of Industrial Engineers Transactions III, 32-36

Sanders, M.S., 1977. Anthropometric survey of truck drivers and bus drivers: anthropometry, control reach and control force. Westlake Village, CA: Canyon Research Group. In Sanders, M.S. and McCormick, E.J., 1987. Human Factors in Engineering and Design, McGraw-Hill Book Co., New York.

Sanders, M.S. and McCormick, E.J., 1987. Human Factors in Engineering and Design. McGraw-Hill Book Co., New York.

Seeman, J.S., Smith, F.H., and Meuller, D.D. A Technique to investigate space maintenance tasks. Report No. AMRL-TR-66-32, Aerospace Medical Research Laboratories, WPAFB, Ohio.

Singleton, W.T. (ed.), 1982. The body at work, Cambridge University Press, Cambridge.

Smith, R.L., Westland, R.A. and Crawford, B.M., 1970. The status of maintainability models: a critical review. Human Factors, 12, 271-282.

Stramler, J.H., 1986. A comparison between space suited and unsuited reach envelopes. In Proceedings of the Human Factors Society, 30th Annual Meeting, Santa Monica, California.

Sudhakar, L.R., Schoemarklin, R.W., Lavender, S.A., Marras, W.S., 1988. The effects of gloves on grip strength and muscle activity. Proceedings of the Human Factors Society, 32nd Annual Meeting, pp 647-650, Santa Monica, California

Swanson, A.B., Matev, I.B, and deGroot, G., 1976. The strength of the hand. Bulletin of Prosthetics Research, 9, 387-396.

Terrell, R. and Purswell, J.L., 1976. The influence of forearm and wrist orientation on static grip strength as a design criterion for hand tools. Proceedings of the 20th Annual Meeting of the Human Factors Society, pp 28-32, Santa Monica, California.

Tichauer, E.R., 1966. Some aspects of stress on forearm and hand in industry. Journal of Occupational Medicine, 8(2):63-71.

Tichauer, E.R., 1973. Ergonomic aspects of biomechanics: the industrial environment - its evaluation and control. U.S. Department of Health, Education and Welfare, NIOSH, Washington, D.C.

Tichauer, E.R., 1978. The biomechanical basis of ergonomics, John Wiley and Sons, New York.

U.S. Department of Defense, 1981. Human factors engineering design criterion for military systems, equipment and facilities (MIL-STD-1472C), Washington, D.C.

Van Cott, H.P. and Kinkade, R.G. (eds.), 1972. Human engineering guide to equipment design. U.S. Department of Defense, Washington, D.C.

Woodson, W.E., 1981. Human factors design handbook, McGraw-Hill Book Co., New York.

PART II
EQUIPMENT AND TOOL DESIGN

Workspace, Equipment and Tool Design
edited by A. Mital & W. Karwowski

WORKSPACE CLEARANCE AND ACCESS DIMENSIONS AND DESIGN GUIDELINES

A. MITAL

Ergonomics Research Laboratory (ERL), Mechanical and Industrial Engineering Department, University of Cincinnati, Cincinnati, Ohio 45221-0072 (U.S.A.)

ABSTRACT

In designing efficient workspaces it is necessary not only to accommodate the variations in the physical size of the work force but also to ensure that operators have adequate clearances for free movement of their limbs. Proper operation of controls, equipment, and tools and maintainability requirements further dictate consideration of accessibility. In this chapter, the problems associated with working in confined spaces are briefly discussed and design recommendations for various clearance and access dimensions are reviewed and summarized. It is concluded that existing design recommendations are scanty and, occasionally, contradictory.

INTRODUCTION

Inadequate clearances for movement of limbs and lack of accessibility, visual or anthropometric, to parts, controls or tools are among the most severe constraints that prevent operators from working efficiently within a workspace. In the absence of adequate clearance for body movement and apertures for operating or accessing the equipment and controls, workspace becomes restricted and operators are forced to function within this confined environment.

Restricted or confined space has profound effect on performance of tasks that require movement. Generally speaking, if it is possible at all to perform a task in a restricted space, the task performance improves with additional space. However, beyond a certain amount, increase in space produces no improvement in performance. Using performance as the criterion, Drury (1985) proposed dividing workspaces into three ranges: **Range 1** - task spaces sc small that the operator will not fit into the space and/or the task cannot be completed; **Range 2** - task spaces are sufficient for task performance but additional space will improve performance and reduce operator stress; **Range 3** - task spaces are so large that any additional space will not improve performance. Our main concern here is with range 2 spaces. The goal is to provide clearances and access apertures of dimensions that will bring the workspace to the upper limit of range 2. That is, any additional increase in clearance or aperture dimension will not result in any further improvement in performance.

In order to understand the importance of providing adequate clearances and access space, it is necessary to understand: (1) the relationship between the size of the working space and its effect on performance and (2) the concept of personal space (PS).

Space restrictions and performance

Studies dealing with a variety of tasks are briefly reviewed in this section. This review is necessary to appreciate the need to provide adequate space. The list of studies reviewed, however, should not be considered exhaustive.

Control tasks. Daniels et al. (1976) and Drury (1971) have shown that when steering a track (with paper and pencil) within fixed boundaries, the speed of movement is directly proportional to the clearance between the track width and the course width. This relationship is similar to that proposed by Beggs and Howarth (1970) and Rashevsky (1970) for automobile driving and tends to indicate that speed (performance) continues to improve as the clearance increases. This, however, is not true.

As the space exceeds the upper limit of range 2, performance no longer increases with space. Drury and Dawson (1974) found that in a fork-lift-truck driving task, as the course width increased the speed increased and errors decreased. However, beyond about 500 mm of clearance, speed started to level out and the errors decreased to zero (Fig. 1) indicating that the total space (with clearance) was now in range 3. Similar results have been reported for car driving (Williams, 1969) and tractor driving (Bottom, 1980), as quoted by Drury (1985), and manoeuvering a hospital cart in hospital corridors (Drury et al., 1975).

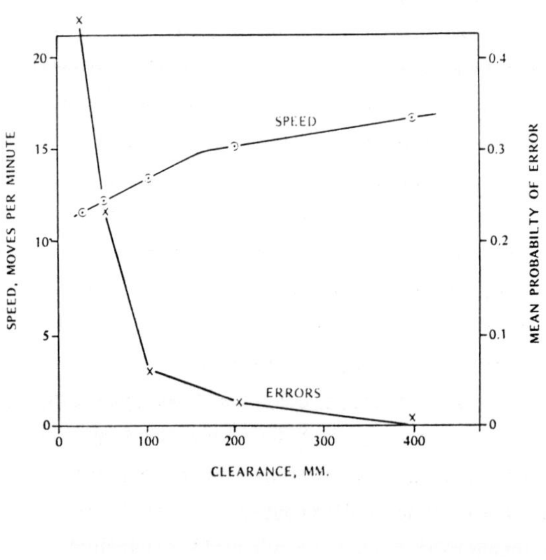

Fig. 1. Speeds and errors as a function of lateral space clearance in fork-lift-truck driving (Drury and Dawson, 1974).

<u>Precision tasks</u>. Kama (1963, 1965) found that as the clearance around the wrench increased, the speed of mounting components in holes first increased and then levelled off (Fig. 2). These findings again show that in region 2, performance is influenced by the amount of space available; in region 3, workspace has no effect on performance.

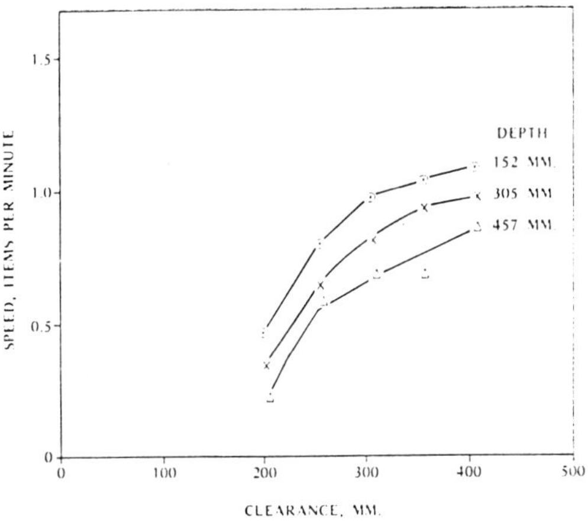

Fig. 2. Variation in the speed of component insertion with hole clearance (Kama, 1963; 1965).

<u>Manual materials handling</u>. The majority of industrial manual materials handling tasks are performed in spaces with some sort of restriction (Drury et al., 1982). Of the various materials handling activities, manual lifting is considered most strenuous and hazardous. When such tasks are performed in confined spaces, individual's capacity to handle materials declines. Ridd (1985) studied the effects of limited headroom, frontal and lateral barriers, and fixed and trapped foot positions on acceptable manual-handling capacity. Limited headroom was found to have the most profound effect on acceptable handling capacity. The acceptable capacity declined almost linearly for reductions in the available headroom of up to 10% of the individual's stature. Further restrictions had little significant effect on handling capacity.

Mital (1986) reported that load carriage in narrow and confined spaces forces individuals to exercise unusual methods. Workers constantly reorient the load while maneuvering it through the confined passage. The effect of spatial restriction on acceptable load carriage capacity is also very significant (Table 1).

TABLE 1
Average acceptable load carrying capacity (kg) in open and confined (0.56 m wide) passages (Mital, 1986).

Space	Males	Females
Open	28.7	18.6
Confined	24.9	16.1

Mital and Wang (1989) investigated lifting and placement of loads in shelves with restricted and unrestricted openings. They found that acceptable lifting capacity declines as the shelf opening clearance (shelf width - container width) declines (Fig. 3). When the shelf clearance is 30 mm or more, the capacity is the same as when the shelf opening has no spatial restrictions.

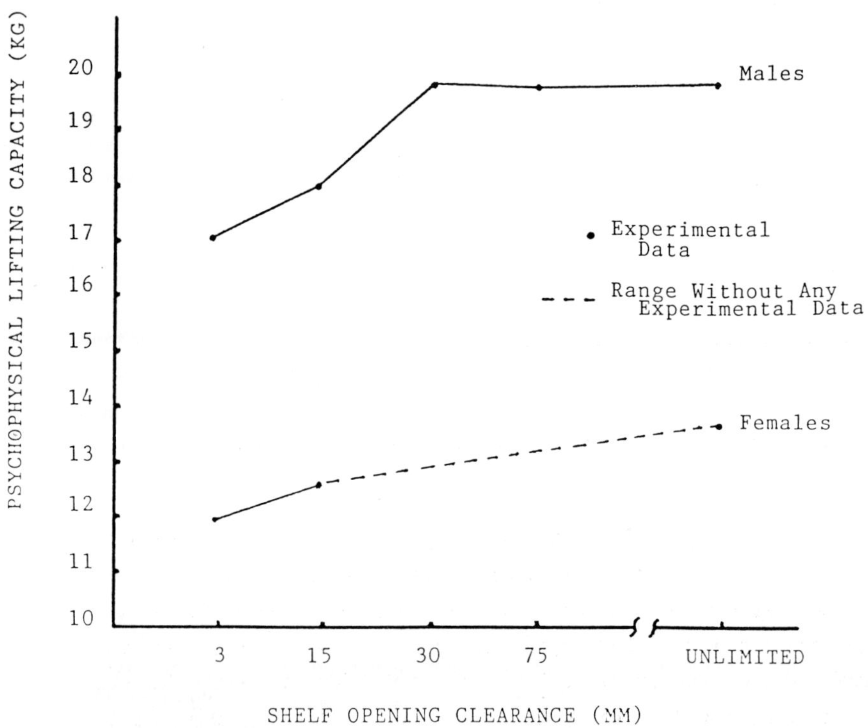

Fig. 3. Change in acceptable lifting capacity with shelf opening clearance (Mital and Wang, 1989).

For pushing manually operated hospital vehicles, Drury et al. (1975) found that the speed of manoeuvering first increases with clearance (corridor width - vehicle width) and then levels off (Fig. 4).

In general, it is clear that manual materials handling task performance improves with additional space, but only up to a point.

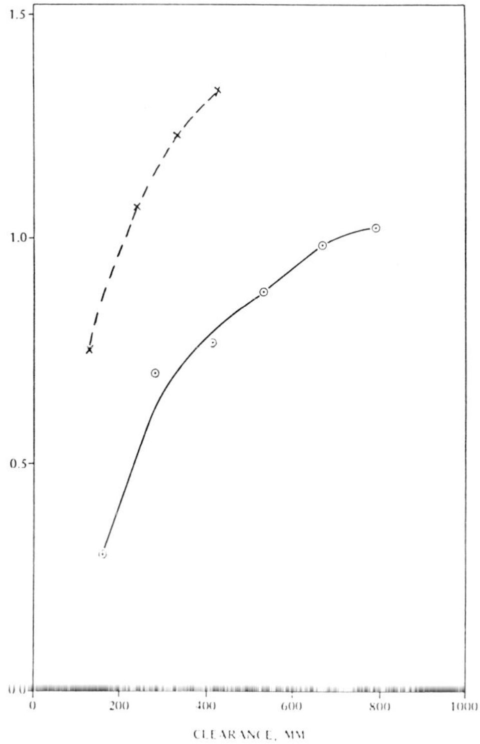

Fig. 4. Variation in cart manoeuvering speed with clearance (Drury et al., 1975).

Entryways and exits. The speed of entry or exit is used to measure performance of doors and corridors. In general, as is the case with other tasks, performance improves as the door and corridor size increase. For walking without colliding with other objects or unusual body motions, a minimum corridor width of approximately 635 mm is needed (Jones, 1970). The speed of walking increases with corridor width up to 915 mm (Drury, 1985) and then levels off (Fig. 5).

Bottom et al. (1979) determined that the speed of entry in a tractor cab increased linearly for door areas up to 0.75 m^2 and then levelled off (Fig. 6). For aircraft emergency exits, Roebuck and Levedahl (1961) determined that the speed of exit increases for exit width up to 508 mm and then levels off (Fig. 7). The speed of forward movement of crowds also increases for areas up to 0.56 m^2 per person and then levels out (U.S. General Service Administration, 1972).

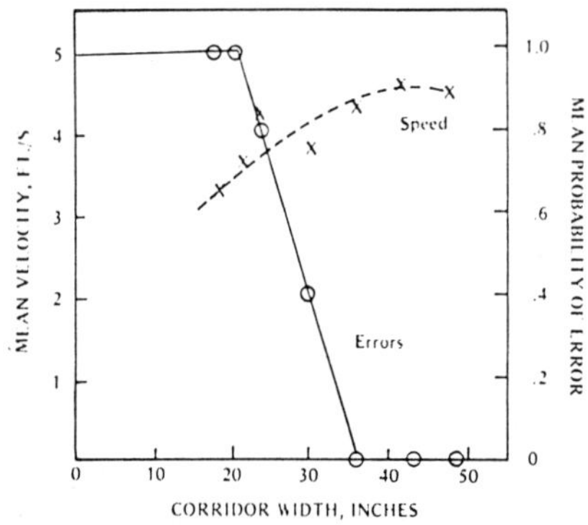

Fig. 5. Speed and errors of traversing a doorway as a function of doorway width (Drury, 1985).

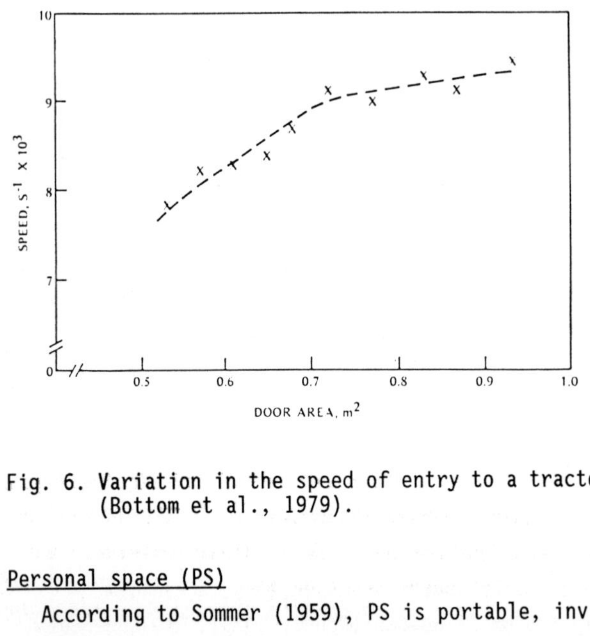

Fig. 6. Variation in the speed of entry to a tractor cab with doorway area (Bottom et al., 1979).

Personal space (PS)

According to Sommer (1959), PS is portable, invisibly bounded, person-centered space that is characterized by withdrawal upon intrusion. Bennett (1978) divides PS into the following three categories: intimate space (interpersonal distance between 0.46 m and 0.61 m), friendly space (interpersonal distance between 0.69 m and 1.22 m), and formal space (interpersonal distance between 2.13 m and 3.66 m). For males, this means at least 0.856 m^2 floor area per person (lightly dressed) for comfort in intimate spaces. For bulky clothing, the minimum area must be 0.948 m^2 per person (calculated from MIL-STD-1472B recommendations).

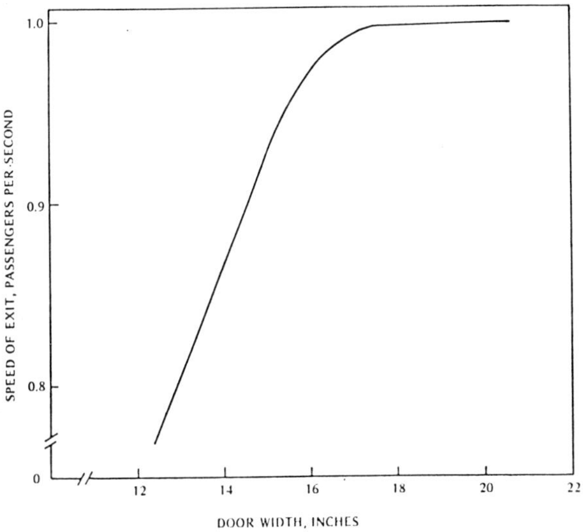

Fig. 7. Aircraft emergency exit width and the speed of exit (Roebuck and Levedahl, 1961).

The minimum PS is significantly influenced by the nature of the task and various psychological, social and environmental factors (Evans, 1973). For example, handicapped individuals, introverts or individuals with some form of social stigma have a greater interpersonal distance and, therefore, need more PS. A hostile or stressful environment also requires larger PS. For detailed review of studies dealing with these factors, the reader is referred to the extensive bibliography provided by Evans (1973).

CLEARANCES BETWEEN CONTROLS

Controls used for different functions must be separated to prevent accidental activation. The spacing between controls is particularly critical when the same type of controls are used for different functions. The U.S. Air Force (1977) recommends that spacing considerations should be based on eight factors: (1) the size of the control and the extent of its displacement, (2) the body member used (controls operated with foot require greater separation), (3) one-hand versus two-hand activation (controls activated with two hands require additional spacing), (4) sequential or simultaneous activation (controls requiring simultaneous activation should be placed closer to each other), (5) sequential or random activation (controls activated in a random order require additional spacing), (6) visual feedback (controls that are activated without any visual feedback require additional spacing), (7) use of gloves and heavy boots (gloves and shoes change the effective dimensions of the extremity and, therefore, greater spacing is needed when activating controls with gloves or shoes), and (8) criticality of the control (critical controls require greater separation to prevent accidental activation). Space availability and standardization and packaging needs further dictate clearances around the controls.

Table 2 summarizes the minimum and preferred clearances between controls as provided by the United States Air Force. If the controls are located at arm's length, below the shoulder level, and grasped by the hand without any visual feedback, the separation should be at least 152 mm; for above shoulder and behind the back locations, separation distance should be at least 305 mm (Damon et al., 1966).

TABLE 2
Minimum and preferred clearances between controls (U.S. Air Force, 1977).

		Separation(mm)	
Control	Condition	Minimum	Preferred
Push button	One finger-random order	12.7	50.8
	-sequential order	6.4	25.4
	Different fingers-random or sequential order	12.7	12.7
Toggle switch	One finger-random order	19.1	50.8
	-sequential order	12.7	25.4
	Different fingers-random or sequential order	15.9	19.1
Rotary selector	One hand-random order	25.4	50.8
	Two hands-simultaneous	76.2	127.0
Knob (continuous setting)	One hand-random order	25.4	50.8
	Two hands-simultaneous	76.2	127.0
Crank	One hand-random order	50.8	101.6
	Two hands-simultaneous	76.2	127.0
Hand wheel	Two hands-simultaneous	76.2	127.0
Thumbwheel		6.35	-
Lever	One hand-random order	50.8	101.6
	Two hands-simultaneous	76.2	127.0

	Maximum simultaneous operation one hand span	-	152.4
Pedal	One foot-random order	101.6	152.4
	sequential order	50.8	101.6

ACCESSIBILITY FOR MAINTENANCE

Accessibility is a major consideration in designing for maintainability. For proper and prompt maintenance, in addition to arranging most frequently replaced parts such that they are reached first, equipment, tools, and workspaces must have visual as well as anthropometric accessibility. Visual accessibility means designing such that parts and items can be viewed quickly. Visibility may be enhanced by mirrors, windows, auxiliary lighting, remote cameras, etc. Snap-in retainers, spring-loaded hinges, latches, etc. may be used to remove and replace components quickly. Major units and assemblies should be in removable housings to permit complete visual inspections. Anthropometric accessibility, on the other hand, means providing sufficient space around the parts and items so that they may be grasped, removed and replaced quickly. This means physically accessing the components with or without tools.

Visual accessibility for displays

Clearances for adequate vision to major displays and other personnel affect the arrangement of equipment (Thomson, 1972). Clear lines of sight to auxiliary displays and between personnel, for communication purposes, may also be necessary. This requires an area of unobstructed vision in several directions. Equipment over 1.4 m in height should be eliminated from the work area in directions requiring unobstructed vision (Fig. 8).

When a display must be viewed by several operators located within a work area, operators should be stationed from the display at a distance such that their lines of sight to the display surface form an angle between 60° and 90°, but never less than 45° (Fig. 9 - Thomson, 1972). To achieve this, the front row of operators should be seated no closer than the width of the screen; preferably at a distance of 1.5 times the width of the display.

When operators are standing, variation in operators' eye height, due to varying leg and torso lengths, creates viewing problems (Fig. 10). These problems can be minimized by seating operators and, thereby, eliminating leg length differential (Thomson et al., 1958).

When a situation involves both seated and standing operators, the required elevation, above the floor, of the operator standing directly behind another seated or standing operator may be determined by the following relationship (see Fig. 11):

Elevation = D - E + ((D - A) * C) / B

Anthropometric accessibility

Among the various forms of access are doors, aisles and corridors, latches, apertures, catwalks, tunnels, etc. In designing access to the equipment,

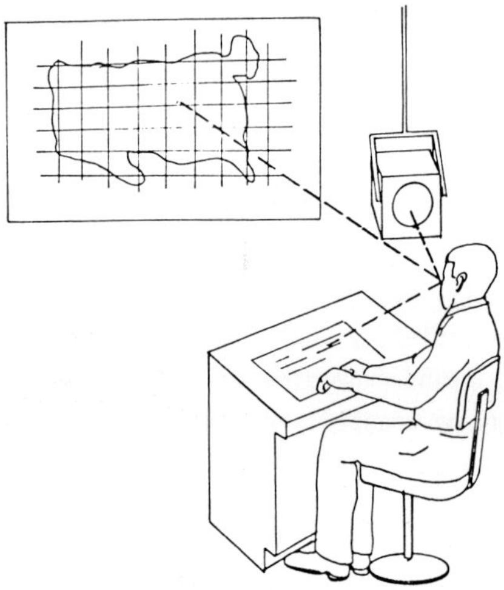

Fig. 8. Unobstructed lines of sight to view several displays (Thomson, 1972).

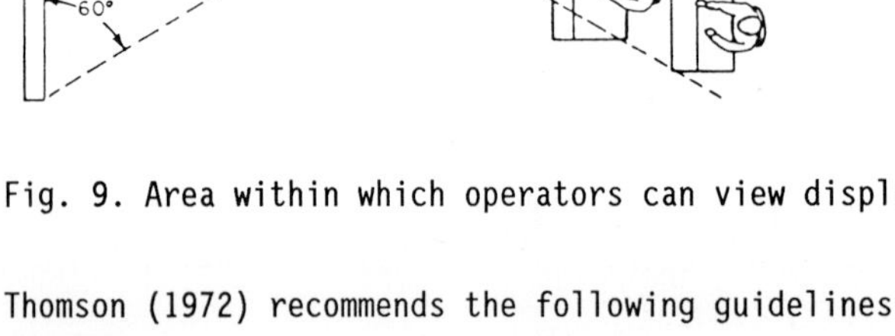

Fig. 9. Area within which operators can view display (Thomson, 1972).

Thomson (1972) recommends the following guidelines: (1) Bulkheads, brackets, and other units should not interfere with removal or opening of cover of units. Unobstructed clearance for opening or removal of covers should be provided, (2) Maintenance checks should be possible without the use of special rigs and harnesses, (3) Units should be laid out to allow a minimum of place-to-place movement by the operator during checkout, (4) Removal of any replaceable unit should involve opening a minimum number of covers or panels, (5) Units should be so located that no other equipment must be removed to gain access, and (6) when one unit is placed behind another, the unit requiring most frequent access should be most accessible.

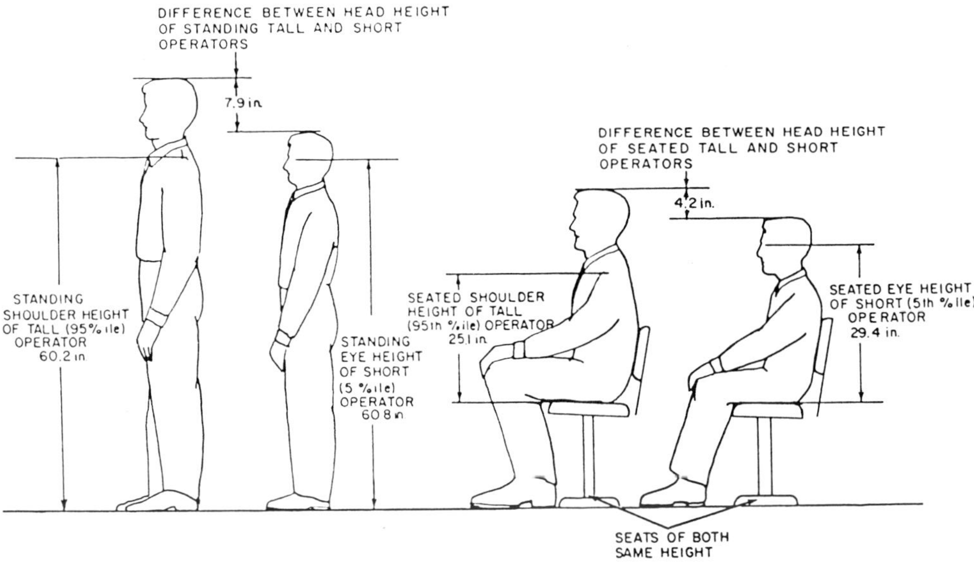

Fig. 10. Seating to reduce viewing problems caused by operator height variations (Thomson et al., 1958).

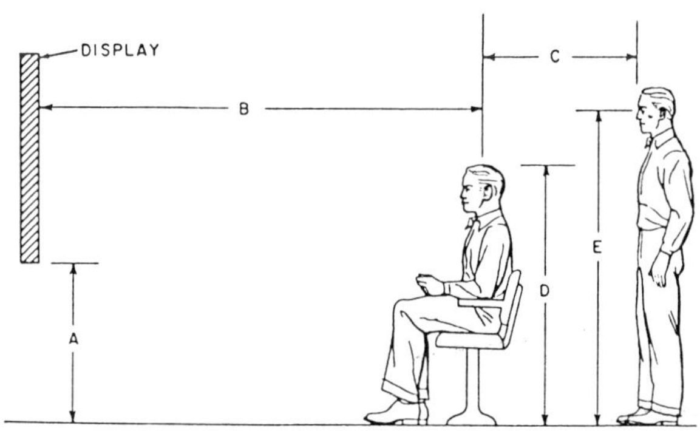

Fig. 11. Dimensions to compute positions of standing and seated operators (Thomson, 1972).

Access openings

Hatches. The minimum dimensions for hatches that will allow one person to pass through are given in Table 3. These recommendations are based on MIL-STD-1472B (1974) and MIL-STD-1472C (1981) and may also be used for emergency exits on aircrafts and ships. For fire exits a minimum width of 915 mm is recommended (Behrens, 1981).

TABLE 3

Minimum acceptable dimensions for hatches (MIL-STD-1472B and MIL-STD-1472C).

Rectangular hatches on a vertical surface (side hatches)	
Light clothing	660 mm high x 760 mm wide
Bulky clothing	740 mm high x 860 mm wide
Rectangular hatches on top or bottom surfaces	
Light clothing	330 mm deep x 580 mm wide
Bulky clothing	410 mm deep x 690 mm wide
Circular hatches	
Minimum diameter	760 mm

It is worth noting that Roebuck and Levendahl (1961) found that the speed of exit levels off when the exit door opening is about 510 mm wide. The recommendations given in Table 3 are also significantly different from those given by Thomson (1972) and Damon et al. (1966). Since Table 3 recommendations are more liberal, other values are not mentioned.

For bulkhead-mounted hatches, the minimum dimensions should be 1524 mm (height) x 508 mm (width). The same dimensions should also be used for watertight doors (Thomson, 1972).

Tunnels and catwalks. These are passages that provide limited access and may be used in high-bays and assembly areas. Woodson (1965) and Damon et al. (1966) have provided minimum dimensions for catwalks and tunnels. Table 4 shows these dimensions (suitable for 95th percentile male). The minimum floor width of a tunnel should be 305 mm (Fig. 12). However, for emergency evacuation or when carrying loads, a minimum floor width of 406 mm is preferable. The minimum width at the shoulder height should be 635 mm. For crawling, the minimum width should be 430 mm (crawling prone - Fig. 13). The minimum height should be 635 mm. For crawling on hands and knees, the minimum height should be 815 mm.

Aisles and corridors. These are passageways that permit appreciable amount of traffic between two points during normal operations. The aisle width dimensions for human traffic have been provided in MIL-STD-1472B (1974), MIL-

STD-1472C (1981) and by Noble (1982). These recommendations, however, do not always agree. For instance, the aisle width recommendation for one person walking, according to MIL-STD, is 610 mm. Noble (1982), on the other hand, recommends 650 mm. Similarly, the recommendations for one person walking while another is pressed against the wall are 914 mm and 1000 mm, respectively. The recommendations in Table 5, therefore, include the most liberal recommendations.

TABLE 4
Minimum dimensions for tunnels and catwalks (Woodson, 1965; Damon et al, 1966).

Activity	Minimum height (mm)	Minimum width (mm)
Walking upright	1955	635
Walking stooped	1600	635
Walking sideways		380
Crawling prone	430	635
Crawling supine (pulling ahead)	635	635
Crawling on hands and knees	810	635

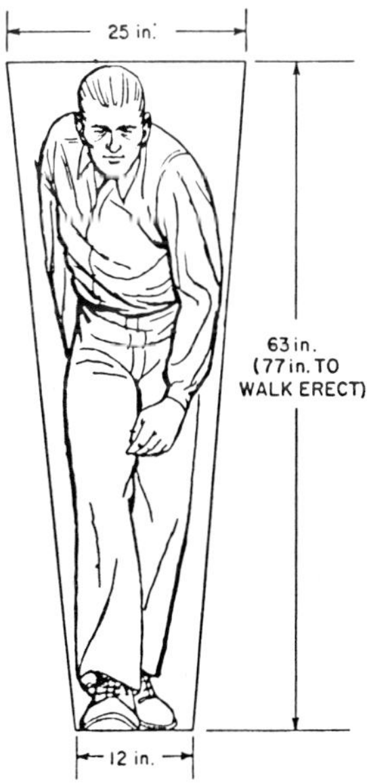

Fig. 12. Minimum tunnel dimensions (Thomson, 1972).

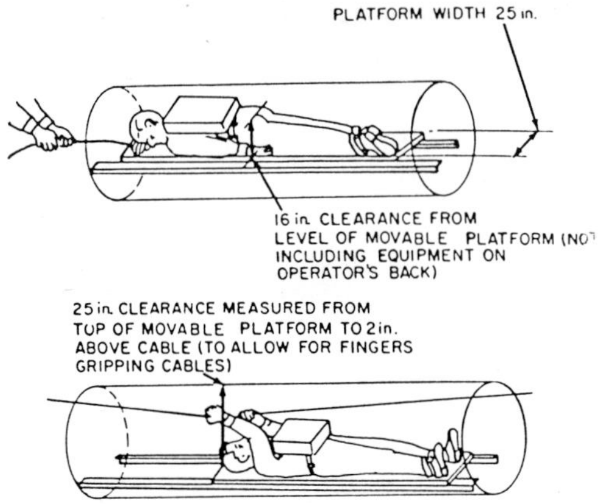

Fig. 13. Minimum crawl tunnel dimensions (Thomson, 1972).

TABLE 5

Aisle and corridor widths (MIL-STD-1472B,C; Noble, 1982).

Activity	Width (mm)
One person walking	650
One person passing, one standing against the wall	1000
One person carrying a tray or similar load	900
One person with a walking frame	1000
Two persons walking abreast	1350
Two persons passing	1372
Two persons passing sideways	900
Three persons walking abreast	1829

When walking between obstacles, such as between two walls, either straight ahead or sideways, the aisle widths could be slightly narrower. Pheasant (1986) summarizes these recommendations from a number of sources (Table 6).

In the design of aisles and corridors, the following design guidelines should be considered: (1) Avoid blind corners, (2) Locate paths for minimum distances, (3) Mark traffic guides on floors, walls, or ceilings, (4) Make intersections converge at 90° to minimize floor space waste, (5) Doors should never open into corridors (consider using folding doors; sliding doors have a tendency to jam), (6) Keep aisles clear (do not use aisle or corridor space for storage), (7) Avoid locating aisles against a blank wall as this allows access only from one side, and (8) Avoid one-way traffic aisles (it is practically impossible to enforce).

TABLE 6

Clearance required for walking between obstacles (Pheasant, 1986).

Activity	Straight ahead (mm)	Sideways (mm)
Between two walls or obstacles higher than 1000 mm in height	650	400
Between one wall or obstacle higher than 1000 mm and one lower than 1000 mm	600	400
Between two obstacles lower than 1000 mm	550	350

Doorways and doors. Open doorways should be at least 1956 mm high; the doorways in most modern houses are about 2007 mm high. Values given in Table 5 should be used to determine the width of open doorways.

Doorways with doors should be at least 2032 mm high. The preferable height of doors is 2134 mm. The clearance between the wall and the moving edge of the door should be 305 mm, but no less than 102 mm (Fig. 14). The clearance between the door opening and equipment should be 1828 mm, but no less than 1524 mm.

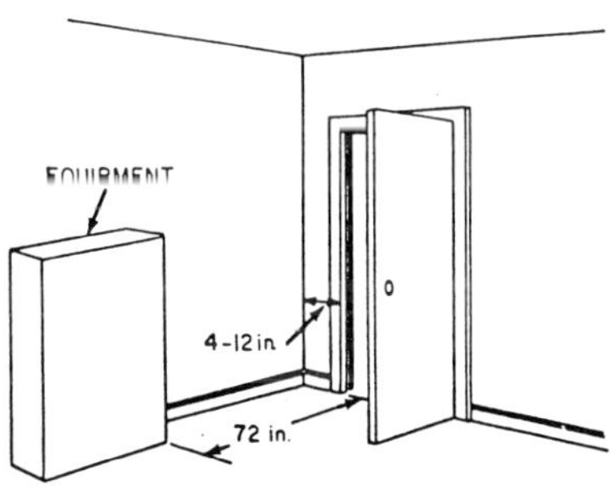

Fig. 14. Door-wall clearance and relation of equipment to door movement (Thomson, 1972).

Ladders and stairs. Vertical ladders should have a clearance of 152 to 203 mm from the wall (or the surface on which it is mounted). The spacing between the underside of the handrails and non-vertical ladders should be between 553 mm and 610 mm. The width of a vertical ladder should be between 305 mm and 533 mm to permit the person to climb up or down. For non-vertical ladders, 533 mm to 610 mm width is desirable. Moreover, non-vertical ladders should preferably rise at an angle of 60^0 from the horizontal.

For stairs, the spacing between the stair and handrail should be between 864 mm and 914 mm. The handrail clearance from the wall should be at least 45 mm. The overhead clearance should be at least 1930 mm (MIL-STD-1472B, 1974). For one-way stairs, a width of 508 mm to 559 mm is needed. If the traffic is both ways, allow between 1219 mm and 1295 mm for width.

Wheelchairs. For one-way passage, aisle clearance should be at least 915 mm (General Service Administration, 1977). For two wheelchairs to pass, 1320 mm is the minimum passage (aisle width of 1830 mm is desirable). The spacing between two fixed tables with back-to-back occupants should be at least 1650 mm for wheelchair passage between them. For turning, a minimum diameter of 1500 mm to 1700 mm is needed (Pheasant, 1986).

The cafeteria line rails should be at least 864 mm wide. The spacing between stacks in libraries should be at least 813 mm wide (1067 mm width is highly desirable).

A person on crutches will need at least 826 mm wide aisle (a width of 914 mm is preferable (Jones, 1978) - contrast it to a width of 1000 mm for one person with a walking frame (Noble, 1982 - Table 5).

CLEARANCE DIMENSIONS IN VARIOUS WORKING POSTURE

Since a variety of body postures are assumed during work, designers must provide appropriate clearances to accommodate the body in these configurations. Figs. 15 through 29 describe some of the most common working postures encoun-

Fig. 15. Stature.

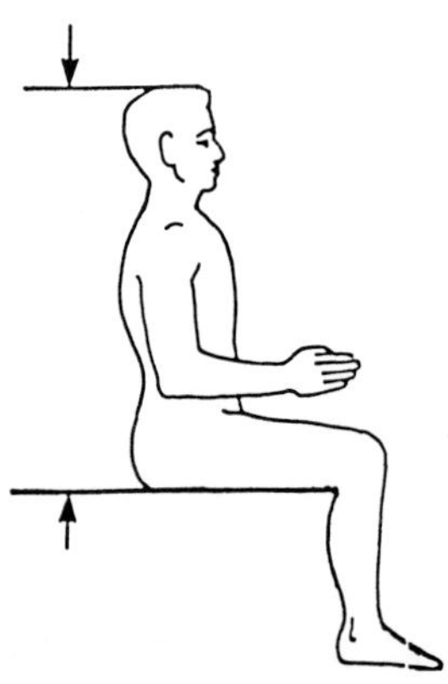

Fig. 16. Sitting height.

tered in industrial work (Hertzberg, 1972). The minimum clearances needed by males and females for these postures are given in Table 7. Certain allowances must be added to these values for clothing. For ordinary clothing, it is recommended that 25 mm be added all round (Pheasant, 1986). This allowance should be doubled for winter clothing. For stature and other heights, approximately 60 mm should be added (Hertzberg, 1972).

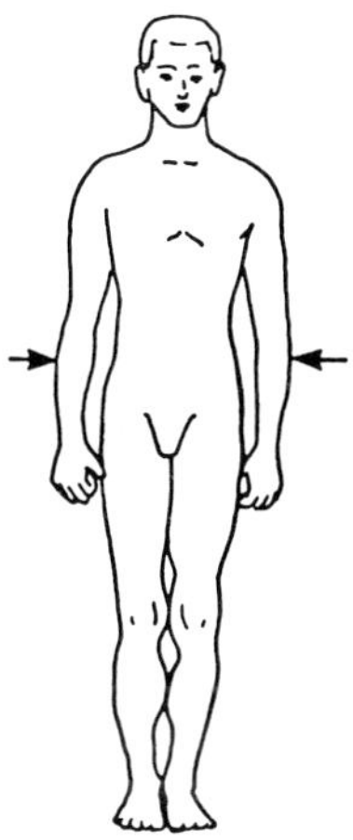

Fig. 17. Maximum body depth.

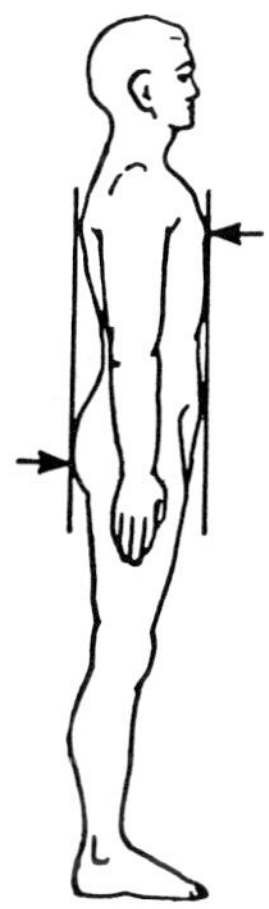

Fig. 18. Maximum body breadth.

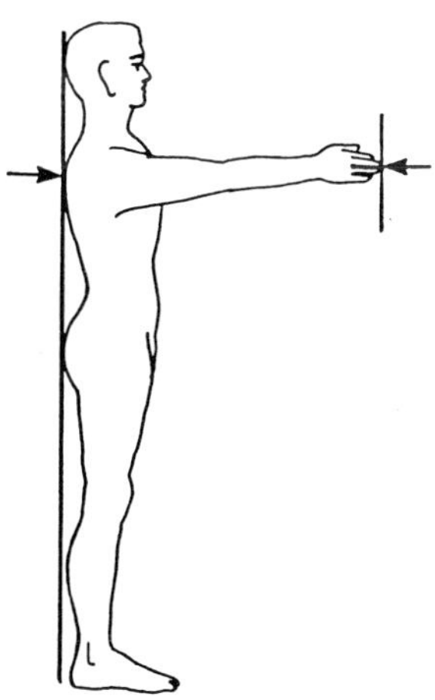

Fig. 19. Arm reach.

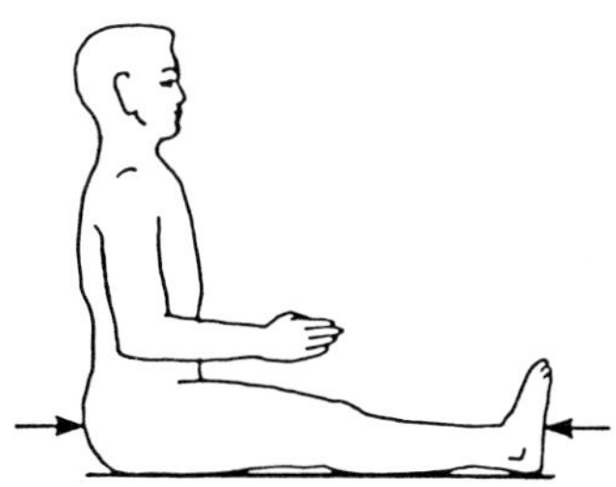

Fig. 20. Buttock-leg length.

Fig. 21. Prone length (A) and prone height (B).

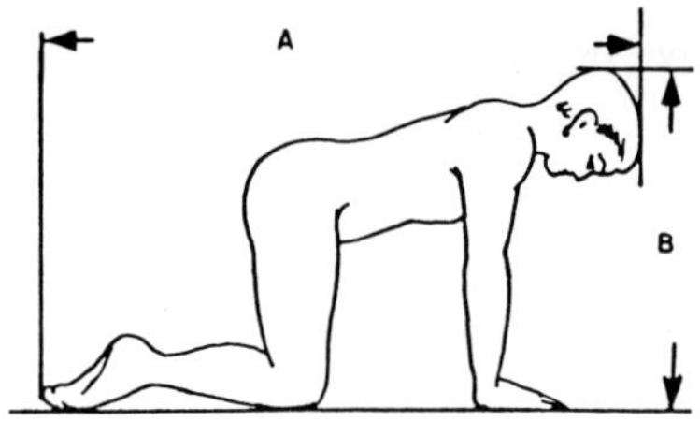

Fig. 22. Crawling length (A) and crawling height (B).

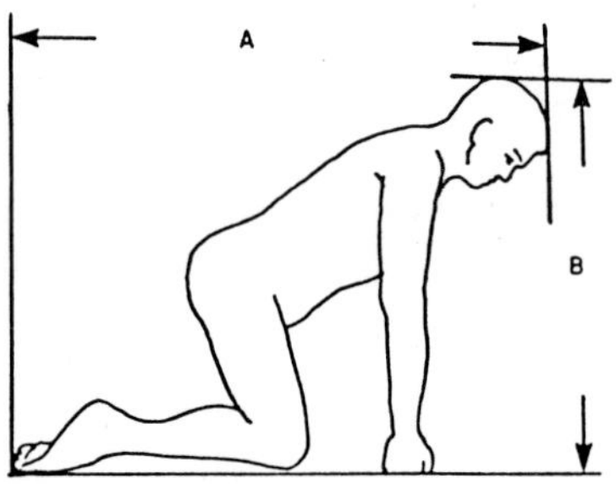

Fig. 23. Keeling length (A) and kneeling height, crouched (B).

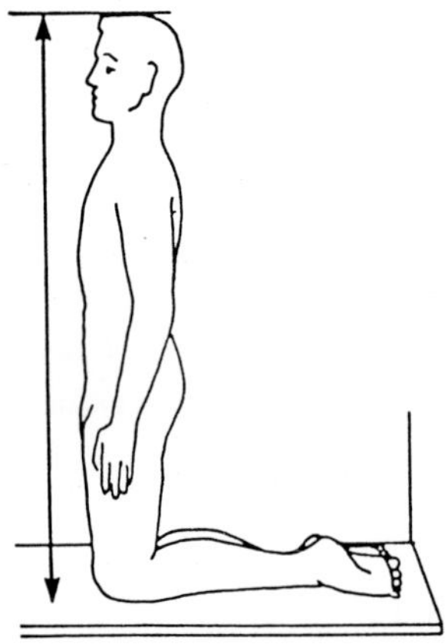

Fig. 24. Kneeling height, upright.

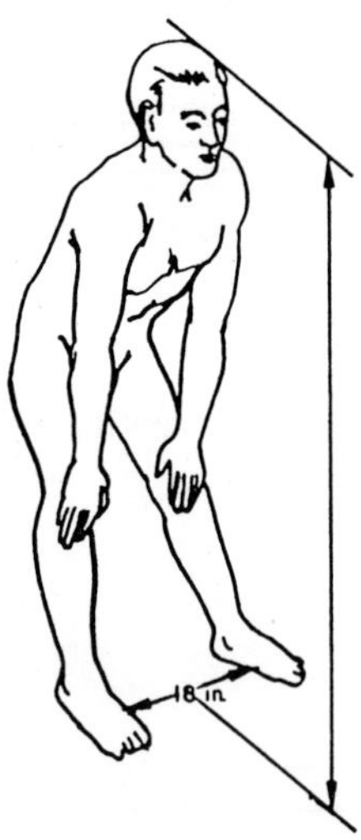

Fig. 25. Bent torso height.

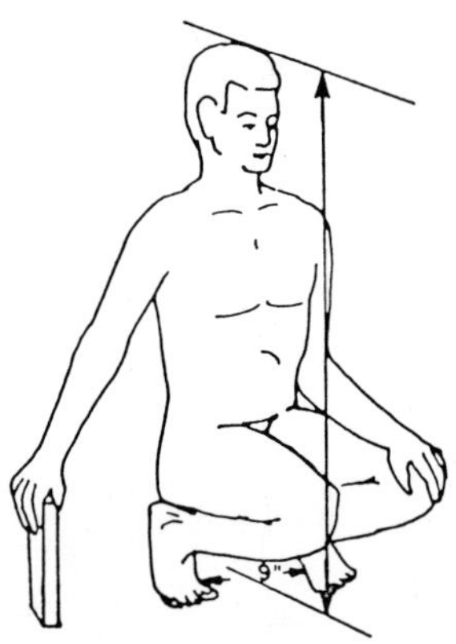

Fig. 26. Squatting height.

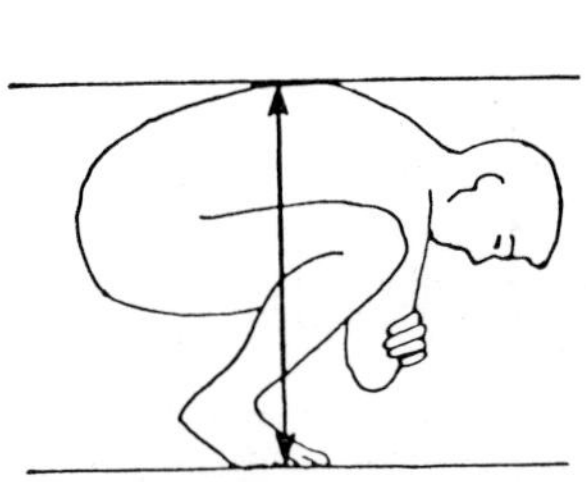

Fig. 27. Minimum squatting height.

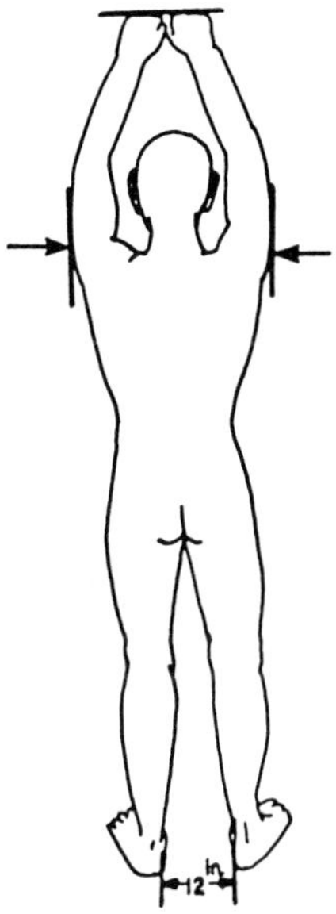

Fig. 28. Overhead reach breadth.

Fig. 29. Vertical reach height, sitting.

TABLE 7

Clearance dimensions (mm) in various working postures (Figs. 15 through 29 - Hertzberg, 1972).

Dimension	Men Percentile 5th	50th	95th	Women Percentile 5th	50th	95th
Stature	1615	1720	1849	1499	1600	1701
Sitting height	843	906	964	785	848	907
Maximum body depth	257	292	330	-	-	-
Maximum body breadth	478	531	579	-	-	-
Arm reach	836	907	975	729	787	851
Buttock-leg length	965	1054	1140	875	965	1055
Prone length (A)	2151	2289	2433	-	-	-
Prone height (B)	312	368	417	-	-	-
Crawling length (A)	1252	1351	1478	1130	1240	1350
Crawling height (B)	665	721	775	605	660	715
Kneeling length (A)	995	1092	1222	-	-	-
Kneeling height, crouched (B)	754	813	876	-	-	-
Kneeling height, upright	1224	1295	1382	1130	1205	1285
Bent torso height	1176	1321	1420	-	-	-
Squatting height	1036	1107	1194	-	-	-
Minimum squatting height	546	630	711	-	-	-
Overhead reach breadth	345	378	404	-	-	-
Vertical reach height, sitting	1311	1397	1499	-	-	-

Note - A variety of sources were used to compile data given in this table. Refer to Hertzberg (1972) for original source. For design purposes, clearances for males should be used.

The clearances for different types of reaches are given in Table 8. This table is reproduced from Pheasant (1986).

TABLE 8

Clearance dimensions (mm) for various kinds of reach (Pheasant, 1986).

Reach type	Men Percentile 5th	50th	95th	Women Percentile 5th	50th	95th
Pinch grip (to the thumbtip)	755	820	875	680	740	795
Fingertip	825	895	960	745	810	870
Forward thrust of the shoulder	835	910	985	745	820	895
10^0 trunk inclination	800	865	930	725	790	850
20^0 trunk inclination	875	950	1020	800	875	940
30^0 trunk inclination	950	1030	1105	875	950	1025

Figs. 30 and 31 show clearance and access dimensions for a standing and seated worker, respectively. The values of these dimensions are given in Tables 9 and 10 (Clark and Corlett, 1984).

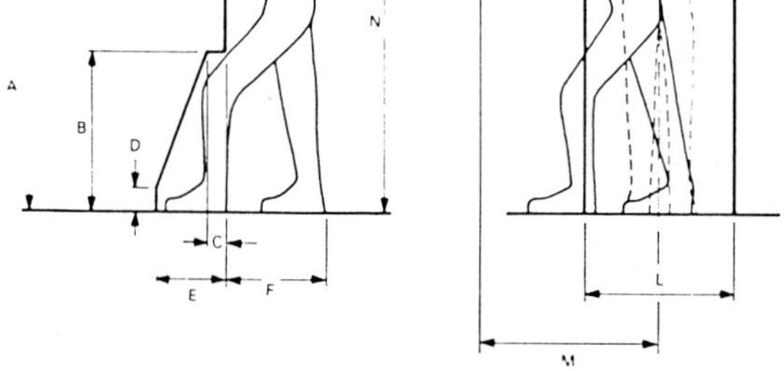

Fig. 30. Clearance and access dimensions for a standing worker (Clark and Corlett, 1984 - see Table 9 for values).

TABLE 9

Clearance and access dimensions (mm) for a standing worker (Clark and Corlett, 1984 - see Fig. 30).

	Men			Women		
	Percentile			Percentile		
Dimension	5th	50th	95th	5th	50th	95th
A Work surface height	970	1070	1190	905	995	1065
B Knee clearance height			560	(minimum required)		
C Knee clearance depth			125			
D Toe clearance height			100			
E Toe clearance depth			250			
F Back clearance			915			
G Max. reach access height	380			380		
H Max. reach access distance	915			890		
J Upright reach access height	685			635		
K Upright reach access distance	455			405		
L Crouch space for side pick-up			1145			
M Extent of max. side reach	1145	(M is less for a twisted trunk)				
N Standing worker's height	1600	1730	1880	1505	1605	1695

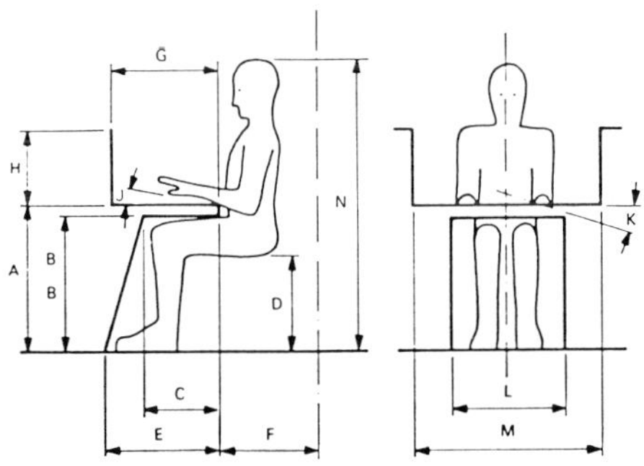

Fig. 31. Clearance and access dimensions for a seated worker (Clark and Corlett, 1984 - see Table 10 for values).

TABLE 10

Clearance and access dimensions (mm) for a seated worker (Clark and Corlett, 1984 - see Fig. 31).

Dimension	Men Percentile 5th	Men Percentile 50th	Men Percentile 95th	Women Percentile 5th	Women Percentile 50th	Women Percentile 95th
A Work surface height	660	710	760	635	685	735
B Knee clearance height (cross legged)			750 (min. needed)			725
B′ Knee clearance height (uncrossed legs)			625 (min. needed)			600
C Knee clearance depth			350			350
D Seat height (horizontal)	447	483	523	424	460	495
E Foot clearance depth			650			650
F Clearance for getting up			635			635
G Grasp objects depth	610			610 (assumes leaning)		
H Grasp objects height	380			380 (assumes leaning)		
J Keyboard/switch panel angle		15°			15°	
K angle for one-handed oper.		20°			20°	
L Knee-well depth			650			650
M Dist. apart of side panels for max. reach to grasp height H	1650			1500		
N Max. head height			1450			

CLEARANCES FOR SAFEGUARDING MACHINERY

The moving parts of machinery constitute an inherent hazard (cutting, shearing, crushing, electrocution, etc.) to operators. In order to minimize the risk of injury, operator access to moving parts should be prevented while the power is on or the machine is running. This can be effectively accomplished by placing moving parts beyond the reach of operators so that the hazard points are inaccessible. The minimum distances at which parts are placed to prevent operator contact are generally referred as safety distances.

A number of recommendations for safety distances have been made in a German standard (DIN 31001, Pt. I, 1983). Table 11 shows the safety distances to hazard points when reaching through different types and sizes of apertures. The safety clearances to avoid squeezing various body parts are shown in Table 12. The safety distances when pivoting movements about various body joints are shown in Table 13. The safety distances for whole-body when reaching over a guard are given in Table 14 (Fig. 32 - Thompson and Booth, 1982).

The safety distances given by Thompson and Booth (1982 - Table 14) are based on experimental work and are much different from those recommended in DIN 31001 (1983). In their experiment, Thompson and Booth found that safety distances recommended in DIN 31001 were almost always exceeded by 95th percentile males.

Therefore, safety distances should be based on 1st or 99th percentile, whichever is applicable.

TABLE 11
Safety distances (mm) to hazard points when reaching through various types and sizes of apertures (DIN 31001, Pt. I, 1983).

Body part at risk	Aperture size (mm) Square/circle	Lateral slot	Safety distance
Fingertip	4 - 8	4 - 8	15
Finger	8 - 25	8 - 20	120
Hand (to root of thumb)	25 - 40	20 - 30	200
Whole upper limb	40 - 250	30 - 135	900*
Whole body	> 250	> 135	see Table 14

* Recommended by Pheasant (1986).

TABLE 12
Safety clearances (mm) to avoid squeezing various body parts (DIN 31001, Pt. I, 1983).

Body part	Safety clearance
Finger	25
Hand, wrist or fist	100
Arm	120
Foot (vertical clearance)	120
Leg	180
Whole body	500

TABLE 13
Safety distances (mm) when pivoting movements about various body joints (DIN 31001, Pt. I, 1983).

Body joint	Safety distance
Root of fingers	120
Wrist	230
Elbow	550
Shoulder	850

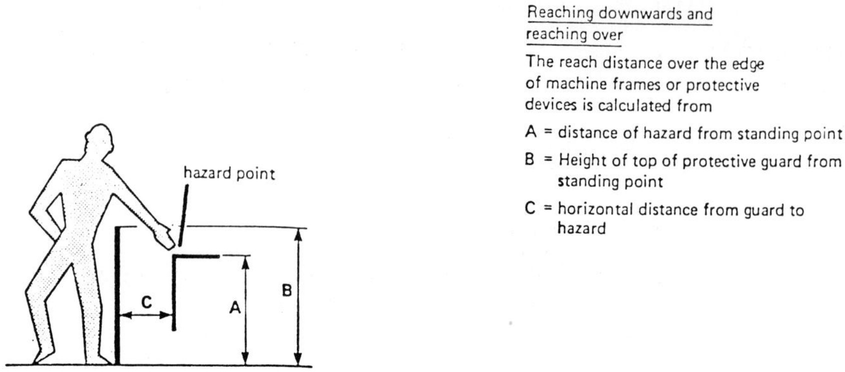

Fig. 32. Safety distances C (mm) for whole-body reach over a barrier (Thompson and Booth, 1982 - see Table 14).

TABLE 14

Safety distance C (Fig. 32 - Thompson and Booth, 1982).

Distance of hazard point from floor A (mm)	Height of protective guard B (mm)						
	2200	2000	1800	1600	1400	1200	1000
2400	308	425	534	486	591	701	761
2200	321	528	699	718	794	920	994
2000		537	785	824	920	1071	1171
1800		477	826	890	993	1189	1295
1600		271	835	898	1026	1267	1397
1400			764	869	1026	1318	1456
1200				788	976	1322	1484
1000				637	889	1310	1482
800					740	1273	1449
600					422	1160	1373
400						1116	1313

For maintenance work, Kennedy and Filler (1966) have recommended the dimensions of vertical rectangular apertures and reach of depth through such apertures. The goal, in this case, is reversed - to be able to reach the component rather than to prevent access to hazard points. Table 15 shows the recommended aperture sizes and the reach depth (Fig. 33) for shirt-sleeved technicians through such apertures.

TABLE 15

Recommended aperture sizes (mm) and depth of reach (mm) for shirt-sleeved technicians in standing and sitting positions (Kennedy and Filler, 1966 - see Fig. 33).

	Standing				
	Forward reach, both arms	Forward reach, preferred arm	Lateral reach, preferred arm	Sitting Normal, both arms	Sitting Cross-legged both arms
A. Depth of reach, 5th %ile	489	514	559	381	349
B. Breadth of aperture	495	305	254	464	451
C. Floor to top of aperture, 95th %ile	1753	1753	1746	1181	711
D. Floor to bottom of aperture, 5th %ile	1327	1327	1327	870	432
E. Vertical dimen. of aperture (C-D) Fig. 33	425	425	419	311	279

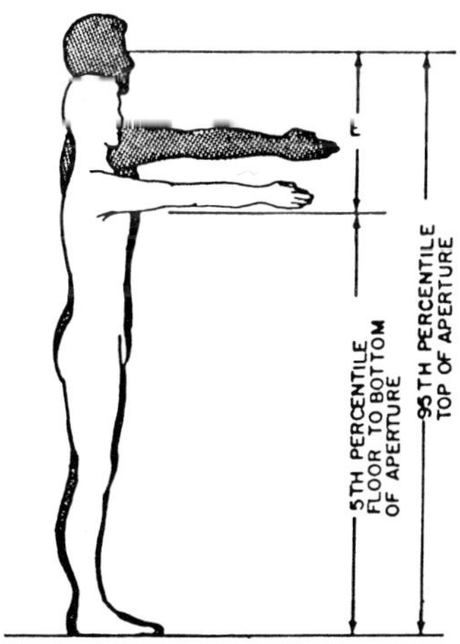

Fig. 33. Reach through vertical aperture (Kennedy and Filler, 1966 - see Table 15 for values).

ACCESS CLEARANCE FOR OPENINGS, HOLES AND HANDLES

Clark and Corlett (1984) have summarized, from various sources, the access clearance dimensions for various openings, holes and handles. The recommendations are summarized in Table 16.

TABLE 16

Access clearance dimensions (mm) for openings, holes, and handles (Clark and Corlett, 1984).

Openings, holes	Height	Width
Empty hand, flat	60	105
Hand with screwdriver	100	100
Fist, clenched	110	110
Inserting box with hands*	Box width + 30	
Thumb plus 2 fingers	50	50
Hole for reaching to 150 mm (hands clasped)	105	115
Hole for reaching to 645 mm	105	460
Full arm reach	see Table 15	
Handles	Clearance for finger thickness - 50 mm (65 mm with gloves)	
	Clearance for hand thickness - 110 mm for one hand (120 mm with gloves) 220 mm for two hands (240 mm with gloves)	
	Separation between handles - 50 mm minimum	

* From Mital and Wang (1989). This value is 15 mm lower than Clark and Corlett's recommendation.

FINAL REMARKS

This chapter presents access and clearance dimensions for designing work spaces. As pointed out in several places, many of the recommendations provided in the published literature are contradictory. Moreover, most design recommendations are based on the United States Department of Defense work. Generally, the work situations encountered in industry are quite different from those encountered on a ship or plane. Therefore, the actual clearance and access dimensions needed are expected to be somewhat different from those given here. However, until such time the actual industrial work situations are simulated and data collected, these recommendations must suffice. The dearth of scientific investigations of clearance needs for the industrial work space is obvious and efforts should be made to generate the relevant data.

REFERENCES

Beggs, W.D.A. and Howarth, C.I., 1970. Movement control in a repetitive motor task. Nature, 225: 725-753.
Behrens, J.F., 1981. For architects and builders. Fire Journal, May, pp. 14-15.
Bennett, C.A., 1978. Spaces for People. Prentice-Hall, Englewood Cliffs, New Jersey.
Bottom, D.J., 1980. The tractor driver's steering control task. Paper presented at the Joint Conference of the Ergonomics Society and the Agricultural Manpower Society, September, 1980.
Bottom, D.J., Barber, T.S., and Chisholm, C.J., 1979. Improving access to tractor cab: an experimental study. Journal of Agricultural Engineering Research.
Clark, T.S. and Corlett, E.N., 1984. The Ergonomics of Workspaces and Machines: A Design Manual. Taylor & Francis, London, United Kingdom.
Damon, A., Stoudt, H.W., and McFarland, R.A., 1966. The Human Body in Equipment Design. Harvard University Press, Cambridge, Massachusetts.
Daniels, E.B., Kobas, G.V., and Drury, C.G., 1976. Monetary and non-monetary incentives in motor performance. Ergonomics, 19: 61-68.
DIN 31001, Part I, 1983. Safety distances for adults and children. Deutsches Institut fur Normung, Berlin.
Drury, C.G., 1971.Movements with lateral constraint. Ergonomics, 14: 293-305.
Drury, C.G., 1985. Influence of restricted space on manual materials handling. Ergonomics, 28: 167-175.
Drury, C.G. and Dawson, P., 1974. Human factors limitations on fork-lift truck performance. Ergonomics, 17: 447-456.
Drury, C.G., Barnes, R.E., and Daniels, E.G., 1975. Pedestrian operated vehicles in hospital. Proceedings of the Institute of Industrial Engineers. Atlanta, Georgia, pp. 184-191.
Drury, C.G., Law, C.H., and Pewenski, C.S., 1982. A survey of industrial box handling. Human Factors, 24: 553-565.
Evans, G.W., 1973. Personal space: research review and bibliography. Man-Environment Systems, 3: 203-215.
Hertzberg, H.T.E., 1972. Engineering anthropology. In Human Engineering Guide to Equipment Design (Editors: H.P. Van Cott and R.G. Kinkade). U.S. Government Printing Office, Washington, D.C., pp. 467-584.
Jones, J.C., 1970. Design Methods. John Wiley, New York, New York.
Jones, M.A., 1978. Accessibility Standards Illustrated. Capital Development Board, State of Illinois, Springfield, Illinois.
Kama, W.N., 1963. Volumetric workspace study: part I. optimum workspace configuration for various screwdrivers. AMRL Technical Report No. AMRL-TDR-63-68(I). Aerospace Medical Research Laboratory, Wright-Patterson Air Force Base, Ohio.
Kama, W.N., 1965. Volumetric workspace study: part II. Optimum workspace configuration for the use of wrenches. AMRL Technical Report No. AMRL-TDR-63-68(II). Aerospace Medical Research Laboratory, Wright-Patterson Air Force Base, Ohio.
Kennedy, K.W. and Filler, B.E., 1966. Aperture sizes and depths of reach for one- and two-handed tasks. Technical report no. AMRL-TR-66-27. Aerospace Medical Research Laboratories, Wright-Patterson Air Force Base, Ohio.
MIL-STD-1472B, 1974. Human Engineering Design Criteria for Military Systems, Equipment, and Facilities. United States Department of Defense, Washington, D.C.
MIL-STD-1472C, 1981. Human Engineering Design Criteria for Military Systems, Equipment and Facilities. United States Department of Defense, Washington, D.C.
Mital, A., 1986. Subjective estimates of load carriage in confined and open spaces. In Trends in Ergonomics/Human Factors III (Editor: W. Karwowski). North-Holland, Amsterdam, The Netherlands, pp. 827-833.
Mital, A. and Wang, L.W., 1989. Effects on load handling of restricted and unrestricted shelf opening clearances. Ergonomics, 32: 39-49.

Noble, J., 1982. Activity and Spaces, Dimensional Data for Housing Design. The Architectural Press, London.
Pheasant, S., 1986. Bodyspace: Anthropometry, Ergonomics and Design. Taylor & Francis, London, United Kingdom.
Rashevsky, N., 1970. Mathematical biology of automobile driving, IV. Bulletin of Mathematical Biophysics, 32: 71-85.
Ridd, J.E., 1985. Spatial restraints and intra-abdominal pressure. Ergonomics, 28: 149-166.
Roebuck, J.A. and Levedahl, B.H., 1961. Aircraft ground emergency exit design considerations. Human Factors, 3: 174-209.
Sommer, R., 1959. Studies in personal space. Sociometry, 22: 247-260.
Thompson, D. and Booth, R.T., 1982. The collection and application of anthropometric data for domestic and industrial situations. In Anthropometry and Biomechanics: Theory and Application (Editors: R. Easterby, K.H.E. Kroemer, and D.B. Chaffin). Plenum Press, New York, New York, pp. 279-291.
Thomson, R.M., 1972. Design of Multi-Man-Machine Work Areas. In Human Engineering Guide to Equipment Design (Editors: H.P. Van Cott and R.G. Kinkade). U.S. Government Printing Office, Washington, D.C., pp. 419-466.
Thomson, R.M., Jacobs, H.H., Covner, B.J., and Orlansky, J., 1958. Arrangement of groups of men and machines. ONR Report No. ACR-33. Office of Naval Research, Washington, D.C.
United States Air Force, 1977. Design of controls. AFSC DH 1-3, DN2D. Air Force Systems Command.
United States General Service Administration, 1972. Building Fire Safety Criteria. PBS P5920.9, CHGE. Government Printing Office, Washington, D.C.
Williams, J.C., 1969. Automobile driving control task. Unpublished Masters Project Report. Department of Production Engineering, University of Birmingham, United Kingdom.
Woodson, W.E., 1965. Human Engineering Guide for Equipment Designers. University of California press, Berkeley, California.

Workspace, Equipment and Tool Design
edited by A. Mital & W. Karwowski

ANTHROPOMETRY APPLICATIONS FOR MACHINE SAFEGUARDING

John R. Etherton
Division of Safety Research, National Institute for Occupational Safety and Health, 944 Chestnut Ridge Road, Morgantown, West Virginia 26505-2888 (USA)

ABSTRACT

Anthropometry is the study of human body measurements. One area of occupational safety which depends upon anthropometry information is the dimensioning of reach distances for machine safeguarding. This paper discusses sources of anthropometric data for designing safeguarding methods which prevent workers from contacting hazards by putting them at distances where the hazards cannot be reached. This information is particularly valuable for U.S. ergonomists, safety engineers, and machine workstation designers who should understand how anthropometric information relates to requirements in their OSHA and ANSI standards for machine guarding.

INTRODUCTION

A good way to protect people from machine hazards is to provide barriers or workstation designs that keep them at a safe distance from the hazard. This chapter will examine how anthropometry can guide ergonomists, safety engineers, and machine workstation designers in making safe distance determinations.

An illustration of lack of attention to safe distance needs in the workplace is presented in the following case. A horizontal milling machine's start/stop handle was located above cutters which were supposedly safeguarded by distance. The start/stop handle for the machine was located at a height of 162.6 cm (64 in) above the floor. A 152.4 cm (60 in) tall female operator reached up to use the handle. There was the possibility for imbalance when this worker reached up and made a horizontal push on the handle. It is easy to imagine the unintentional reach toward the cutters that this woman would have made with her free hand to support herself. On one occasion, this free hand was, in fact, placed in the cutter path and she suffered a hand amputation.

Part of the machine workstation designer's problem in preventing further injuries of this kind is finding human reach data to evaluate the ease of functional reaches to normally used controls and the extended distance of unintentional reaches (see Figure 1). What should the parameters be for one hand making a convenient reach overhead to a control while the other hand reaches forward toward a hazard but remains a safe distance from it?

For this example, the available anthropometric data for evaluating the interaction between these two reaches is unsatisfactory. To get data on the overhead reach component of this situation, the wrist to floor distance is sought because a bent wrist grasp would be an efficient way to move such a handle. For a woman of small stature (2.5 %ile), the wrist to floor distance with the arm held overhead is given in one anthropometric data source as 169.2 cm (66.6 in) (Diffrient, Tilley, and Harman, 1981a). In a different data source, a similar dimension, 2 %ile female grip reach overhead to center line of grip, is given as 182.4 cm (71.8 in) (Clauser et al, 1972). There is a 13.2 cm (5.3 in) difference between these two values. This may be due to the first source being based on composite body dimensions, whereas the second source represents actual measurements of this dimension. Neither source represents the functional situation which was not a reach directly overhead, but a reach up and increasingly forward as the handle moved. Reach distance

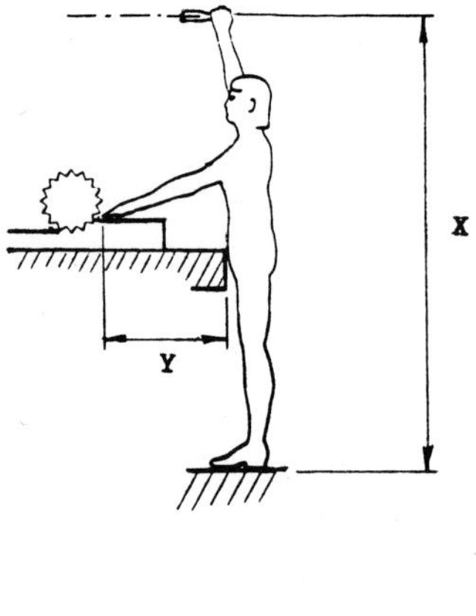

X - Reach overhead, grasp, and push/pull: Is this a suitable reach distance for this kind of control?
Y - Reach forward for balance: If this reach is made to maintain balance while reaching overhead, can the hazard be contacted?

Figure 1 Anthropometric data on these concurrent reaches is needed to evaluate the safety of this machine workstation.

data on the hand reaching directly toward the hazard in combination with an overhead reach was not found. From an injury prevention perspective, anthropometric measurements for these reach interactions for small women is needed if this workstation's design is to be changed to improve safety, and if reach distance is to remain the safeguarding method.

In performing machine safety analysis services for an industrial company, ergonomists and safety professionals may identify a machine action similar to the one illustrated in the injury account above which could amputate a hand if someone should inadvertently reach into it. Whether anyone could actually reach part of a hand or even the finger tips into the hazard from their normal work position has usually not been determined, although assumptions probably have been made that no one would normally reach in that direction. Looking at this problem as an ergonomist, rather than as an inspector for safety code compliance, how do you decide whether to recommend a more positive form of safeguarding than safe location for this hazard? Human reach capabilities must be considered.

THE SCIENCE OF ANTHROPOMETRY

Industrial Anthropometry is the science concerned with determining the dimensions of the human body and defining how these dimensions apply to work tasks. This chapter presents only a few concepts of anthropometry and how anthropometry can be applied to machine design and safeguarding. Other references should be consulted for a more complete description of this discipline (Pheasant, 1986; NASA, 1978; Bennet, 1977; Roebuck et al., 1975; Diffrient et al., 1974; Jones, 1974).

Some of the more important terms and concepts of anthropometry are:

Percentile (%ile)

A convenient way of indicating how much of the range of human variability is accommodated by a given dimension, or the fraction of the population who measure less than this dimension. For example the 50 %ile U.S. civilian male height is 173.6 cm (5ft 7in) indicating that 50% of the U.S. civilian male population will be 173.6 cm or less in stature.

The average man fallacy

For large populations of workers very few or none at all will have average measurements in each of several body part dimensions. In other words, there are not many people who are average height, average weight, have average size hands, average size feet, and average size chest circumference.

Percentile values are not additive

There is little correlation between the relative size of different body dimensions. Tall people can have short arms and short people can have long fingers. Combining 95 %ile values for different body dimensions to get a composite dimension does not represent the dimension which would be determined if a population were actually measured for that dimension.

Static dimensions

Measurements taken with the body in a rigid posture.

Functional dimensions

Measurements taken with the body in a working posture such as crawling or reaching.

Comfortable dimensions

Measurements taken so that the body does not assume a fatiguing posture such as twisting of the trunk, extended reaches, or forced reach into narrow openings. Comfortable dimensions are commonly needed on machine workstations to accommodate for easy maintenance access and to provide operators with comfortable seating.

Anthropometric data sources include: Humanscale 1/2/3 (Diffrient et al., 1974), Humanscale 4/5/6 (Diffrient et al., 1981a), Humanscale 7/8/9 (Diffrient et al., 1981b), and Anthropometric Source Book (NASA, 1978). The Center for Anthropometric Data (CARD) provides an on-line data base which can be accessed free-of-charge by writing the U.S. Department of the Air Force, AAMRL/HEG, Wright Patterson AFB, OH 45433. Anthropometric design aids include physical manikins and computer aided design (CAD) packages with built-in anthropometric figures.

If a body dimension of interest for the population of interest is not found in tables of anthropometric measurements, then it may be necessary to carry out your own measurements. Computation methods have also been devised to determine unknown dimensions based on other body dimensions. Robinette and McConville (1981) suggest a method using regression equations. Pheasant (1982a) describes a technique for determining 1st and 99th percentile values for body measurements if one knows the mean value of the dimension of interest and the mean stature and its standard deviation for the population. Haslegrave (1986) proposes that to compute any body dimension for a so-called 5 %ile woman or a 95 %ile male (remember that people who are large or small in one body dimension cannot be expected to also be large or small in other body dimensions), one should obtain subjects whose stature and weight are at these extremes, take measurements of the body dimension of interest, calculate the median value, and use this value in your design. He reports fourteen values for characterizing these anthropometric extremes.

REVIEW OF RECENT RESEARCH

Table 1 presents a review of recent research in which anthropometry was used to evaluate conditions affecting the safety and health of workers at machine workstations. Useful results from recent research include analyses of the anthropometric acceptability of workstations used by workers of different nationalities (Courtney, 1984; Courtney and Wong, 1985; Ong et al., 1988) and of specific kinds of workstations (engine drivers' stations: Hedberg et al., 1981; laser scanner checkout stations: Wilson and Grey, 1984; simple farm implements: Gite and Yadav, 1989). Other recent research should be useful for developing values for body dimensions not included in tables of measurements (Pheasant, 1982b; Haslegrave, 1986).

SAFE DISTANCE OR SAFEGUARDING BY LOCATION

Distance can be an effective protection from a hazard when there is no need to access the area near the hazard. The Occupational Safety and Health Administration (OSHA) publication 3067 (U.S. Department of Labor, 1980) states this concept as follows:

> "To safeguard a machine by location, the machine or its dangerous moving parts must be so positioned that hazardous areas are not accessible or do not present a hazard to a worker during the normal operation of the machine."

The advantages of using safe distance as a safeguarding technique are that it is usually inexpensive and provides good visibility of work in process (i.e., additional, physical safeguards are normally not needed.) Its limitations are that safe distance does not protect against ejected materials nor does it apply when access is necessary, for example during maintenance.

To provide safety by using a safe distance criterion, what anthropometric information on reach capabilities is available? The British Standards Institute's BSI 5304 (1988) provides one reference for U.S. machine workstation designers. The reach tables in this source are taken from the West German standard DIN 31001 (1983). This standard describes unforced reach over barriers for 95% of the West German population over the age of 14 years.

Research published by Booth and Thompson (1980) evaluated the use of these West German population reach values for an English population made up of 95 %ile stature males. Booth and Thompson collected their data by permitting

Table 1
Recent anthropometry literature related to machine safety.

Author	Population	Dimension
Davies et al, 1980	92 female workers	5th and 95th percentile hand dimensions for safe guard openings
Haslegrave, 1980	2000 male and female car drivers	Seventeen dimensions required in designing vehicular workstations
Booth and Thompson, 1981	500 Asian, European, and West Indian women	Forced reach through variously shaped, rigid openings
Hedberg et al, 1981	150 engine drivers and 91 driver trainees	5th and 95th percentile data for sitting and standing postures, particularly reach to controls distance
Pheasant, 1982b	British civilian adults	Estimated measurements for dimensions for which anthropometric data has not been collected
Das and Grady, 1983	U.S. military	Adjusted dimensions for workplace layouts for standing and seated operators
Courtney and Ng, 1984	100 Hong Kong females	Hand measurements for the design of machine guards
Courtney, 1984	Hong Kong, Japan, U.S. and U.K. females	Hand measurement comparisons for design of machine guarding
Wilson and Grey, 1984	Laser-scanner checkout operators	Reach dimensions for laser-scanner checkout workstations
Courtney and Wong, 1985	Hong Kong Chinese bus drivers	Evaluation of seat and pedal dimensions on buses used
Lee and Wheway, 1986	70 males and 70 females	Reach dimensions between two horizontal photoelectric beams

Table 1 (continued)
Recent anthropometry literature related to machine safety.

Author	Population	Dimension
Haslegrave, 1986	British civilian adults	Composite body dimensions for 95 %ile males and 5 %ile females
Boussena and Davies, 1987	Employment rehabilitation clients	13 body dimensions
Ong et al, 1988	Singapore female VDU operators	Seated body dimensions
Gite and Yadav, 1989	39 Indian farm workers	52 body dimensions for the design of farm equipment

subjects to perform a dynamic reach by placing a hand on the barrier and straining to reach as far as possible. They found that the BS 5304 values would not in all cases protect the population they measured. In reporting their results they sought to protect 99% of the population and therefore they made reach distance recommendations which added a safety factor to the actual reach distances they measured. Their recommendations are shown in Table 2. A followup study by Thompson (1989) using 99 %ile stature subjects produced results which still exceed the BS5304 values but are within the limits of Table 2.

Table 2 Safety distance when reaching over barriers.

Barrier height (mm)	Safety distance (mm)
1000	1600
1200	1400
1400	1170
1600	940
1800	870
2000	590
2200	370

Data from Thompson and Booth (1982), as stated by Pheasant (1986)

APPLICATION OF ANTHROPOMETRICS TO SAFE DISTANCE

The ANSI standard for the safety of press brakes (ANSI B11.3, 1982) provides that safe distance is an acceptable method of safeguarding. However, this standard later states that "A dimension value has not been assigned to the minimum safe distance." A press brake is shown in Figure 2 and the configuration for proposed safe distance is illustrated in Figure 3.

As shown in Figure 4, the ergonomist would find a range of reach values to consider for a standing, working adult who attempts to reach into the die area of a press brake. The height is based on a normal working height. Following are comments on the use of each of these values for safe distance.

Dynamic reach over a horizontal workpiece

This distance could not be reached by a very tall worker (99 %ile) who was able to bend forward and assist his reach by supporting himself with his other hand and strain to reach the die. It seems that this distance would protect all press brake operators, although this kind of forced reach would seem an unlikely occurrence.

Reach over a horizontal workpiece

This distance could be reached by a very tall (99 %ile) male worker who was able to bend forward and reach toward the die. This distance would protect about 95% of the workers who would make such a reach. Again this kind of reach seems unlikely in the workplace.

Fingertip reach for 95 %ile male

Using a method developed by Das and Grady (1983), a reach distance of 807 mm could be achieved if the worker kept a straight back.

Convenient reach for 95 %ile male

This distance could be reached without undue strain and without bending forward.

Maximum easy reach

This distance is one which is easy for any standing worker to reach if they are intended to reach forward to operate controls or otherwise work easily in front of themselves.

OVERHEAD REACH

Another condition in which U.S. standards do not account for a large segment of human capability is the OSHA rule (1910.219) that guarding is not

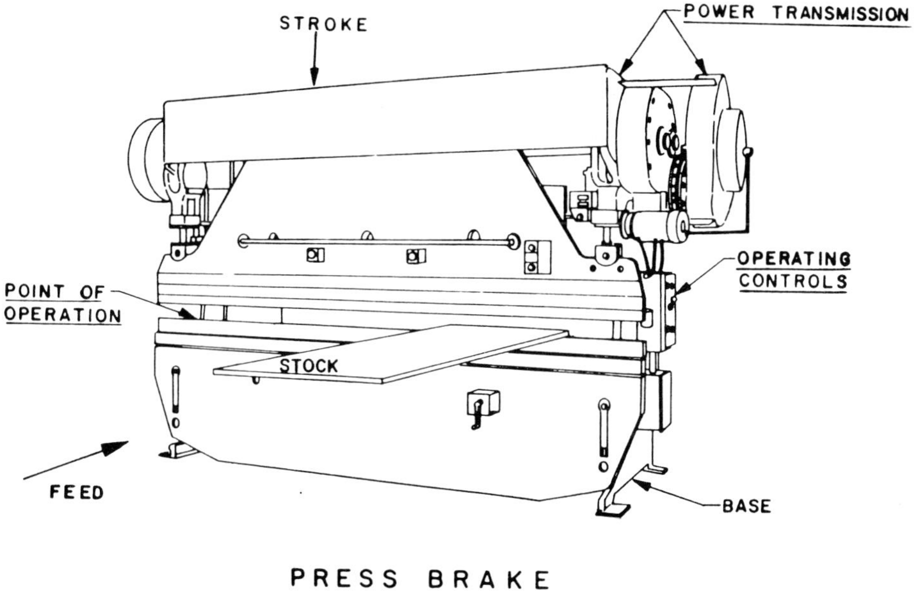

Figure 2 A press brake. The stock must be held as it swings up due to the pressing action of the dies.

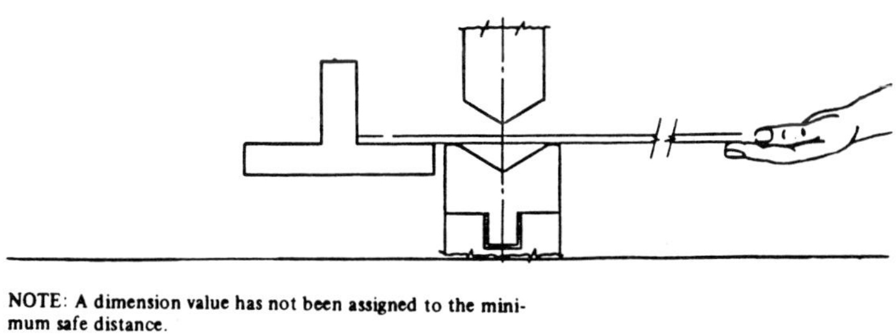

Figure 3 Safe distance concept applied to the press brake.

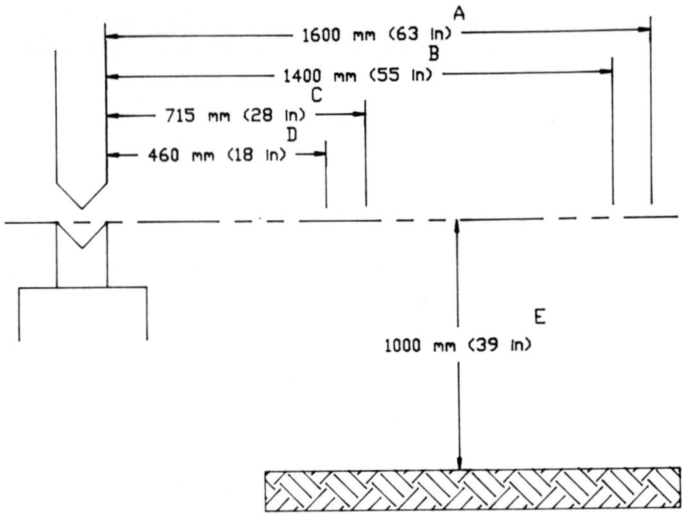

A-Dynamic reach over horizontal workpiece (Pheasant, 1986)
B-Reach over horizontal workpiece (BS 5304, 1988)
C-Convenient reach 95%ile male (Pheasant, 1986)
D-Maximum easy reach to controls (Diffrient et al., 1981b)
E-Normal work height for average standing adult (Diffrient et al., 1981b)

Figure 4 Range of possible safe reach distances for a task of holding a workpiece in a press brake.

needed if a hazard overhead is higher than seven feet (84 inches). Common examples of overhead hazards would include overhead conveyors and mechanical power transmission equipment such as belts and flywheels. This requirement can be viewed as an obvious compromise when we look at the anthropometric tables for overhead reach and find that a 95 %ile male, standing flat-footed in his shoes, can reach 240.0 cm (94.5 in), fully 25.4 cm (10 in) higher than the value in the OSHA standard.

RECOMMENDATIONS

Correctly applying anthropometric data to the safety of a machine operator's task should involve using measurements which reflect the functional nature of the task. For this reason, safe distance in press brake work may differ from the dynamic forced reach evaluated by Thompson and Booth (1982) because the workpiece does not constitute a rigidly constructed, fixed barrier. If an operator were to reach toward the die they would have to use one hand to support the work and also position their feet for stability. A practical method for determining safe distance is to have the employee with the longest reach (probably the tallest employee) make a dynamic reach with one hand with the press brake not running. Three trials would be made and the furthest horizontal reach used in subsequent analysis. Subjects should be instructed to reach as far as they can, but always to support the workpiece. An alternative method would be to randomly sample a minimum of 30 people from each gender and ideally each racial group which commonly uses the equipment.

It is clear that anthropometric information provides only one of the criteria which have thus far been included in voluntary and mandatory standards for the safety of machine operators in the U.S. Tradeoffs are being made on what percentile of the population will be protected by standards. In making recommendations to industry, ergonomists must make these tradeoffs clear. The safest reach distance to use will be the one which protects the greatest proportion of the machine operator population. Task/machine designers who are developing tasks or workstations need to survey the workers who will be performing those tasks and using those workstations and find out how far those workers can reach. The safest dimensions to use, if safe location or safe distance is to be the primary safeguarding method, are the dimensions which protect workers by not allowing them to reach the machine hazard.

Another way to express this approach is to suggest that when designing or evaluating machine workstations and machine operating tasks, one should use the smallest openings and the longest distances needed for the workers who are being protected. The best way to prevent reaching injuries is to use safe dimensions in the initial design of machines and machine operating tasks. There is a need for further research on safe distance and safe location to provide designers with these dimensions. Such research should consider the needs of the workers from the many national backgrounds who make up the machine operating workforce in the U.S.

REFERENCES

ANSI B11.3, 1982. Safety requirements for construction, care, and use of power press brakes. American National Standards Institute, 1430 Broadway, New York.

Booth, R. and Thompson, D., 1980. Anthropometry and machine guarding: an investigation into forced reach over barriers and through openings - part I reach over barriers. Journal of Occupational Accidents, 3: 45-56.

Booth, R. and Thompson, D., 1981. Anthropometry and machine guarding: an investigation into forced reach over barriers and through openings - part II reach through openings. Journal of Occupational Accidents, 3: 87-99.

Boussena, M. and Davies, B., 1987. Engineering anthropometry of employment rehabilitation centre clients. Applied Ergonomics, 18: 223-228.

BS 5304, 1988. British Standards Institution, Safeguarding of Machinery. Global Engineering Documents, Suite 400, 1990 M. St NW, Washington, D.C.

Clauser, C., Tucker, P. McConville, J., Churchill, E., Laubach, L., and Reardon, J., 1972. Anthropometry of Air Force Women. Technical Report AMRL-TR-70-5, Aerospace Medicine Research Laboratory, Wright Patterson AFB, OH (Available as document AD 743113 from National Technical Information Service, Springfield VA).

Courtney, A., 1984. Hand anthropometry of Hong Kong Chinese females compared to other ethnic groups. Ergonomics, 27: 1169-1180.

Courtney, A. and Ng, M., 1984. Hong Kong female hand dimensions and machine guarding. Ergonomics, 27: 187-193.

Courtney, A. and Wong, M., 1985. Anthropometry of the Hong Kong Male and the design of bus driver cabs. Applied Ergonomics, 16: 259-266.

Das, B. and Grady, R., 1983. Industrial workplace layout design: an application of engineering anthropometry. Ergonomics, 26: 433-437.

Davies, B., Abada, A., Benson, K., Courtney, A., and Minto, I., 1980. Female hand dimensions and guarding of machines. Ergonomics, 23: 79-84.

Davies, B. and Moustafa, A., 1986. A low-cost, accurate electronic anthropometer. Ergonomics, 29: 713-717.

DIN 31001, Part I, 1983. Safety distances for adults and children. Deutsches Institut fur Normung, Berlin.

Gite, L. and Yadav, B., 1989. Anthropometric survey for agricultural machinery design. Applied Ergonomics, 20: 191-196.

Halstrom, F., 1987. Horizontal milling machine: lack of proper guarding, improper location of control lever. Product Liability Law Review, October, p. 120.

Haslegrave, C., 1980. Anthropometric profile of the British car driver. Ergonomics, 23: 437-467.

Haslegrave, C., 1986. Characterizing the anthropometric extremes of the population. Ergonomics, 29: 281-301.

Hedberg, G,. Bjorksten, T., Ouchterlony, K., and Jonsson, B., 1981. Anthropometric study of Swedish engine drivers. Ergonomics, 24: 257-264.

Lee, J. and Wheway, G., 1986. Reach distance between two horizontal photoelectric beams. Ergonomics, 29: 87-97.

Ong, C., Koh, D., Phoon, W., and Low, A, 1988. Anthropometrics and display station preferences of VDU operators. Ergonomics, 31: 337-347.

Pheasant, S., 1982a. A technique for estimating anthropometric data from the parameters of the distribution of stature. Ergonomics, 25: 981-992.

Pheasant, S., 1982b. Anthropometric estimates for British civilian adults. Ergonomics, 25: 993-1001.

Robinette, K. and McConville, J., 1981. An alternative to percentile models. SAE Technical Paper 810217, Society of Automotive Engineers, Warrendale, PA.

Thompson, D., 1989. Reach distance and safety standards. Ergonomics, 32: 1061-1076.

Thompson, D. and Booth, R., 1982. The collection and application of anthropometric data for domestic and industrial standards, In: Anthropometry and Biomechanics: Theory and Application, (Editor R. Easterby, K. Kroemer, and D. Chaffin), Plenum, New York, 279-291.

US Department of Labor, 1980. Concepts and Techniques of Machine Safeguarding. OSHA, Document No. 3067, Washington, D.C.

Wilson, J. and Grey, S., 1984. Reach requirements and job attitudes at laser-scanner checkout systems. Ergonomics, 27: 1247-1266.

SELECTED ANTHROPOMETRY REFERENCES

Bennet, C., 1977. Spaces for People: Human Factors in Design. Prentice Hall, Englewood Cliffs, NJ.

Clark, T. and Corlett, E., 1984. The Ergonomics of Workspaces and Machines: A Design Manual, Taylor and Francis, Philadelphia.

Diffrient, Tilley, Bardagjy, 1974. Henry Dreyfuss Associates, Humanscale 1/2/3. MIT Press, Cambridge, Massachusetts.

Diffrient, Tilley, Harman, 1981a. Henry Dreyfuss Associates, Humanscale 4/5/6. MIT Press, Cambridge, Massachusetts.

Diffrient, Tilley, Harman, 1981b. Henry Dreyfuss Associates, Humanscale 7/8/9. MIT Press, Cambridge, Massachusetts.

Etherton, J., 1987. The use of safety devices and safety controls at industrial machine workstations. In Handbook of Human Factors, (Editor G. Salvendy), John Wiley & Sons, New York.

Huchingson, R., 1981. New Horizons for Human Factors. McGraw Hill, New York.

Jones, J., 1974. Design Methods: Seeds of Human Futures. Wiley-Interscience, New York.

Kennedy, K., 1978. Reach Capability of Men and Women: A Three-dimensional Analysis. AMRL TR-77-50, Aerospace Medical Research Laboratory, Wright Patterson AFB, OH.

NASA, 1978. Anthropometric Source Book, 3 Vols. NASA Reference Publication 1024, Houston, TX.

Pheasant, S., 1986. Bodyspace. Taylor and Francis, London.

Roebuck, J., Kroemer, K., and Thompson, W., 1975. Engineering Anthropometry Methods. John Wiley and Sons, New York.

Tisserand, M., 1980. Anthropometric Basis for Evaluating Safety Distances. Cahiers de Notes Documentaires, No. 1288-101-80, INRS, Paris.

Workspace, Equipment and Tool Design
edited by A. Mital & W. Karwowski

HAND TOOLS: INJURIES, ILLNESSES, DESIGN, AND USAGE

A. MITAL

Ergonomics Research Laboratory (ERL), Mechanical and Industrial Engineering Department, University of Cincinnati, Cincinnati, Ohio 45221-0072 (U.S.A.)

ABSTRACT

Hand tools have been in use for almost a million years. Over the years, hand tools have become indispensable to humans as they are the primary means to extend the capabilities of the hand. In this chapter, the problems associated with the use of hand tools are reviewed. The guidelines for designing hand tools are provided. Effects of tool and worker factors influencing human force/ torque exertion capability with human powered non-impact hand tools are also discussed to assist designers in selecting or designing fastener devices.

INTRODUCTION

Historical development

Hand tools have been in existence for perhaps as long as the human race. There is evidence that pre-human primates used stone tools for digging, pounding, cutting, and scrapping almost a million years ago (Leaky, 1960; Washburn, 1960). About 35 000 years ago, handle was added to the stone tool (Freivalds, 1987). This was a major development in tool design. The specialized development of hand tools began in earnest around 10 000 years ago and led to the development of a screw system, around 300 BC, and, later, the screwdriver (Fraser, 1980).

With changing tasks, a change in the anatomy of the hand also occurred (Napier, 1962, 1963). The increasing differentiation of human activities, from agricultural activities to industrial activities, also led to the development of specialized tools. Even though changes in tool design are continually being made, many tools, for instance axe and hammer, have survived their basic design without any major changes and are still widely used.

The lack of design changes in many basic hand tools may be attributed to the absence of motivation for improvements. The motivation for developing new and improved tools was provided by the need to develop specialized tools for mass production. Today, one can find a wide variety in basic tools. Numerous different styles of hammers, screwdrivers, pliers or wrenches, for example, can be easily obtained in the market. Such variety in tool design indicates several things: (1) these tools have yet to reach the optimal form, (2) these tools possibly do not fulfil industry's needs and tool manufacturers come up with new designs in the hope of meeting technological demands, and (3) there is insufficient testing and evaluation of these tools which causes problems when such tools are used. As a matter of fact, health and productivity problems associated with the use of hand tools have become well known in recent years (Putz-Anderson, 1988; Mital and Channaveeraiah, 1988). It is now widely recognized that improper design and use of hand tools may lead to disabilities and cost billions of dollars in lost work days, medical costs, etc. The design, evaluation, selection, and use of hand tools, therefore, is a major concern of ergonomists.

This chapter is divided into five major sections. The remainder of this section contains a short introduction to the basic anatomy of the hand and forearm, a description of the "operator-hand tool" system, and a statement of the goal of ergonomists interested in improving the efficiency of the operator-hand tool system. The second and third sections deal with hand tool injuries and illnesses, respectively. Hand tool design is considered in the fourth section. In the fifth section, the impact of various task factors on human force/torque exertion capabilities with nonpowered wrenches and screwdrivers is discussed.

Basic anatomy of the hand and forearm

Hand tools are devices that extend capabilities of the hand. In order to design better tools it is essential that we understand the basic anatomy and functioning of the human hand and the forearm.

Bones of the hand and forearm. The bones of the forearm and hand are shown in Fig.1. Radius and ulna are the bones of the forearm. The radius is the lateral bone of the forearm and has a small flat and round head proximally. The distal extremity of the radius is expanded and is the major contribution to the wrist joint. The Ulna is large proximally but narrows distally into a small round head. Both bones are united through synovial joints proximally and distally. The proximal joint shares an articular capsule with the elbow joint; the distal joint shares one with the wrist. In the supine position, both bones are parallel. The radius connects to the thumb side of the wrist; the ulna to the little finger side of the wrist. When the hand is pronated, the radius is wrapped around the ulna.

The hand includes 8 carpal bones, 5 metacarpals and 14 phalanges. The bones of the carpus are arranged in two rows. Each of the digits is composed of three phalanges except the thumb, which has two. These phalanges are named by their position (proximal, middle or medial, and distal).

Nerves in the hand and forearm. There are four major nerves in the forearm and the hand: median nerve, musculocutaneous nerve, ulnar nerve, and radial nerve. The primary function of these nerves is to innervate the skin and

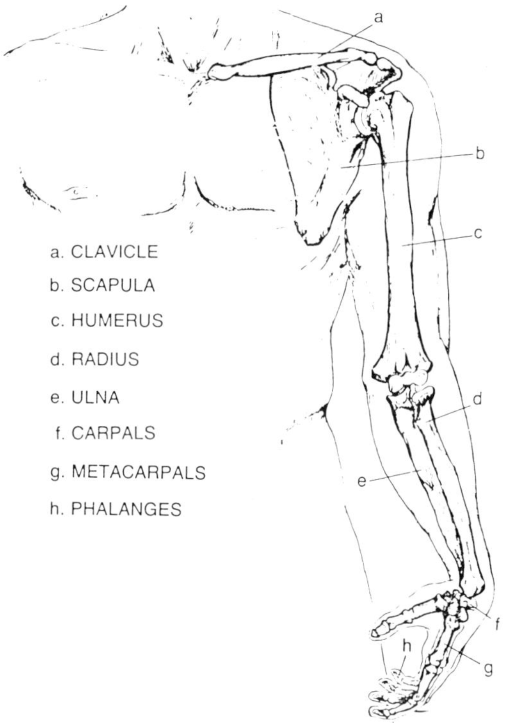

Fig. 1. Bones of the hand and arm (Putz-Anderson, 1988).

control the muscular contraction in the hand and the forearm.

Motions in the wrist and hand. The muscles that produce motions in the area of the wrist are: flexor carpi radialis and flexor carpi ulnaris (flexion); extensor carpi radialis longus, extensor carpi radialis brevis, and extensor carpi ulnaris (extension); extensor carpi radialis longus, flexor carpi radialis, and abductor pollicis longus (abduction or radial deviation); flexor carpi ulnaris and extensor carpi ulnaris (adduction or ulnar deviation). The muscles that produce motion in the hand are: lumbricals, dorsal and ventral interossei, flexor digitorum profundus, and flexor digitorum superficialis (metacarpophalangeal (MP) flexion except for the thumb); extensor digitorum (MP extension except for the thumb); dorsal interossei (digital abduction); ventral interossei (digital adduction); lumbricals, dorsal and ventral interossei (proximal interphalangeal (IP) extension); lumbrical, dorsal, and ventral interossei (distal IP extension); flexor digitorum superficialis and flexor digitorum profundus (proximal IP flexion); flexor digitorum profundus (distal IP flexion); flexor pollicis longus and brevis (thumb flexion); extensor pollicis longus and brevis (thumb extension); abductor pollicis longus and brevis (thumb abduction); adductor pollicis (thumb adduction); opponens pollicis, abductor pollicis brevis and flexor pollicis brevis (opposition of the thumb).

The muscles are connected to the bones by tendons. The tendons that are connected to the fingers pass through the wrist channel (carpal tunnel). The various nerves and blood vessels also pass through this channel (Fig. 2). The wrist joint motion is possible only in two planes. Movement in one plane allow ulnar and radial deviations. Movement in the second plane, at 90^{0} to the first plane, allow palmer flexion and dorsiflexion.

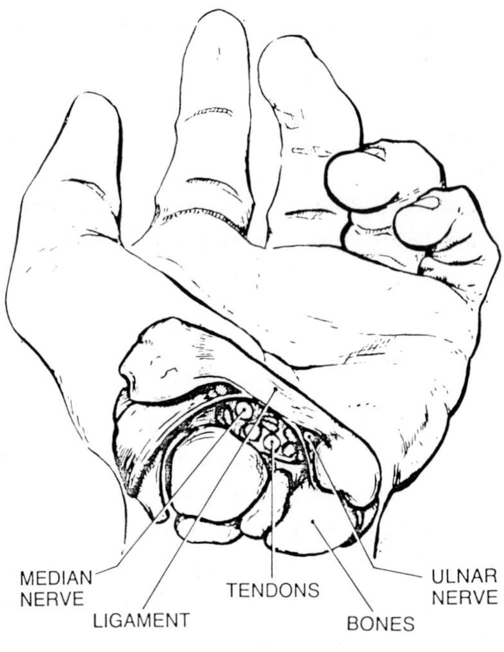

Fig. 2. A cross section of the wrist showing the carpal tunnel (Putz-Anderson, 1988).

Grips. The dexterity of the hand permits various kinds of grips. The two most basic grips have been defined by Napier (1956) as the power grip and the precision grip. In a power grip, tool axis is perpendicular to the forearm axis and the hand makes a fist with four fingers on one side and the thumb on the other side. This grip is used when large forces are to be exerted. Depending upon the direction of the line of force, three subcategories of the power grip are defined: (1) force parallel to the forearm (e.g., saw), (2) force at an angle to the forearm (e.g., hammer), and (3) torque about the forearm (e.g., corkscrew).

In a precision grip, the tool is pinched between the thumb and fingers. This grip is primarily used for precision work which requires control rather than exertion of large forces. The precision grip is classified into two categories: internal precision grip and external precision grip. In an internal precision grip, the shaft of the tool is internal to the hand (e.g., knife). In the external precision grip, the shaft of the tool is external to the hand (e.g., pencil).

There are some situations that require a combination of precision and power grip (e.g., serving in tennis). In industrial work, such combination grips are not recommended because the use of control muscles, which are smaller, to produce power can accelerate fatigue (Greenberg and Chaffin, 1977). In addition, a hook grip is also sometimes defined. This grip is used in situations when loads are held by the fingers and the fingers form a hook (e.g., holding a suitcase). Such grips should be avoided if the hand is to be used for precision tasks in the immediate future (within hours). According to Patkin (1969), surgeons should avoid carrying suitcases up to 24 hours before surgery to avoid fatiguing fingers.

The "operator-hand tool system"

The operator-hand tool system has three major components: the worker, the tool, and the task (Kriefeldt and Hill, 1975). All three components interact with each other and involve several factors which influence the efficiency of the system. Among the worker-related factors are factors such as age, gender, strength, body size, technique, experience, and posture. The tool and task-related factors include factors such as type of tool, grip/handle shape and size, gloves, wrist orientation, reach distance, force/torque requirements, and duration of application/use. In order to optimize the efficiency of the system, the effect of all these factors on system performance should be known and accounted for in tool designs.

Goal of ergonomists

Since the usage of hand tools affects productivity as well as the occupational safety, the overall goal of ergonomists should be to optimize the relationship between the worker and the hand tool. This means: (1) designing jobs and/or hand tools which alleviate common physiological problems and musculoskeletal complaints associated with the design and usage of hand tools and (2) proper selection and usage of hand tools.

The selection and usage of a hand tool is generally a function of the workpiece design and force/torque requirements of the job. The selection and usage will be satisfactory if the force/torque generated by the equipment, with or without external power source, is greater, but not excessive (excessive force/torque might damage the workpiece), than that required (most problems occur when human-powered tools are used). Therefore, force/torque exertion capabilities with various hand tools should be known and jobs (couplings) should be designed keeping a particular hand tool in mind.

ACUTE INJURIES

Acute injuries are the injuries that occur instantaneously causing immediate pathological damage. Common classifications of types of accidents used by statisticians (e.g., OSHA, U.S. Department of Labor) include struck by or struck against, caught in or between and burned by chemicals, caustics or acids (Bureau of Labor Statistics, 1982).

Acute injuries include burns, cuts, lacerations, scratches, abrasions, punctures, fractures, sprains, strains, hernias and dislocations. More insidious consequences of hand tool related accidents are arm, hand or finger amputations. Occasionally, such injuries cause fatality (Aghazadeh and Mital, 1987).

In 1984, the Ergonomics Research Laboratory at the University of Cincinnati mailed a questionnaire to various federal and state agencies in the United States to determine the frequency, severity and annual cost of hand tool-related injuries in industry and to identify problem areas with regard to tool type, accident type, nature of injury, parts of body affected, type of industry and characteristics of the injured worker. The findings of this survey were reported in detail by Mital and Aghazadeh (1985) and Aghazadeh and Mital (1987). For the purpose of the survey, the hand tools were classified as shown in Table 1.

Response data from the questionnaire were tabulated and grouped and then analyzed. Tables 2 through 9 show the relationships between injuries and hand tool type, accident or exposure type, nature of the injury, part of body affected, industry type, and age and occupation of the injured worker. The results of the questionnaire showed that in the United States hand tool injuries account for 9% of all industrial injuries and cost approximately $ 10 billion annually.

Questionnaire results

Frequency. Hand tool injuries comprise about 9% of all work-related compensable injuries (approximately 265 570 injuries per year).

TABLE 1

Hand tool classification (Fraser, 1980).

Type	Example
Manually driven tools	
Percussive tools	Axe, Hammer
Scraping tools	Saws, Files, Chisels, Planes
Drilling and boring tools	Awl, Gimlet, Borer, Drill
Screwdriver and wrenches	
Holding tools	Tongs, Pliers, Pinchers
Cutting tools	Knives, Scissors, Shears, Tinsnips
Power driven tools	
Electric power tools	Power saws, Drills, Screwdrivers, Percussion drills, Electric hammers, Sanders, Grinders, etc.
Compressed air tools	Percussive tools, Rotary tools
Internal combustion tools	Chain saws
Explosive drive tools	Bolt guns, Cutters, Splicers, Riveters

Severity. The responses to the questionnaire revealed that not only are hand tool injuries numerous, but they can also be severe. Approximately 3.9% of all amputations were caused by non-powered hand tools; 5.1% of all amputations were caused by powered tools (similar percentages have also been reported by McCaffrey (1981)). The most severe of the injuries were caused by knives among non-powered hand tools and saws among powered hand tools.

Costs. Table 2 gives an indication of some of the costs associated with hand tool injuries. It is apparent that indemnity and medical costs associated with both types of hand tools are very close. The annual cost (indemnity cost plus cost of work day losses, wage losses, insurance and other direct and indirect costs) was estimated to be $ 10 billion.

TABLE 2

Cost of hand tool injuries (Aghazadeh and Mital, 1987).

Tool type	Average cost indemnity/case($)	Average cost medical/case ($)	Total cost ($)
Non-powered	978	563	1541
Powered	995	496	1491

Tool type. The percentages of injuries by tool type are depicted in Table 3. The majority of injuries (80%) are inflicted by non-powered hand tools. This does not mean that non-powered hand tools are more hazardous to operate than powered hand tools. It is, however, an indication of the widespread use of non-powered hand tools. The four highest injury causing non-powered hand tools were knife, hammer, wrench, and shovel. Among the powered hand tools, saws, drills, grinders, and hammers were involved in more injuries than any other hand tool. Drills and hammers have also been identified as the tools that have the majority of lost time associated with them in the mining operations (Marras et al., 1988a, b).

TABLE 3
Percentage distribution of cases by
hand tool type (Aghazadeh and Mital, 1987).

Hand tool type	Percent
Non-powered tools	
Axe	0.8
Crowbar	3.0
Hammer	10.0
Knife	44.3
Pliers, tongs	1.0
Rope, chain	3.4
Scissors	2.4
Screwdriver	2.1
Shovel	5.7
Wrench	8.9
Others	Remainder
Total	100.0
Powered tools	
Grinder	8.3
Drill	17.6
Hammer	8.0
Saw	33.6
Others	Remainder
Total	100.0

Type of accident/exposure. Table 4 presents the distribution of type of accident or exposure. The information gathered indicated that in the case of non-powered hand tools, 71.2% of all injuries were caused by "striking by" or "striking against" the tool. Overexertion was the second leading cause. "Caught in" or "betweens" and "falls" contributed the least. The same types of accidents were responsible for most injuries in the case of powered hand tools.

TABLE 4
Percentage of cases by type of accident or exposure (Aghazadeh and Mital, 1987)

Type of accident/exposure	Percent
Non-powered tools	
Struck by or struck against	71.2
Fall	0.7
Caught in or between	2.0
Over-exertion	24.8
Other	1.4
Total	100.0
Powered tools	
Struck by or struck against	62.6
Fall	1.5
Caught in or between	4.4
Over-exertion	27.9
Other	3.7
Total	100.0

"Struck by" and "struck against" types of accidents were caused by saws, drills, hammers, and grinders more than any other powered hand tool.

Nature of injury. Table 5 presents the distribution of injuries. As shown in the table, cuts and lacerations, sprains and strains are the most frequent injuries for both powered and non-powered hand tools.

Part of body affected. According to Table 6, the parts of the body most injured are the upper extremities. Thirty percent of all body parts affected are fingers. Head, neck, and the eyes are the least injured parts.

Industry type. As shown in Table 7, hand tool related injuries are numerous in manufacturing, mining and construction, and wholesale and retail groups. The service group (building maintenance workers, automotive and other repair shops, garden and health care services) also contributed significantly as the hand tools are also used frequently in these areas.

Occupation of the injured worker. Table 8 shows the distribution of hand tool injuries and worker occupation. Laborers, craftsmen and kindred workers, and operatives are the three groups of workers who are most frequently injured.

Age and hand tool injuries. Table 9 shows the distribution of injury data with the age of the worker. Workers between ages of 18 and 35 years suffer more from hand tool injuries than any other age group. The relative safety of older workers may be attributed to their experience, conservative behavior, and relatively less demanding jobs. Older workers generally do not take unnecessary risks and are usually assigned to softer jobs due to their seniority.

TABLE 5
Percentage distribution of cases by nature of injury (Aghazadeh and Mital, 1987).

Nature of injury	Percent
Non-powered tools	
Contusions, bruises	10.0
Cuts, lacerations	61.0
Fractures	4.0
Sprains, strains	19.0
Other	7.0
Total	100.0
Powered tools	
Contusions, bruises	8.0
Cuts, lacerations	53.0
Fractures	5.0
Sprains, strains	25.0
Other	9.0
Total	100.0

TABLE 6
Percentage distribution of cases by part of body affected (Aghazadeh and Mital, 1987).

Part of body affected	Percent
Non-powered tools	
Eyes	0.7
Head and neck	3.9
Upper extremities	59.3
Trunk	8.5
Back	16.7
Other	10.9
Total	100.0
Powered tools	
Eyes	0.3
Head and neck	4.0
Upper extremities	51.0
Trunk	5.6
Back	15.4
Lower extremities	22.1
Other	1.6
Total	100.0

TABLE 7

Percentage distribution of cases by industry type (Aghazadeh and Mital, 1987).

Industry	Percent
Agriculture	5.5
Mining and construction	24.1
Manufacturing	26.4
Wholesale plus retail	20.5
Service and government	18.2
Other	5.3
Total	100.0

TABLE 8

Relationship of occupation to hand tool injuries (Aghazadeh and Mital, 1987).

Occupation	Percent
Laborers, except farm	32.3
Craftsmen and kindred workers	23.0
Industrial operatives	22.1
Transport equipment operatives	5.7
Service workers	9.3
Other	7.6
Total	100.0

TABLE 9

Relationship between age and injury (Aghazadeh and Mital, 1987).

Age	Percent
17 or less	2.7
18-25	36.8
26-35	27.5
36-45	15.0
46-55	10.3
56-65	6.0
66 and above	1.8
Total	100.0

Prevention of acute injuries

Worker training in proper and safe operating procedures is perhaps the most critical action that can be taken to prevent acute injuries. Specific safe practice rules for using many types of hand tools have been established and are available (Lahey, 1981). Important design and operational guidelines for safety include eliminating pinching hazards, sharp corners and edges. Putting guards over pinch points and using stops to prevent handles from closing completely and pinching the palm are example of simple solutions. Brake devices on power tools should be installed so that in case of emergency the operation can be stopped instantly. Proper placement of power and emergency switches is also critical.

The maintenance of tools is also important for safety. Tools should be inspected periodically and defective tools should be replaced immediately. Workers should be instructed not to leave tools on the stairways or in positions where they can fall off elevated positions. Provisions should be made for storing tools safely. Use of protective equipment, such as gloves and safety glasses, should be encouraged.

ILLNESSES (CHRONIC INJURIES)

Illnesses or chronic injuries (also called cumulative trauma, repetitive trauma, repetitive strain, or cumulative trauma disorder) are the result of (1) frequency, (2) extreme joint movement, and (3) force. These disorders can be a major problem, particularly for ages from 18 to 64. Armstrong et al. (1982) found that the overall incidence of cumulative trauma in a poultry processing plant to be almost 13 cases/200000 work hours. Armstrong (1983) further reported an incidence rate of 25.6 in an automobile upholstery plant and 35.8 in an athletic products plant. Clearly, the incidence rate varies from job to job. Hymovich and Lindholm (1966) and Armstrong et al. (1986) provide a detailed breakdown of various medical incidents resulting from repetitive strain.

Common cumulative trauma disorders (CTD)

CTDs include disorders such as tenosynovitis, carpal tunnel syndrome, tendonitis, ischemia, De Quervain's disease, Raynaud's syndrome (white finger or dead finger syndrome), epicondylitis (tennis elbow), and ganglionic cyst.

Carpal tunnel syndrome. Carpal tunnel syndrome (CTS) is a special case of cumulative trauma--it refers to injury of the median nerve of the wrist. Primarily CTS is due to a decrease in the effective cross-section of the tunnel, either by decreasing its diameter or by increasing the volume of the tunnel contents, resulting in increased tunnel compartment pressure. The hand is kept small and flexible by locating the muscles which power it in the bulky forearm. The fingers are moved by tendons leading from the fingers through the wrist (wrist=carpal in Latin) tunnel to the muscles. The median nerve goes through the tunnel. If the soft, fragile nerve is injured by contact with a hard surface within the tunnel (such as a tendon or ligament or bone) the result is carpal tunnel syndrome.

It is important to note that the majority of CTS cases are not occupationally related. CTS can occur from recreational activities and even can be idiopathic--that is, occur from no known cause. Females, especially those pregnant and breast feeding, have more CTS. People over 40 have more CTS. Diabetes and rheumatoid arthritis also are associated with CTS.

Early symptoms of CTS include numbness or tingling and burning sensations in the fingers. More advanced problems involve pain, wasting of the muscles at the base of the thumb, dry or shiny palms, and clumsiness. Many symptoms are noticed first at night and may be confined to part of the hand. Later the pain radiates to the elbows and shoulders.

CTS is established by tests. Reduced median nerve conductivity is a clinical test. Two non-clinical tests are Phalen's test and Tinsel's test. Phalen's test involves prolonged palmer flexion of the wrist in a vertical position; CTS is indicated if there is a loss of feeling (parathesia); bending the wrist in the opposite direction (hyperextension) is called the reverse Phalen's test.

Tinsel's test is a light tap over the median nerve of the wrist; CTS is indicated if there is a loss of feeling or tingling in the appropriate digits. Tinsel's test is not positive in all CTS cases.

Tendonitis. It is a form of tendon inflammation that occurs when a muscle/tendon unit is repeatedly tensed. Further exertions may result in torn tendon The tendon becomes thickened, irregular and bumpy and may be permanently weakened (Curwin and Stanish, 1984).

Tenosynovitis. It is a general term for a repetition induced tendon injury involving the synovial sheath. The most recognized stenosing tenosynovitis is the De Quervain's Disease. This disorder affects the tendons on the side of the wrist and at the base of the thumb.

If the tendon sheath of a finger is sufficiently swollen so that the tendon becomes locked in the sheath, attempts to move that finger will cause a snaping and jerking movements. This condition is called trigger finger (Thompson et al., 1951).

Epicondylitis. This disorder is commonly known as the tennis elbow. It is an inflammatory reaction of the tissues in the lateral elbow region and occurs when the hand is exposed to radial deviation, particularly combined with pronation and dorsiflexion (Boyd and McLeod, 1973; Charash, 1982).

Ischemia. This is a condition when blood supply in a tissue is lacking. Symptoms include numbness, tingling, and fatigue depending on the degree of ischemia or blockage of peripheral blood vessels. Undesirable compressive forces on the palm may cause ischemia.

Vibration syndrome. Also known as the white finger or dead finger syndrome or Raynaud's phenomenon. It is characterized by recurrent episodes of finger blanching due to complete closure of the digital arteries. Thermoregulation of fingers during exposure to cold, under such circumstances, is impaired. Prolonged use of pneumatic hand tools cause this condition.

Control of CTS

Of all the CTDs, CTS is most prevalent (Phalen, 1972). Konz and Mital (1989) have outlined the various ways to collect and analyze data for recognizing the problem and controlling it. Following is a brief description of data collection, analysis and interpretation, and solution procedures.

Data Collection. Health records - In order to determine how widespread CTS is, the employee health records should be analyzed. This analysis will also identify problem jobs that need ergonomic evaluation and possible redesign; the need to maintain records of medical visits, workers' complaints, first aid logs, workman's compensation, health related absenteeism, OSHA logs, etc. can not be overemphasized. From the health records, incidence rates (number of cases per 200,000 work hours or per 1000 workers) can be calculated for: (1) each affected job classification, (2) each affected work station, (3) age (up to 40 years and above 40 years), and (4) both genders.

Photographic data - Recording of work posture and motions photographically is very helpful for later frame-by-frame analysis. A simple video camera recorder will serve the purpose and also is quick and economical.

Vibration data - Accelerometers can be used to quantify vibration caused by power tools or vibrating surfaces.

Frequency - The number of times a job cycle is performed over several days and over several shifts should be determined. If there is a large variation in output over shifts and days, health data should be classified by shift or day of the week to isolate shift and work day effects.

Data Analysis/Interpretation. The incidence rates, computed from health records, should be tabulated to provide: (1) an indication about which job classifications and work stations are problems, (2) a relative ranking of jobs and work stations in terms of severity, (3) a priority list for ergonomic evaluation of jobs and/or work stations, (4) an idea about how widespread the CTS problem is, and (5) an indication if the problem is influenced by gender or age.

The photographic record of the job should be subjected to a frame-by-frame analysis using a VCR with slow motion or freeze-frame. The work cycle should be broken into sequential elements performed by left and right hands, using therbligs or predetermined time nomenclatures (such as MTM). This analysis will identify job elements that require use of excessive force, cause extreme joint deviation, or are frequently performed and thus must either be eliminated or redesigned. The redesign alternatives, discussed in the next section, should be applied to reconfigure or eliminate these elements.

Repeated vibration (between 40 and 130 Hz) exposure may cause CTS. The vibration exposure time should be reduced, ideally to zero. In any event, the transmission of vibration to the hand should be avoided. The vibration should be reduced by: (1) using damping material between the vibrating surface and the surrounding areas, (2) reducing the radiating surface area, (3) reducing the driving force, (4) isolating the vibrating surface from the rest of the surrounding areas, (5) proper equipment maintenance, and (6) using materials that absorb vibration (nonmetallic materials).

The frequency data provides an indication of cumulative fatigue or exposure to elements that appear to cause CTS. If the incidences of CTS show a strong correlation with the working shift or work day, production volume should be made more uniform (which may not be possible, due to fluctuating market demand) or additional work stations and/or workers should be added during periods of high output. Options given in the next section should be considered.

Solution Recommendations/Implementation. There are ergonomic and non-ergonomic approaches. Ergonomic approaches are divided into frequency, angle and force.

CTS appears to be more influenced by repetition than force. The goal is to reduce the number of cycles for a specific wrist. Within frequency there are four approaches: automation, mechanization, job enlargement and job rotation. The automation approach is to eliminate the person. For example, a robot (or fixed-purpose machine) might be used to load boxes or insert parts into an assembly. In addition, robots can hold a tool (such as paint gun, welding gun). A scanner reading bar code can eliminate the need for manual keying of data.

The next three approaches retain the individual but reduce the number of cycles on a specific wrist. Mechanization requires a machine to do part of the overall task while the operator does the rest. A slight variation on mechanization is to replace hand operated controls by foot or knee operated controls.

The number of cycles of a specific motion can be reduced for a specific person if the job is not as specialized. This can be done by adding one or more people but each performing part of the job for all cycles. From the organization viewpoint, output/employee is the same but CTS stress is spread over more wrists, giving less strain.

Another alternative is to rotate wrists rather than redesign jobs. One technique is to have an individual alternate hands; that is, a task is split for the right wrist and for the left wrist. A key here is a tool which can be used well by either hand. The job can also be rotated among people. A common rotation technique is to use part-time workers. A high number of repetitions may occur when overtime is worked; this may dictate using part-time workers to do overtime work instead of full-time workers.

The wrist should be kept in the neutral (handshake) position (right thumb at "11 o'clock"). This should be accomplished by changing the job or changing the object grasped. One approach is to change the job posture temporarily. For example, the worker can alternate standing and sitting. Another approach is to change worker posture permanently. The chair position should be adjusted to minimize awkward wrist deviations. The work orientation should be modified by tilting the worksurface or moving the object on the worksurface.

In addition to work height, thumb orientation should be considered. Close observation may show that the right thumb may be at 7 o'clock (e.g., grasping horizontal handle from above) or 12 o'clcok (e.g., grasping vertical handle) or even 3 o'clock (e.g., horizontal handle grasped from below) instead of the preferred 11 o'clock.

The object grasped may be a hand tool or a control or an item to be assembled. These objects should be redesigned to put the bend in the object, not the wrist. Controls should be such that large push buttons ("palm buttons") are avoided in repetitive operations as the operator may actuate them by repeated blows of the palm (using the palm as a hammer). A foot or knee control is preferable. For keyboard work, a wrist support below the keyboard is recommended for use in the pauses between entries. The keyboard orientation on the worksurface should be adjustable by the user for all three axes. The controls should be touch-sensitive. The vibration should be dampened by having worksurfaces with rubber-backed low-pile carpet (also refer to the section on hand tool design).

Both force amount and force duration should be reduced. A reduction in load may increase the permitted work time by several fold. Reducing the duration has an additional benefit as it also increases recovery time. Force duration can be reduced by designing a control so it only needs to be activated, not held continuously.

The non-ergonomic approaches to controlling CTS are medical. One recommendation is to have the person with CTS wear a wrist splint while sleeping. Salt-restricted diets are recommended, based on the concept of reducing edema and thus, potentially, pressure in the carpal tunnel. Vitamin B6 has been recommended by some. There is considerable controversy whether it reduces CTS; it may just reduce nerve conduction speed (which is used as an index of CTS). A final alternative is surgery. Surgery usually works but there is no guarantee. In addition, surgery does not prevent CTS from recurring if the work does not change.

Measurement of wrist deviation

It has been shown by a number of people that wrist deviation can cause cumulative trauma disorders (Tichauer, 1976; Tichauer and Gage, 1977, 1978; Armstrong, 1983). It is, therefore, important to be able to measure the wrist deviation. Schoenmarklin and Marras (1987) have developed a wrist monitor for measuring workers' wrist positions during hand tool usage. This monitor (Fig. 3), however, can only be used in tasks where the worker is holding the tool with a power grip.

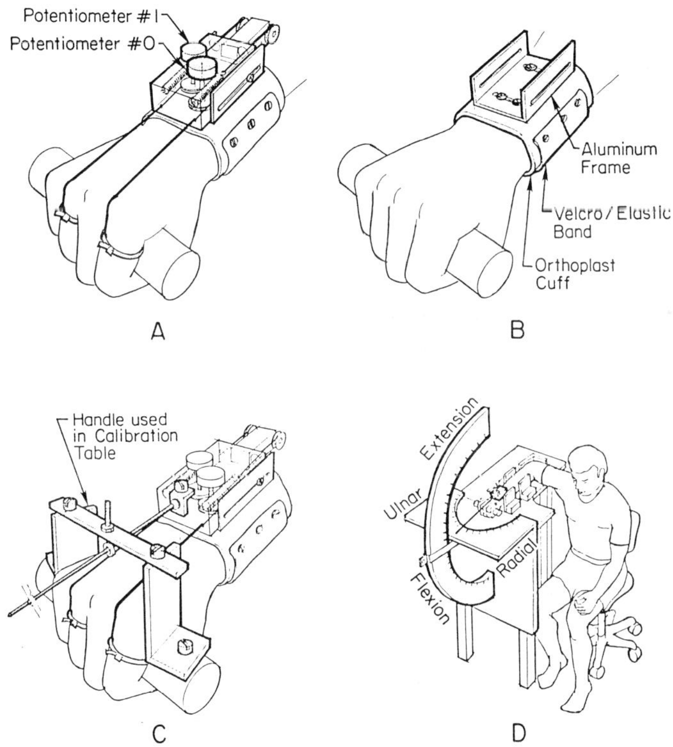

Fig. 3. Wrist monitor (A, B, and C show the wrist monitor and cuff on a subject's right hand; C shows the handle and pointer used in the calibration table D - Schoenmarklin and Marras, 1987).

DESIGN OF HAND TOOLS

Hand tools can be broadly classified into two categories (Freivalds, 1987): (1) agricultural and forestry tools and (2) industrial and technological tools. The emphasis in this book is on industrial work and, therefore, the discussion in this chapter is limited to industrial and technological tools. The recommendations made in various sections should be used in designing hand tools. For discussion pertaining to agricultural and forestry tools readers should refer to Freivalds (1987).

Principles of hand tool design

Based on the results of the research into hand tool design and hand tool injuries, Konz (1983) has outlined the following 8 principles for designing or selecting hand tools:

1. Use special-purpose tools.
2. Design tools to be used by either hand.
3. Power with motors more than with muscles.
4. Use a power grip for power and a precision grip for precision.
5. Make the grip the proper thickness, shape, and length.
6. Design the grip surface to be compressible, nonconductive, and smooth.
7. Consider the angles of the forearm, grip, and tool.
8. Use the appropriate muscle group.

The first three of these principles are general in nature and emphasize the use of special-purpose hand tools. The remaining five deal with the grip and geometry of the tools. Properly designed tools should be efficient and meet the six basic requirements outlined by Drillis (1963):

1. The tool or implement must perform effectively the function for which it is intended. Striking tools, for example, should convert a maximum amount of their kinetic energy into useful work (Gorjatchkin, 1925; Drillis et al., 1963).
2. The tool should be properly proportioned to the body dimensions of the operator. In suitably adjusted impact tools, the ratio of the kinetic energy at the instant of contact with the material to the total energy expenditure during the full working cycle characterizes the efficiency of the movement and should approach a maximum value (Contini et al., 1959, 1963; Zheligovsky, 1925). In non-impact tools the efficiency would be would be determined by the ratio of the force transmitted to the work piece to the force generated in the muscles.
3. The tool should be suitably adjusted to the strength and work capacity of the operator. The body build, gender, and training of the operator should be taken into consideration.

4. The tool should not cause premature fatigue.
5. The tool must provide feedback (pressure, force, surface texture, temperature, etc.) to the operator.
6. The tool should be inexpensive and easy and economical to maintain.

For selecting or evaluating pneumatic screwdrivers and nut runners, based on the review of published scientific information, the Technical Research Center of Finland (1988) has provided guidelines. These guidelines consider a variety of factors:

1. Torque - The torque delivered on to the workpiece should be approximately 60% of the maximum torque that can be generated by the tool.
2. Tool type - Straight and angled tools are best for downward screwing; pistol grip type tools (angle between the shaft and the handle should be 78°) are best for screwing horizontally.
3. The level of noise generated by the tool must not exceed 85 dBA for 8 hours of exposure.
4. The vibration level must not exceed 126 dB for 8 hours of exposure.
5. The tool should be light, well balanced, proportionately sized and shaped.
6. The tool should be supported for reaction force if the torque exceeds 6Nm for straight type tools, 12Nm for pistol type tools, and 50Nm for angle type tools.
7. Handle length should be at least 100 mm. Circular handles should be between 28 and 38 mm in diameter. Rectangular handles should be 40x25mm in size.

Use special-purpose tools. In general, special-purpose tools, even though costly, are faster. The savings in labor cost, due to shorter time per use, over the life of the tool more than offsets the high cost of the tool.

Design tools to be used by either hand. Approximately 10% of the population is left-handed (Konz, 1974). Yet tools are mostly designed for right-handed users. Since the preferred hand is stronger (about 7% to 20% - Shock, 1962; Miller, 1981), more dexterous (Kellor et al., 1971), and faster (Konz and Warraich, 1985), a left-handed user is at a disadvantage when using a right-handed tool. Besides the strength disadvantage, certain right-hand tools require a different action when used by the left hand (Capener, 1956). For instance, the right-hand scissors when used by a left-handed person requires a reversal of pressure upon the rings of the scissors. This leads to a tendency to twist it. The shearing surfaces thus lie in the vertical plane instead of the horizontal plane. In many instances, tools designed for right handers can not be used by left handers (Laveson and Meyer, 1976). Tools which are designed for use by either hand avoid such problems.

Power with motors more than with muscles. Mechanical energy is cheaper than human energy. Besides, the humans tend to fatigue much faster than machines.

The efficiency of human power generation is also much lower than machine power generation.

Use a power grip for power and a precision grip for precision. Precision grips, on the average, provide only 20% strength of a power grip (Swanson et al., 1970). This is primarily due to the fact that a precision grip involves small muscles whereas a power grip involves larger muscle groups (also see earlier discussion on grips). Tools designed primarily for exertion of force, such as a hammer, should use a power grip; tools designed for manipulation, such as a surgical knife, should use a precision grip.

Make the grip the proper thickness, shape, and length. Grip thickness - Rubarth (1928) reported that with screwdrivers larger the diameter (between 18 and 40 mm) greater the force for a power grip. Pheasant and O'Neill (1975) also reported similar findings. Hertzberg (1973) reported an increase in grip strength with grip diameter up to 65 mm and then a decrease. Ayoub and LoPresti (1971), on the basis of minimum electromyographic (EMG) activity, found a grip diameter of 51 mm to be the best. However, on the basis of grip force to EMG ratio and the number of work cycles completed before fatigue, they recommended a grip diameter of 38 mm. Rigby (1973) also made similar recommendation.

Greenberg and Chaffin (1977) recommended that a power grip diameter should be between 50 and 85 mm; preferably 50 mm. If the handles move towards each other, the maximum span should be 100 mm; the minimum span should be 50 mm. The grip strength is maximized when the span is between 75 and 80 mm (Fig. 4). As shown in fig. 4, a grip force of approximately 9 kg (about 88 N) can be produced by 95% of the female population and, therefore, should be the maximum force required to close the grip span.

Replogle (1983) recommended a grip diameter twice the size of the grip span as it maximized the torques. Eastman Kodak Company (1983) recommended that a power grip diameter should be 40 mm. Petrofsky et al. (1980) concluded that the optimum hand grip span should be between 50 and 60 mm. Moreover, they stated that an increase or decrease in the grip span of even 6 mm results in loss of strength. For T-handles, Saran (1973) recommended a diameter of 25 mm. In general, the diameter of a power grip should be between 25 and 50 mm.

For a precision grip, Hunt (1934) reported doubling of time to drive a screw (from 1.9 seconds to 3.6 seconds) when the screwdriver handle diameter increased from 8 mm to 16 mm. Kao (1976, 1977) recommended preferred diameter for pens to be 13 mm. According to Konz (1983), diameters less than 6 mm should be avoided for precision grips as they tend to cut into the hand if force is required.

For powered screwdrivers, Johnson and Childress (1988) recommended a grip diameter of 38 mm. This recommendation was based on EMG activity which was lower for the 38 mm grip diameter compared to the 28 mm grip diameter screw-driver.

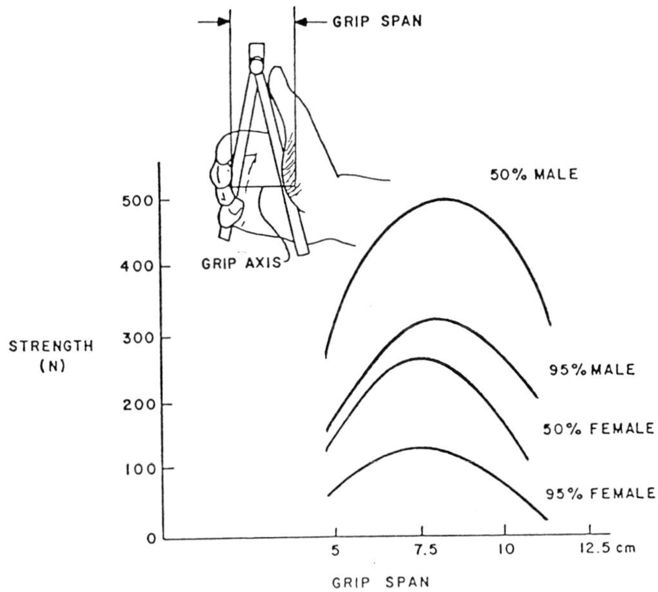

Fig. 4. Maximum grip strengths for various handle openings (Greenberg and Chaffin, 1977).

Gloves Often, the work requires that workers use gloves for safety. In general, gloves interfere with a person's grasping ability and hand movements. A fraction of force/torque generated by muscular contraction also is wasted in maintaining the grip. This results in reduced torque production (Swain et al., 1970). Sudhakar et al. (1988) also found no significant difference in muscle activity between glove and no glove conditions and yet the peak grip strength (output) with rubber and leather gloves was 10 to 15% lower compared to grip strength without gloves. According to Hertzberg (1973), grip strength declines by 20% when gloves are used. Weidman (1970) studied the influence of different kinds of gloves on manual performance and observed that performance times decreased by 12.5%, as compared to no gloves, when Neoprene gloves were worn. With terry cloth, leather, and PVC gloves, performance decreased by 36%, 45%, and 64%, respectively. Reduction in strength by as much as 30% or more have also been reported by Lyman (1957) and Sperling (1980).

In holding tasks gloves may enhance friction and reduce the task strength requirements (Riley et al., 1985). In some cases gloves improve occupational speed (Bradley, 1969a, b). Friction between the handle and fingers is not always sufficient to prevent slipping. This is particularly true when the surfaces are wet. In such cases, gloves enhance friction.

Gloves also change the effective dimensions of the hand. The hand thickness may increase anywhere from 5 mm to 25 mm when gloves are worn; the hand width at the thumb may increase from 8 mm to 40 mm (Damon et al., 1966). The increased size of the hand should be taken into consideration in designing hand tools.

Overall, gloves are a mixed blessing. They provide safety and comfort but at the cost of reduction in manual performance.

Grip length - Hand width ranges from approximately 79 mm for a 1st percentile female (Garrett, 1969a; 1971) to approximately 99 mm for a 99th percentile male (Garrett, 19699b; 1971). Using these dimensions as the basis, Konz (1983) recommended a minimum grip length of 100 mm; 125 mm grip length is more comfortable. Eastman Kodak Company (1983) recommended a grip length of 120 mm. If gloves are to be worn or the grip is enclosed (e.g., hand saw), the minimum grip length should be 125 mm (Konz, 1983). Lindstrom (1973), on the other hand, recommended a handle length of 110 mm for men and 100 mm for women. In reality, it is impractical to provide different handle length tools for the same work done by males and females. In general, the grip length should be such that it would not limit the tool head opening and at the same time avoid excessive compressive forces or pressure concentration on the tender parts of the palm. The pliers redesigned by Tichauer (Tichauer and Gage, 1977), because of an extended upper handle (Fig. 5), thus, do not cut into the palm.

For external precision grip, the tool shaft must be at least 100 mm in length and must be long enough to be supported at the base of the first finger or thumb. For internal precision grip, the tool should extend past the tender palm but not so far as to hit the wrist (Konz, 1983).

Magill and Konz (1986) reported that torque exerted by a screwdriver was proportional to the handle grip length. The stem length of screwdrivers, however, do not influence human torque exertion capability (Mital et al., 1983, 1985; Huston et al., 1984; Mital and Sanghavi, 1986).

Grip shape - The grip shape should be such that maximizes the area of contact between the palm and the grip. Maximizing this area of contact leads to better pressure distribution and reduces the chances of forming pressure ridges or pressure concentration points. This is particularly important for tools that require a power grip.

Generally, the tools available in the market have a cylindrical shape grip. Rubarth (1928) reported that cylindrical handles with rounded end maximized force. For a turning action, exerted on tools with cylindrical grips, Pheasant

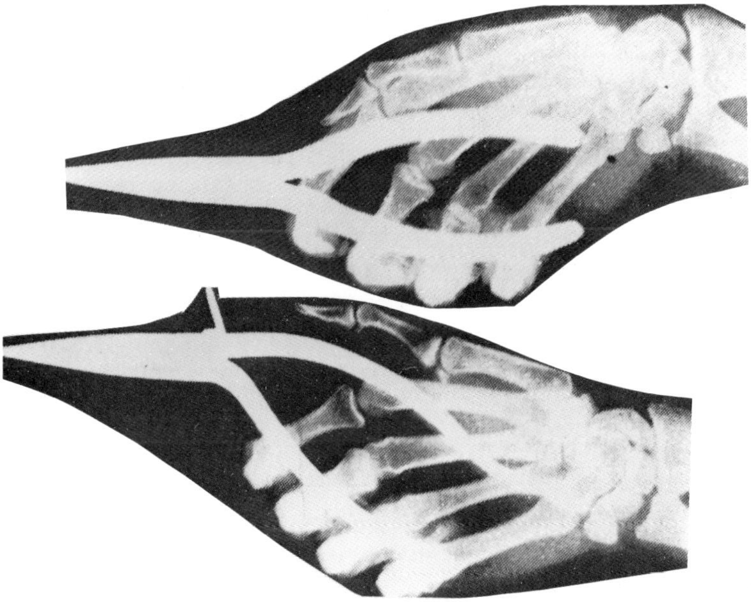

Fig. 5. Conventional and redesigned pliers (Tichauer and Gage, 1977).

and O'Neill (1975) and Grieve and Pheasant (1982) reported that the torque is proportional to the handle diameter. Mital et al. (1985) and Mital and Sanghavi (1986), however, did not observe that; they found the torque increased with the grip diameter but not proportionately.

When there is no turning action on the grip and yet torque is exerted on the workpiece (e.g., wrenches), torque varies with the lever arm (Mital et al., 1985, Mital and Sanghavi (1986); Mital and Channaveeraiah, 1988).

Pheasant and O'Neill (1975) reported that the shape of the handle is not relevant as long as the hand did not slip around it. The chances of hand slippage are reduced when a non-circular handle, such as rectangular or triangular, is used. The edges in such grips resist the slippage. Mital and Channaveeraiah (1988) compared cylindrical, triangular, and rectangular handles (Fig. 6) and reported that the torque exertion capability of individuals with triangular handle screwdrivers was greater than with cylindrical handle screwdrivers (Manufacturers in West Germany have been marketing triangular handle screwdrivers for quite some time (Fraser, 1980). In the last 5-6 years, similar screwdrivers have also become available in the United States). For wrenches, torque exertion capability was maximized when cylindrical handles were used. According to Cochran and Riley (1986a), the thrust forces exerted with straight knives are as much as 10% greater with triangular handles as compared to cylindrical or rectangular handles. A handle perimeter of 110 mm is

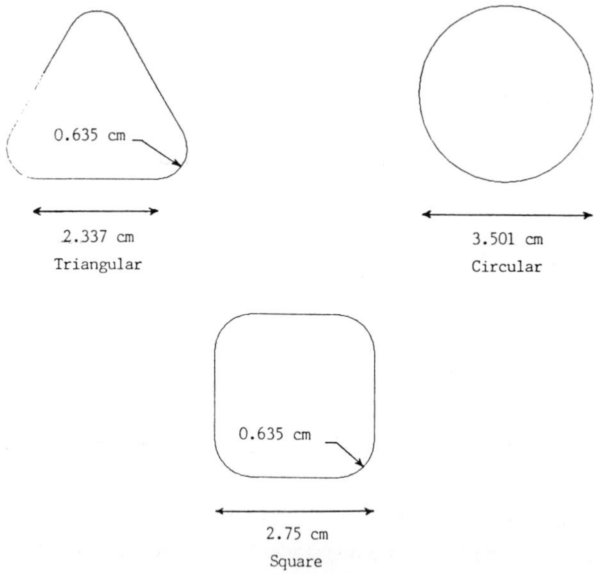

Fig. 6. Grip shapes used by Mital and Channaveeraiah (1988).

recommended for knives.

A T-handle is preferable for screwdrivers than the straight handle (Pheasant and O'Neill, 1975) as it not only prevents wrist deviations but also increases torque exertion by as much as 50%. According to Saran (1973), the T-handle should be slanted (60° angle) to allow the wrist to be straight and should be 25 mm in diameter.

For knives, a pistol grip is preferable to a straight grip (Armstrong et al., 1982; Karlqvist, 1984). The angle of the grip should be 78° from the horizontal (Fraser, 1980).

Guards - The primary purpose of the guard in front of the grip is to prevent injury when the hand slips or the hand and tool collide against a sharp or rigid surface. The guard in such cases prevents the hand from slipping and shields the hand against the impact. According to Cochran and Riley (1986b), guards of 1.52 cm height provide adequate safety. Greater guard heights improve safety only minimally.

Grooves and indentations - Form fitting tools that have grooves for fingers, etc., are not recommended. The main problem with grooves is that they do not fit people; either the grooves are too big or too small. In both cases, pressure ridges are created, leading to nerve compression and impairment in circulation (Fig. 7). Vertical grooves on the handle, provided by the designer to prevent the hand from slipping, also cut into the palm of the hand and

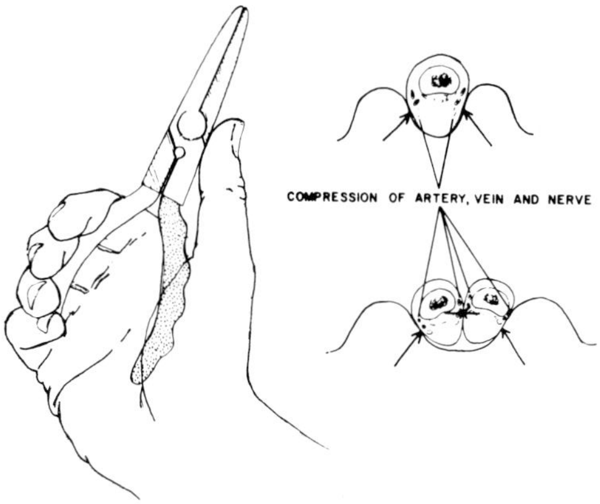

Fig. 7. Pressure concentration from recessed handles (Tichauer and Gage, 1977).

create pressure ridges. Such grooves or flutes should be avoided. According to Bobjer (1984), Ergo in Sweden has developed screwdrivers that have 40 small axial grooves. These grooves are small enough to avoid pressure concentration but large enough to provide better friction between the hand and the grip. For small screwdrivers, used in precision work, regular size handles were used to provide better control.

Pheasant and O'Neill (1975) reported that screwdrivers with cylindrical and knurled surfaces provided greater torque exertion capability than smooth surface cylindrical handles. The 50 mm diameter handle with knurled surface maximized torque exertion.

In general, grooves and indentations are undesirable. Slight and uniform surface indentations, however, are desirable as they allow greater torque exertion capability.

Design the grip surface to be compressible, nonconductive, and smooth. A compressible grip material is easier on the hand (Konz, 1983). Compressible materials dampen vibration and allow better distribution of pressure across the palm-grip contact area. A word of caution - the grip material should not be too soft otherwise sharp objects, such as metal chips, will get embedded in the grip and make it difficult to use. The grip material should not absorb oil or other liquids and should not permit conduction of heat or electricity. Wood or plastic are desirable handle materials (Wu, 1975). In general, metallic handles should be avoided or encased in a rubber or plastic sheath (Konz, 1983).

Smooth grip surfaces are desirable. The surface, however, should not be slippery. Sharp indentations, flutes or grooves should be avoided as they cut into the palm, causing nerve compression and impaired circulation (Fig. 7).

Consider the angles of the forearm, grip, and tool. As pointed out in the

"Illnesses" section, wrist deviations lead to various problems. Wrist deviations also lead to loss in productivity (Tichauer, 1976) and grip strength (by as much as 14-27% - Terrell and Pursewell, 1976). To avoid chronic illnesses and loss in productivity, the wrist should be kept straight (handshake orientation). If any bending is required, the tool, rather than the wrist, should bend. Fig. 5 shows a conventional and a redesigned plier (Tichauer and Gage, 1977). By bending the handles and increasing the length of the upper handle, it is possible to keep the wrist straight and avoid nerve, tissue, and blood vessel compression. Yoder et al. (1973) also used bent-handle pliers and successfully reduced complaints of wrist fatigue and stress. Similar bends have also been included in the designs of soldering irons (Tichauer, 1966) and hammers (Emanuel et al., 1980; Granada and Konz, 1981; Krohn and Konz, 1982; Knowlton and Gilbert, 1983; Konz and Streets, 1984; Konz, 1986; Schoenmarklin and Marras, 1988a, b).

For hammers, Konz (1986) concluded that a 10^0 bend is preferable to subjects. Both, Konz (1986) and Schoenmarklin and Marras (1988a), concluded that the angle of bend does not influence performance. Vertical hammering, however, was reported to be more difficult, less accurate, and more stressful and fatiguing (Schoenmarklin and Marras, 1988b). In order to reduce fatigue, the tool should not weigh more than 2.3 kg (Greenberg and Chaffin, 1977; Eastman Kodak Company, 1983). This is also corroborated by Johnson and Childress (1988) who observed that powered screwdrivers weighing about 1.12 kg or less do not produce significantly different magnitudes of EMG activity.

<u>Use the appropriate muscle group.</u> Power grips are used when force is the prerequisite. In such cases, larger and stronger forearm muscles, instead of the smaller and weaker finger muscles, should be used. Fingers should not be used to squeeze the tool. For triggering, trigger strips are preferable to trigger buttons (Konz, 1983).

Since the muscles closing the hand are stronger than the muscles opening the hand (Radonjic and Long, 1971), a spring should be used to open the handles (Tannen et al., 1986).

USE OF HAND TOOLS

In the previous section, guidelines for designing new hand tools or redesigning existing hand tools to alleviate health and safety problems and increase human torque exertion capabilities were discussed. In bits and pieces, much of this information has been available for several years (Khalil, 1972; Tichauer and Gage, 1977; Greenberg and Chaffin, 1977; Fraser, 1980; etc.). Most hand tools in use today, especially those in maintenance and repair work, however, do not conform to these guidelines. Also, ergonomically designed hand tools are not easily available in the market. The common outlets for marketing hand tools generally do not carry ergonomically designed hand tools.

Since common nonpowered hand tools, such as those shown in Fig. 8, are widely used, it is necessary to know what worker, task, and tool factors influence human torque exertion capability and how this capability can be maximized. Studies dealing with the determination of human torque exertion capability with common nonpowered hand tools are few (Mital et al., 1983, 1985; Huston et al., 1984; Mital et al., 1985; Mital, 1986; Mital and Sanghavi, 1986; and Mital and Channaveeraiah, 1988). In all these studies, nonpowered screwdrivers and wrenches were included and the effects of various factors defining

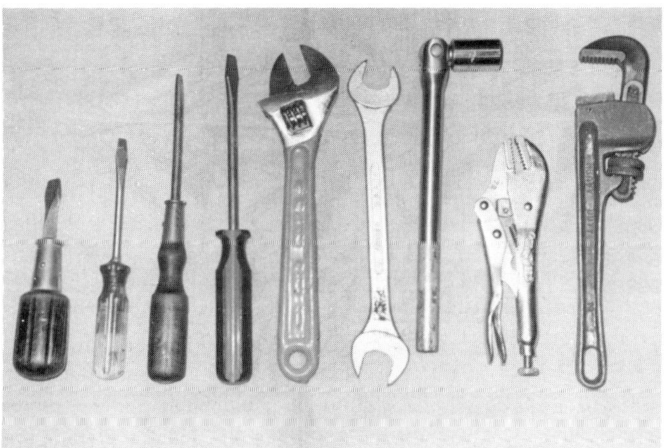

Fig. 8. Common wrenches and screwdrivers available in the market.

the operator - hand tool system on torque exertion capability was investigated. The effects of the relevant factors are summarized in this section.

Measurement of human torque exertion capability

To measure the torque transmitted on to the workpiece, strain gauges are mounted on the experimental workpiece (at 45^{0}), such as the one shown in Fig. 9, and torques exerted are recorded by a strainmeter (Fig. 10 shows a VISHAY-20 digital strainmeter). The entire system needs to be calibrated periodically against known torques.

Fig. 9. Workpiece wired for recording torque exertions.

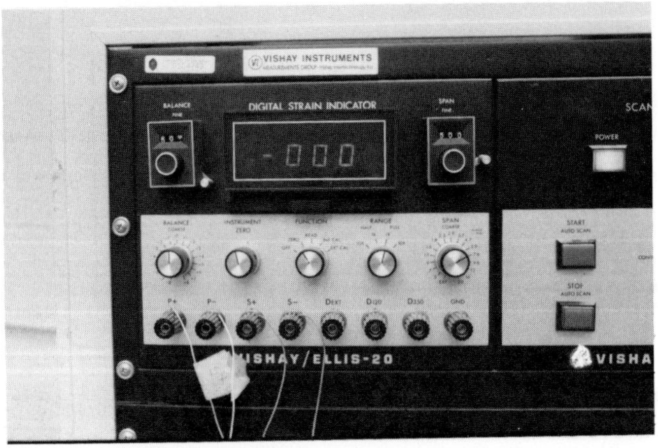

Fig. 10. VISHAY-20 digital strainmeter.

Effect of worker, tool, and task related factors

Worker gender and age. The grip strength of females is approximately 50 to 67% of male grip strength (Lindstrom, 1973; Konz, 1983; Chaffin and Anderson, 1984). Males, on the average can exert about 500N force while females on the average are able to exert 250N force. The difference in force is primarily due to females having smaller grip size and muscles. This means that tools designed on the basis of male strength and anthropometry will not accommodate females (Ducharme, 1975). Mital and Channaveeraiah (1988) also reported that the torque exertion capability of males with screwdrivers and wrenches is 10% to 56% greater than females. Thus, the tools designed for male dimensions and force capabilities will not permit easy integration of females into the workforce.

The grip strength is also significantly affected by age (Shock, 1962). By age 65, the decline in strength is about 20 to 40%. No study of the effect of age on torque exertion capability with nonpowered screwdrivers and wrenches has been reported in the literature. However, since human strengths decline with age, it is logical to assume that variation in the age of the workforce would affect their torque exertion capability and should be taken into consideration in selecting tools and designing fastener devices.

Isometric strengths and anthropometry. Mital and Sanghavi (1986) observed a significant, but not very large, correlation between isometric shoulder strength and torque exertion capability (0.26 for males and 0.23 for females). In general, heavier and stronger individuals applied more torque. Other isometric strengths did not show appreciable correlations. Thus, it appears that shoulder strength is a limiting factor in torque exertion capability. Body heights, sitting or standing, were unimportant. Similar conclusion about anthropometric characteristics was drawn by Johnson and Childress (1988) in their study of powered screwdrivers.

Body posture. The effect of posture on torque exertion capability was studied in detail by Mital (1986) and Mital and Channaveeraiah (1988). Postures ranging from standing to lying-on-the-side were studied. It was concluded that minor variations in posture are unimportant. Extreme variations, such as between standing and lying on the stomach (Fig. 11) or leaning sideways from a ladder (Fig. 12), however, cause large differences in torque exertion capabilities. Mital and Channaveeraiah (1988) have given detailed tables showing the distributions of torque exertion profiles of males and females in different postures with various wrenches and screwdrivers.

Technique and experience. Marras and Rockwell (1986) studied the force generated by experienced and inexperienced workers during the spike maul use. They observed that experienced workers generated almost twice as much force (136 846N) as the novice workers (64 446N). This large difference was due to the technique used by experienced workers (snapping action as opposed to sustained force application). Many tools, such as screwdrivers, however, do not permit snapping action and, therefore, the difference in the force/torque exerted by experienced and inexperienced workers is not expected to be as large but yet significant.

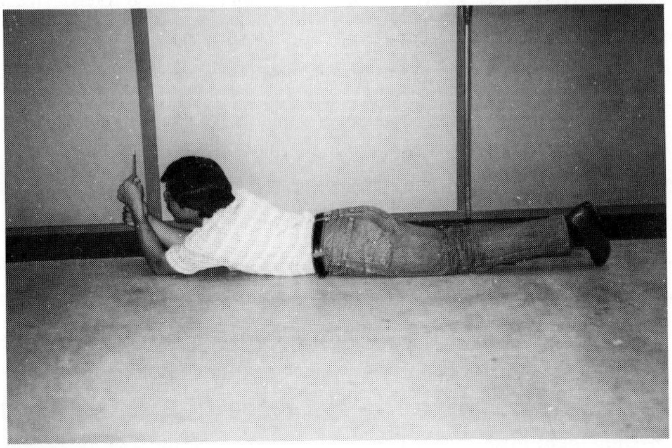

Fig. 11. Lying on the stomach and exerting with a screwdriver (Mital and Channaveeraiah, 1988).

Fig. 12. Leaning sideways from a ladder and exerting with a screwdriver (Mital and Channaveeraiah, 1988).

<u>Type of tool.</u> Considerably greater torque is exerted with wrenches (10 to 20 times more) than with screwdrivers as they use different muscles and lever arms (Mital et al., 1983, 1985; Mital and Sanghavi, 1986; Mital, 1986; Mital and Channaveeraiah, 1988). For screwdrivers, torque exertion capability generally increases with the handle diamter. For wrenches, the torque varies linearly with the lever arm. However, there are exceptions. For instance, more torque is exerted with a socket wrench than with a vise grip or spanner wrench or crescent wrench of about the same lever arm. This is expected to be caused by the type of grip and the nature of coupling between the tool and the workpiece. For the tools shown in Fig. 8, Mital and Channaveeraiah (1988) have provided extensive torque exertion capability data. Table 10 shows the minimum and maximum average torques that can be exerted with these tools and the body posture associated such exertions.

TABLE 10
Average minimum and maximum torques (Nm) that can be exerted with the tools shown in Fig. 8 (Mital and Channaveeraiah, 1988).

Tool	Gender	Torque	Posture(s)
Short screwdriver	Male	2.61	Leaning sideways from a ladder, tool axis horizontal (outward) or vertical (upward); overhead, tool axis horizontal (outward) or vertical (upward); squatting in a confined space, tool axis horizontal (outward)
		5.13	Squatting, tool axis vertical (upward)
	Female	2.02	Same as for males
		3.91	Kneeling on one or two knees, tool axis vertical (upward)
Medium screwdriver	Male	1.98	Lying on the side, tool axis horizontal (outward)
		2.64	Kneeling on one knee, tool axis vertical (upward)
	Female	1.40	Same as for males
		1.95	Squatting, tool axis vertical (downward)
Long screwdriver	Male	2.82	Same as for medium screwdriver
		3.52	Same as for short screwdriver
	Female	1.82	Lying on the side, tool axis horizontal (outward)
		2.85	Sitting, tool horizontal (outward) and at the eye level
Longest screwdriver	Male	3.40	Same as for medium screwdriver
		4.62	Standing, bent at the waist, tool axis vertical (upward)

	Female	2.46	Same as for long screwdriver
		3.58	Standing, knees bent, back supported, tool axis vertical (upward)
Crescent wrench	Male	6.41	Leaning sideways from a ladder, tool axis vertical (upward)
		34.88	Standing, tool axis horizontal (outward)
	Female	5.13	Same as for males
		24.71	Same as for males
Spanner wrench	Male	5.86	Same as for crescent wrench
		37.91	Standing, bent at the waist, tool axis horizontal (inward and outward)
	Female	4.76	Same as for crescent wrench
		25.00	Same as for males
Socket wrench	Male	7.02	Same as for crescent wrench
		43.66	Kneeling on one knee, tool axis horizontal (outward); kneeling on both knees, tool axis horizontal (outward) or vertical (upward)
	Female	5.77	Same as for crescent wrench
		31.13	Standing, tool axis horizontal (outward)
Pipe wrench	Male	6.58	Same as for crescent wrench
		46.79	Standing, tool axis horizontal (outward)
	Female	5.49	Same as for crescent wrench
		31.04	Same as for males

Height of torque application. The height at which torque is applied (eye, shoulder or elbow) has no appreciable influence on torque exertion capability with screwdrivers and wrenches (Mital and Channaveeraiah, 1988).

Handle shape and size, and influence of gloves. These have been discussed in detail in the "tool design" section. With screwdrivers, triangular handles maximized torque. Torque with wrenches was maximized when cylindrical handles were used (Mital and Channaveeraiah, 1988).

Arm longitudinal axis orientation in the horizontal plane. The orientation of the forearm, as indicated by its angle from the sagittal plane in the horizontal plane, has little or no influence on torque exertion capabilities (Mital et al., 1983, 1985; Huston et al., 1984).

Reach distance. The torque exertion capability with wrenches and screwdrivers decreases linearly with the reach distance (distance between the front of the ankles and the workpiece) between 33 cm to 58 cm in the standing posture and between 46 cm to 71 cm in the sitting posture (Mital and Sanghavi, 1986).

Wrist orientation. The maximum torque with wrenches is exerted when the long axis of the wrench is kept horizontal. With screwdrivers, torque exertion is maximized when the wrist is rotated 90° counterclockwise from the prone position (Mital and Channaveeraiah, 1988).

Repeated exertions and fatigue. If repeated exertions are made at a self determined pace, the maximum torque that can be exerted with screwdrivers and wrenches declines. The decline is as much as 38% after 240 seconds. The relationships between torque and duration of repeated exertions for various tools are given in Table 11.

TABLE 11
Relationships between torque (%) and duration of repeated exertions (min) (Mital and Channaveeraiah, 1988).

Tool	Regression equation
Pipe wrench	$\text{Torque} = 1.05 - 0.301\,(\text{Time}) + 0.08\,(\text{Time})^2$
Socket, crescent, and spanner wrenches	$\text{Torque} = 1.056 - 0.304\,(\text{Time}) + 0.104\,(\text{Time})^2$
Screwdrivers	$\text{Torque} = 1.024 - 0.138\,(\text{Time}) + 0.04\,(\text{Time})^2$

FINAL REMARKS

Hand tools have been in use for centuries. While many efficient tool designs have evolved over the years, a large number of common hand tools have remained virtually unchanged. As the technology advances, the need for more sophisticated and efficient hand tools is bound to grow. Lack of compliance with the design guidelines will most certainly create difficulties in using these tools effectively. As discussed in the "Acute Injury" and "Illnesses" sections, improper designs and use of hand tools costs the society billions of dollars. This capital is no longer available for investment and economic growth.

Besides the financial loss, improper designs also lead to permanent or long term disabilities. The humane concern alone requires that design of hand tools be taken seriously. The review of injuries, costs, design principles, and use of hand tools covered in this chapter will hopefully alert designers to the perils of bad tool designs and assist them in developing safer and more efficient hand tools.

REFERENCES

Aghazadeh, A. and Mital, A., 1987. Injuries due to hand tools: results of a questionnaire. Applied Ergonomics, 18: 273-278.
Aghazadeh, F., Latif, N.T., and Mital, A., 1989. Guidelines for preventing hand tool related accidents and illnesses. Ergonomics SA, 1: 2-9.
Armstrong, T.J., 1983. An Ergonomic Guide to Carpal Tunnel Syndrome. American Industrial Hygiene Association, Akron, Ohio.

Armstrong, T.J., Foulke, J., Joseph, B., and Goldstein, S., 1982. Investigation of cumulative trauma disorder in a poultry processing plant. American Industrial Hygiene Association Journal, 43: 103-116.
Armstrong, T.J., Radwin, R.C., Hansen, D.J., and Kennedy, K.W., 1986. Repetitive trauma disorders: job evaluation and design. Human Factors, 27: 325-336.
Ayoub, M.M. and LoPresti, P., 1971. The determination of an optimum size cylindrical handle by use of electromyography. Ergonomics, 14: 509-518.
Bobjer, O., 1984. Screwdriver handle design for power and precision. In Proceedings of the 1984 International Conference on Occupational Ergonomics. Human Factors Association of Canada, Rexdale, Ontario, pp. 443-446.
Boyd, H.B. and McLeod, Jr., A.C., 1973. Tennis elbow. Journal of Bone and Joint Surgery, 55A: 1183-1187.
Bradley, J.V., 1969a. Effect of gloves on control operation time. Human Factors, 11: 13-20.
Bradley, J.V., 1969b. Glove characteristics influencing control manipulability. Human Factors, 11: 21-35.
Bureau of Labor Statistics, 1982. Work related hand injuries and upper extre mity amputations. Bulletin No. 2160.
Capener, N., 1956. The hand in surgery. The Journal of Bone and Joint Surgery, 38B: 128-151.
Chaffin, D.B. and Anderson, G.B.J., 1984. Occupational Biomechanics. John Wiley & Sons, New York, New York.
Charash, R., 1982. Tools need not harm the hands that use them. Industrial Design, 29: 42-45.
Cochran, D.J. and Riley, M.W., 1986a. The effects oh handle shape and size on exerted forces. Human Factors, 28: 253-265.
Cochran, D.J. and Riley, M.W., 1986b. An evaluation of knife handle guarding. Human Factors, 28: 295-301.
Contini, R., Drillis, R.J., and Slote, L., 1959. Development of techniques for the evaluation of high altitude pressure suites. WADC Technical Report 58-641, Wright Patterson Air-Force Base, Ohio.
Contini, R., Drillis, R.J., et al., 1963. A report on efficiency of american clawhammers. Unpublished report, New York University Department of Engineering, New York, New York (as cited in Drillis, 1963).
Curwin, S. and Stanish, W.D., 1984. Tendinitis: Its Etiology and Treatment. Collamore Press, Lexington, Kentucky.
Damon, A., Stoudt, H.W., and McFarland, R.A., 1966. The Human Body in Equipment Design. Harvard University Press, Cambridge, Massachusetts.
Drillis, R.J., 1963. Folk norms and biomechanics. Human Factors, 5: 427-441.
Drillis, R.J., Schneck, D., and Gage, H., 1963. The theory of striking tools. Human Factors, 5: 467-478.
Ducharme, R.E., 1975. Problem tools for women. Industrial Engineering, September: 46-50.
Eastman Kodak Company, 1983. Ergonomic Design for People at Work. Lifetime Learning Publications, Belmont, California.
Emanuel, J.T., Mills, S.J., and Bennett, J.F., 1980. In search of a better handle. In Proceedings of the Symposium on Human Factors and Industrial Design in Consumer Products. Tufts University, Medford, Massachusetts, pp. 34-40.
Fraser, T.M., 1980. Ergonomics principles in the design of handtools. Occupational Safety and Health Series, No. 44, International Labour Office, Geneva, Switzerland.
Freivalds, A., 1987. The ergonomics of tools. International Reviews of Ergonomics, 1: 43-75.
Garrett, J., 1969a. References on hand anthropometry for women. AMRL Technical Report, 69-42. Army Medical Research Laboratory, Wright Patterson Air Force Base, Ohio.
Garrett, J., 1969b. References on hand anthropometry for men. AMRL Technical

Report, 69-26. Army Medical Research Laboratory, Wright Patterson Air Force Base, Ohio.
Garrett, J., 1971. The adult human hand: some anthropometric and biomechanical considerations. Human Factors, 13: 117-131.
Gorjatchkin, W., 1925. Theory of striking hand tools. Herald of Metallindustry, Moscow (in Russian).
Granada, M. and Konz, S.A., 1981. Evaluation of bent hammer handles. In Proceedings of the Human Factors Society 25th Annual Meeting. Human Factors Society, Santa Monica, California, pp. 322-324.
Greenberg, L. and Chaffin, D.B., 1977. Workers and Their Tools. Pendell, Midland, Michigan.
Grieve, D. and Pheasant, S., 1982. Biomechanics. In The Body at Work (Editor: W.T. Singleton). Cambridge University Press, Cambridge, pp. 142-150.
Hertzberg, H., 1973. Engineering Anthropometry. In Human Engineering Guide to Equipment Design (Editors: H. Van Cott and R. Kincaid). Superintendent of Documents, U.S. Government Printing Office, Washington, D.C.
Hunt, L., 1934. A study of screwdrivers for small assembly work. The Human Factor, 9: 70-73.
Huston, T., Sanghavi, N., and Mital, A., 1984. Human torque exertion capabilities on fastener devices with wrenches and screwdrivers. In Trends in Ergonomics/Human Factors I (Editor: A. Mital). North-Holland, Amsterdam, pp. 51-58.
Hymovich, L. and Lindholm, M., 1966. Hand, wrist and forearm injuries. Journal of Occupational Medicine, 8: 573-577.
Johnson, S.L. and Childress, L.J., 1988. Powered screwdriver design and use: tool, task, and operator effects. International Journal of Industrial Ergonomics, 2: 183-191.
Kao, H., 1976. An analysis of user preference toward handwriting instruments. Perceptual and Motor Skills, 43: 522.
Kao, H., 1977. Ergonomics in penpoint design. Acta Psychologia Taiwanica, 18: 49-52.
Karlqvist, L., 1984. Cutting operation at canning bench - a case study of hand tool design. In Proceedings of the 1984 International Conference on Occupational Ergonomics. Human Factors Association of Canada, Rexdale, Ontario, pp. 452-456.
Kellor, M., Kondrasuk, R., Iverson, I., Frost, J., Silberberg, N., and Hoglund, M., 1971. Hand strengths and dexterity tests. Sister Kenny Institute, Manual 721, Minneapolis, Minnesota.
Khalil, T.M., 1972. Design of tools and machines to fit the man. Industrial Engineering, January: 32-35.
Knowlton, R.G. and Gilbert, J.C., 1983. Ulnar deviation and short term strength reductions as affected by a curve handled ripping hammer and a conventional claw hammer. Ergonomics, 26: 173-179.
Konz, S., 1974. Design of hand tools. In Proceedings of the Human Factors Society 18th Annual Meeting. Human Factors Society, Santa Monica, California, pp. 292-300.
Konz, S., 1983. Work Design: Industrial Ergonomics. Grid Publishing Inc., Columbus, Ohio.
Konz, S., 1986. Bent handle hammers. Human Factors, 28: 317-323.
Konz, S. and Mital, A., 1989. Guidelines: carpal tunnel syndrome. International Journal of Industrial Ergonomics, 5: in press.
Konz, S. and Streets, B., 1984. Bent hammer handles performance and preference. In Proceedings of the Human Factors Society 28th Annual Meeting. Human Factors Society, Santa Monica, California, pp. 438-440.
Konz, S. and Warraich, M., 1985. Performance differences between the preferred and non-preferred hand when using various tools. Ergonomics International 85 (Editors: I.D. Brown, R. Goldsmith, K. Coombes, and M.A. Sinclair), Taylor & Francis, London, pp. 451-453.
Kriefeldt, J.G. and Hill, P.K., 1975. Towards a theory of man-tool system design applications to the consumer product area. In Proceedings of the

Human Factors Society 19th Annual Meeting. Human Factors Society, Santa Monica, California, pp. 301-309.

Krohn, R. and Konz, S., 1982. Bent hammer handles. In Proceedings of the Human Factors Society 26th Annual Meeting. Human Factors Society, Santa Monica, California, pp. 413-417.

Lahey, J., 1981. Portable tools, part I: manual tools. National Safety News, 124: 64-68.

Laveson, J.I. and Meyer, R.P., 1976. Left out "lefties" in design. In Proceedings of the Human Factors Society 20th Annual Meeting. Human Factors Society, Santa Monica, California, pp. 122-125.

Leaky, L.S.B., 1960. Finding the world's earliest man. National Geographic, 118: 420-435.

Lindstrom, F.E., 1973. Modern Pliers. BAHCO Vertyg, Enkoping, Sweden.

Lyman, J., 1957. The effects of equipment design on manual performance. In Production and Functioning of the Hands in Cold Climates (Editor: R. Fisher), National Academy of Sciences, National Research Council, Washington, D.C., pp. 86-101.

Magill, R. and Konz, S.A., 1986. An evaluation of seven industrial screwdrivers. In Trends in Ergonomics/Human Factors III (Editor: W. Karwowski), North-Holland, Amsterdam, pp. 597-604.

Marras, W.S. and Rockwell, T.H., 1986. An experimental evaluation of method and tool effects in spike maul use. Human Factors, 28: 267-281.

Marras, W.S., Bobick, T.G., Lavender, S.A., Rockwell, T.H., and Lundquist, R.L., 1988a. Risks of hand tool injury in U.S. underground mining from 1978 through 1983 part I: coal mining. Journal of Safety Research, 19: 71-85.

Marras, W.S., Lavender, S.A., Bobick, T.G., Rockwell, T.H., and Lindquist, R.L., 1988b. Risks of hand tool injury in U.S. underground mining from 1978 through 1983 part II: metal-nonmetal mining. Journal of Safety Research, 19: 115-124.

McCaffrey, D.P., 1981. Work-related amputations by type and prevalence. Monthly Labor Review, 3: 35-41.

Miller, G.D., 1981. Significance of dominant hand grip strengths in hand tools. Unpublished M.S. thesis, Pennsylvania State University, University Park, Pennsylvania.

Mital, A., 1986. Effect of body posture and common hand tools on peak torque exertion capabilities. Applied Ergonomics, 17: 87-96.

Mital, A. and Aghazadeh, A., 1985. A review of hand tool injuries. Ergonomics International 85 (Editors.: I.D. Brown, R. Goldsmith, K. Coombes, and M.A. Sinclair), Taylor & Francis, United Kingdom, pp. 85-87.

Mital, A. and Sanghavi, N., 1986. Comparison of maximum volitional torque exertion capabilities of males and females using common hand tools. Human Factors, 28: 283-294.

Mital, A. and Channaveeraiah, C., 1988. Peak volitional torques for wrenches and screwdrivers. International Journal of Industrial Ergonomics, 3: 41-64.

Mital, A., Huston, T., and Sanghavi, N., 1983. Hand-exertable maximum torque profiles for wrenches and screwdrivers. In Proceedings of the VIIth International Conference on Production Research. University of Windsor, Windsor, Ontario, pp. 843-849.

Mital, A., Sanghavi, N., and Huston, T., 1985. A study of factors defining the "operator-hand tool system" at the work place. International Journal of Production Research, 23: 297-314.

Napier, J., 1956. The prehensile movements of the human hand. Journal of Bone and Joint Surgery, 38B: 902-913.

Napier, J., 1962. The evolution of the hand. Scientific American, 207: 56-62.

Napier, J., 1963. Early man and his environment. Discovery, 24: 12-18.

Patkin, M., 1969. Ergonomic design of a needleholder. Medical Journal of Australia, 2: 490-493.

Petrofsky, J.S., Williams, C., Kamen, G., and Lind, A.R., 1980. The effect of handgrip span on isometric exercise performance. Ergonomics, 23: 1129-1135.

Phalen, G.S., 1972. The carpal tunnel syndrome, clinical evaluation of 598 hands. Clinical Orthpaedics and Related Research, 83: 29-40.
Pheasant, S. and O'Neill, D., 1975. Performance in gripping and turning - a study in hand/handle effectiveness. Applied Ergonomics, 6: 205-208.
Putz-Anderson, V., 1988. Cumulative Trauma Disorders: A Manual for Musculo-skeletal Diseases of the Upper Limbs. Taylor & Francis, London.
Radonjic, D. and Long, C., 1971. Kinesiology of the wrist. American Journal of Physical Medicine, 50: 57-71.
Replogle, J.O., 1983. Hand torque strength with cylindrical handles. In Proceedings of the Human Factors Society 27th Annual Meeting. Human Factors Society, Santa Monica, California, pp. 412-416.
Rigby, L.V., 1973. Why do people drop things? Quality Progress, 6: 16-19.
Riley, M., Cochran, D., and Schanbacher, C., 1985. Force capability differences due to gloves. Ergonomics, 28: 441-447.
Rubarth, B., 1928. Untersuchung zur bestgestaltung von handheften fur schraub-enziehar und ahnlich werkzeuge (Investigations concerning the best shape for handles for screwdrivers and similar tools). Industrielle Psychotechnik, 5: 129-142.
Saran, C., 1973. Biomechanical evaluation of T-handles for a pronation supina-tion task. Journal of Occupational Medicine, 15: 712-716.
Schoenmarklin, R.W. and Marras, W.S., 1987. Measurement of hand and wrist position by a wrist monitor. In Proceedings of the IXth International Conference on Production Research (Editor: A. Mital), University of Cincinn-ati, Cincinnati, Ohio, pp. 410-416.
Schoenmarklin, R.W. and Marras, W.S., 1988a. An investigation into the effects of angled hammers and hammering orientation; part I: wrist motion and hammering performance. Human Factors, In press.
Schoenmarklin, R.W. and Marras, W.S., 1988b. An investigation into the effects of angled hammers and hammering orientation; part II: muscle fatigue and subjective preference. Human Factors, In press.
Shock, N., 1962. The physiology of aging. Scientific American, 206: 100-110.
Sperling, L., 1980. Test program for work gloves. Research report no. 1980:18. Department of Occupational Safety, Division for Occupational Medicine, Labor Physiology Unit, Umea, Sweden.
Sudhakar, L.R., Schoenmarklin, R.W., Lavender, S.A., and Marras, W.S., 1988. The effects of gloves on grip strength and muscle activity. In Proceedings of the Human Factors Society 32nd Annual Meeting. Human Factors Society, Santa Monica, California, pp. 647-650.
Swain, A.D., Shelton, G.C., and Rigby, L.V., 1970. Maximum torque for small knobs operated with and without gloves. Ergonomics, 13: 201-208.
Swanson, A., Matev, I., and Groot, G., 1970. The strength of the hand. Bulletin of Prosthetics Research, pp. 145-153.
Technical Research Center of Finland, 1988. Evaluating and choosing pneumatic screwdrivers and nut runners. Occupational Safety Engineering Laboratory, Tampere, Finland.
Terrell, R. and Pursewell, J., 1976. The influence of forearm and wrist orientation on static grip strength as a design criterion for hand tools. In Proceedings of the Human Factors Society 20th Annual Meeting. Human Factors Society, Santa Monica, California, pp. 28-32.
Thompson, A.R., Plewes, L.W., and Shaw, E.G., 1951. Peritendinitis crepitans and simple tenosynovitis: a clinical study of 544 cases in industry. British Journal of Industrial Medicine, 8: 150-160.
Tichauer, E.R., 1966. Some aspects of stress on forearm and hand in industry. Journal of Occupational Medicine, 8: 63-71.
Tichauer, E.R., 1976. Biomechanics sustains occupational safety and health. Industrial Engineering, February, pp. 46-56.
Tichauer, E.R.and Gage, H., 1977. Ergonomic principles basic to hand tool design. American Industrial Hygiene Association Journal, 38: 622-634.
Tichauer, E.R., 1978. Biomechanical Basis of Ergonomics. John Wiley & Sons, New York, New York.

Washburn, S., 1960. Tools and human evolution. Scientific American, 203: 3-15.
Weidman, B., 1970. Effect of safety gloves on simulated work tasks. AD 738981, National Technical Information Service, Springfield, Virginia.
Wu, Y., 1975. Material properties criteria for thermal safety. Journal of Materials, 7: 575-579.
Yoder, T.A., Lucas, R.L., and Botzum, C.D., 1973. The marriage of human factors and safety in industry. Human Factors, 15: 197-205.
Zheligovsky, W., 1925. Striking hand tools and work with them. Herald of Metallindustry (in Russian).

Workspace, Equipment and Tool Design
edited by A. Mital & W. Karwowski

DISPLAYS AND CONTROLS

S. S. Asfour, V.K. Omachonu, E.L. Diaz, and E. Abdel-Moty

Department of Industrial Engineering, University of Miami, Coral Gables, FL 33124 USA

ABSTRACT

This chapter examines the various types of displays and controls, as well as the design characteristics that make them effective or less so. Since there are many types of displays and controls already in existence, an overview is presented in this chapter to assist the practicing ergonomist in the selection of displays and controls. Important distinctions are made between quantitative and qualitative displays as well as discrete and continuous controls. Population stereotypes are also discussed as they affect the design of displays and controls.

INTRODUCTION

Displays and controls are essential in our everyday living. We use them at home, in cars, and most importantly, at work. Every existing technology today relies, in some form or the other, on the use of displays and controls. As technology takes on added significance in our lives, the design of displays and controls will become a critical component of the effectiveness with which we perform our daily work. Without proper design, displays and controls can constitute sources of annoyance, accidents, errors, fatigue, delays, and general systems failure.

DISPLAYS

Information can generally be transmitted to an intended receiver through visual (sight) and auditory (hearing) stimuli. When information is intended to be received through visual cues, the instrument used to convey the information is known as a visual display. Similarly, information received through auditory cues is transmitted with the aid of an auditory display. A traffic light signal is an example of a visual display, whereas a siren is an example of an auditory display. In addition to these two types of displays, there is also the use of the sense of touch for communicating information such as tactual feedback in control systems or braille for the blind.

In order to design displays that are suitable for industrial applications, it is critical to understand as much as possible about the task to be performed. The selection of the appropriate type of display for a given task is perhaps the most important decision a designer can make. A necessary starting point is to decide what sense to utilize; sight, hearing, etc. Since the majority of industrial applications rely on the use of visual and auditory displays, they are discussed below.

Visual Displays

In general, visual displays can be classified according to the type of information being conveyed. Examples of different kinds of visual displays are given in Figure (1).

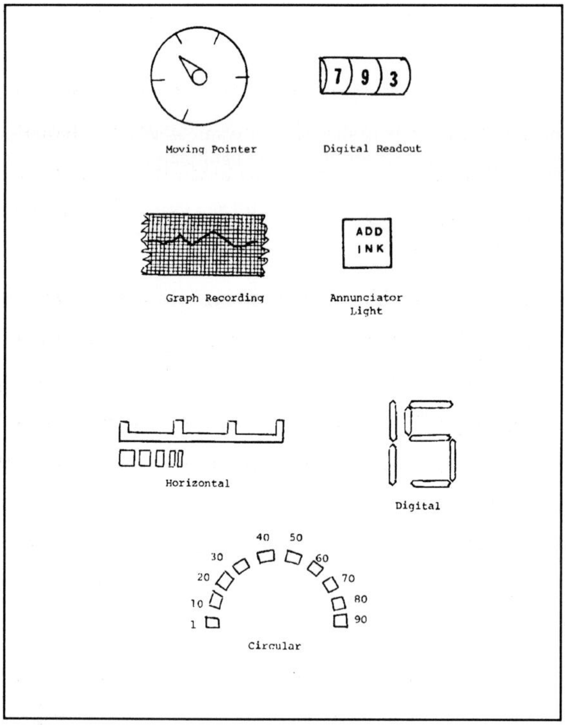

Fig. 1. Examples of Visual Displays

The information displayed could be one or a combination of the following:

1. Quantitative or Qualitative
2. Static or dynamic
3. Representational

Quantitative displays are essential where the user requires numerical information from the instrument. The information may be presented in two forms: analog and/or digital. In analog indicators, the position of the pointer on the scale is analogous to the value it represents (meters, gauges). Analog indicators are suitable for quick check-readings, since the general position of a pointer can be easily checked. This type is of interest whenever the rate and direction of change are of concern. In digital indicators, information is presented directly as a number. This type is more suitable for precise readings.

Qualitative displays are suitable where the user needs to distinguish between a smaller number of different conditions. The main requirement is that the indicators for each of the conditions should be as distinctive as possible. Visual indicators are essential for representing three or more conditions, because they can be made distinctive through differences in position, color, shape and size.

In dynamic displays, the information being presented changes with time (e.g. temperature gauge), and is fed back from some system parameter such as an over-heated engine.

In static displays, the information being presented does not change with time (stationary), and are not connected to system output. Examples are tables and graphs in a book, and conventional road signs.

Representational displays can be pictorial or symbolic. Pictorial displays include TV pictures, road signs with silhouettes of objects, photographs, etc. A clock with a conventional dials is generally considered a symbolic display. Combined symbolic and pictorial displays also exist in the form of maps and radar scopes such as those seen on TV weather reports.

Quantitative visual displays. In general, conventional quantitative displays can be classified under one of three categories:

1. Fixed scale with moving pointer
2. Moving scale with fixed pointer
3. Digital display

The first two are referred to as analog indicators in that the pointer position is analogous to the value it represents. (Examples of the various types of scales utilized in quantitative displays are shown in Figures (2A) and (2B)).

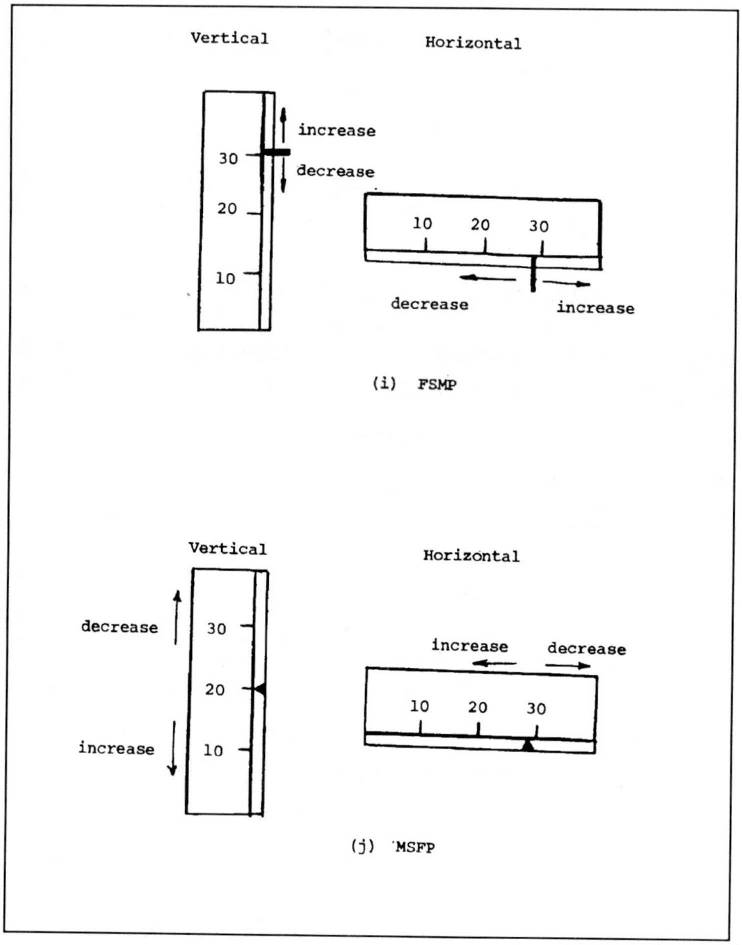

Fig. 2A. Examples of Vertical and Horizontal Scales

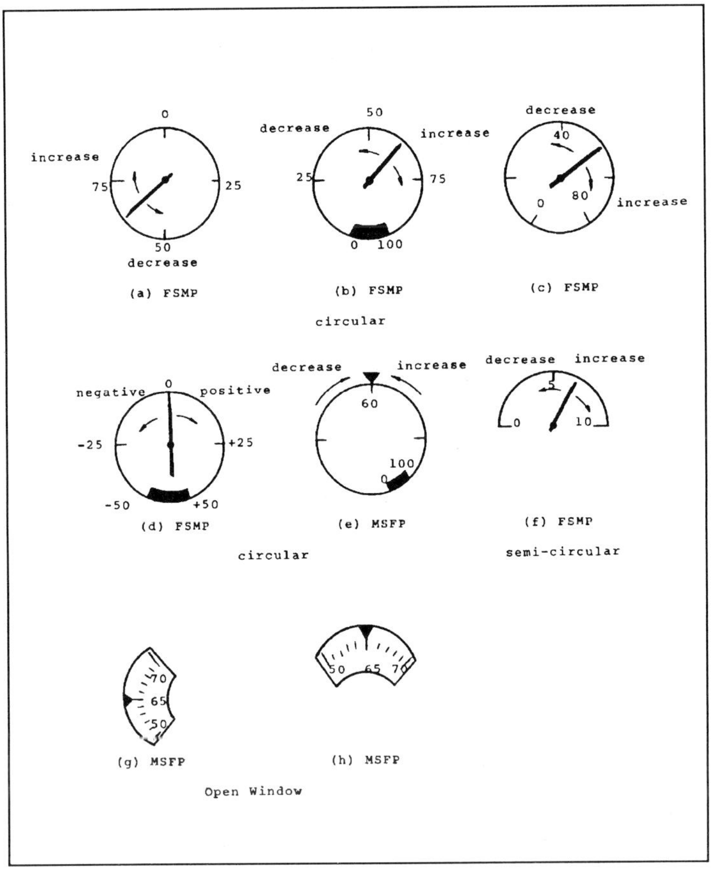

Fig. 2B. Examples of Circular, Curved, and Open-Window Scales

FSMP: Fixed Scale, moving pointer

MSFP: Moving Scale, fixed pointer

<u>Design features of quantitative visual displays</u>. These include the numeric progressions used, length of the scale units, use of scale markers, width (thickness) of markers, design of pointers, and location of scale numbers. The design implications of these features are presented below:

i. Numeric progressions of scales. In general, a progression by 1's (0, 1, 2, 3, etc.) is considered to be the easiest to use. Major markers can then be used on the scale at 0, 10, 20, etc., with intermediate markers at 5, 15, 25, etc., and with minor markers at individual numbers. Decimals are generally difficult to work with. Progression systems such as by 3s, 8s, etc., should be avoided as much as possible.

ii. Length of scale unit. This value represents the smallest numeric value to which the scale is to be read. Length of scale unit should be chosen such that the distinctions between the values can be made easily.

iii. Design of scale markers. It is recommended to include a scale marker for each scale unit to be read. Figure 3 shows an acceptable design for scale markers under conditions of poor and good illumination.

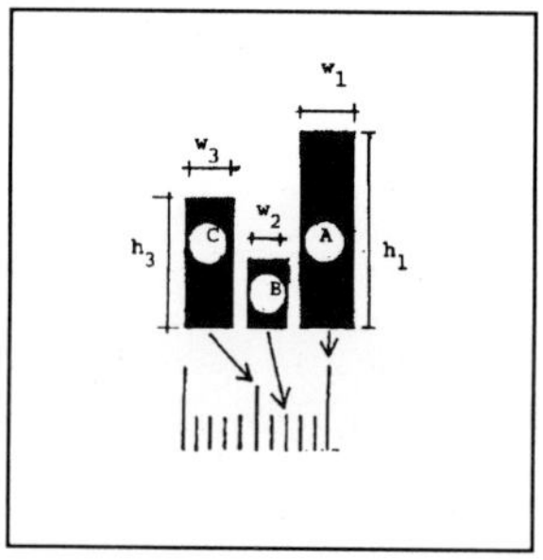

		Low illumination		Normal illumination	
		w	h	w	h
A	Major Marker	0.035	0.100	0.013	0.090
B	Intermediate Marker	0.030	0.160	0.013	0.160
C	Minor Marker	0.025	0.220	0.013	0.220

Fig. 3. Recommended format of scale markers
(Dimensions are in inches)

iv. Scale markers and interpolation. It is generally necessary to interpolate when there is the need to estimate specific values between markers when not all scale units have markers. It is desirable to have a scale marker for each unit to be read. When high accuracy is desired, a marker should be placed at every scale unit, even though this might require a larger scale.

v. Scaled dimensions for normal illumination: The following recommendations made by Bailey (1982) apply to indicators that are reasonably well illuminated. Assuming high contrast between the graduation marks and dial face, adequate illumination levels on the dial face, and reading distances of

13 to 28 inches, the following recommendations should be observed:

a. Minimum width of a major graduation mark should be 0.0125 in.

b. Although graduation marks may be spaced as close as 0.035 in., the distance should not be less than twice the stroke width for black marks on white dial faces.

c. The minimum distance between graduation marks should be 0.05 in.

d. The height of major, intermediate and minor graduation marks should not be less than 0.22 in., 0.16 in. and 0.09 in. respectively.

vi. Numbers, letter size and style. The width of all numbers should be 3/5 of the height, except for the "4", which should be one stroke width wider than the others, and the "1", which should be one stroke width wide. In addition, the stroke width should be from 1/6 to 1/8 of the numeral height. The width of the uppercase letters should be 3/5 of the height except for the "I" which should be one stroke width, and the "m" and "w" which should be about 1/5 wider than the other letters. The stroke width should be from 1/6 to 1/8 of the numeral height.

vii. Scale size. For linear scales, the optimum size for general purpose is a scale length of 2.5 - 3.0 in.; for high accuracy reading 4.0 - 5.0 in. scales are advisable. Floor monitoring and space-saving purposes, linear scales with a minimum length of 2.0 in. are acceptable (Diffrient, 1974)

viii. Design of pointers. It is generally recommended to use pointers (with a tip angle of 20); have the tip of the pointer touch but not overlap, the smallest scale markers; have the color of the pointer extend from the tip to the center of the scale (for circular scales); and have the pointer close to the scale surface to avoid parallax. (Diffrient, 1974)

ix. Pointer size. The pointer should not obscure the scale numbers or scale marking strokes, and it should be mounted so that the possibility of reading errors due to visual parallax is minimized. The optimum pointer, therefore, is one visible over most of its length from tip to pivot; its tip should reach to within 1/16 to 1/64 in. of the scale markings but should not overlap them; it should be about 1/32 to 3/32 in. wide, its tip narrowing to a point, or at least down to the width of the scale marking strokes (Diffrient, 1974).

x. Dial pointers (Diffrient, 1974).

a. Optimum length: 1.12 in.

b. Pointer tip widths: same as major graduation marks

c. Color of pointer: same as graduation marks

d. Pointer shape: straight and simple

e. Distance from pointer to indices: 0.03 in. (max); no min., no overlap

f. Number of pointers on the same dial: one is preferred

xi. Characters for dials and panels (Diffrient, 1974).

a. Letter style: simple and sans serif (ex. future, helvetica, venus or gothic)

b. Capital letters: use for titles (can be read further than lowercase because of their increased size)

c. Lowercase letters: use for instructions (can be read 13.4% faster than caps because of the up-and-down visual cues)

d. Location of titles: above displays or controls (exception: locate below if displays or controls are above eye level)

e. Label color (ample light): black on light

f. Label color (dark adaptation): white on dark

g. Label color (poor light): illuminated on dark

h. Recommended character height for 28 in. viewing distance or less:

1. Main panel title: 0.25 in.
2. Subdivision titles: 0.188 in.
3. Component titles: 0.125 in.

i. If reading distance is 18 inches or less, and panel space is tight, all titles can be 0.09 in. high.

j. Stroke width (black on light): 1/6 character height

k. Stroke width (white on black): 1/7 - 1/8 character height

l. Character spacing: 1 stroke width (min.)

m. Word spacing: 1 character width (min.)

Qualitative visual displays. In the case of qualitative information, the user is mainly interested in the approximate value of some continuously changeable variable (e.g., pressure, temperature, or speed) or in its trend or rate of growth. Qualitative information are generally derived from quantitative data. Quantitative data may be used as the basis for generating a qualitative reading in at least three ways:

a. For determining the status or condition of the variable, e.g., if the temperature gauge of an automobile is cold, normal, or hot.

b. For maintaining some desirable range of values such as driving speed between 55 and 60 mph.

c. For monitoring trends, rate of change, etc., such as the rate of change in altitude of an airplane.

The most common method for making distinctions among readings is color coding. Certain color-coding conventions are fairly well established and should be followed unless it can be demonstrated that changing the color code would enhance performance. Examples of conventional color codes are:

a. Red: Used to alert an operator that the system or any portion thereof is inoperative and successful operation is impossible until appropriate corrective or override action is taken. Indicates no-go, error, failure, malfunction, etc.

b. Flashing Red: Used for conditions in which action must be taken to prevent personnel and/or equipment disaster.

c. Amber: Used to advice an operator that a condition exists which is marginal insofar as system effectiveness is concerned, that an unsatisfactory condition is being approached or exists but that the system can operate, or that preparedness should exist for an emergency or corrective action (e.g. battery approaching replacement/recharging time).

d. Green: Used to imply that a unit is in tolerance or a condition is satisfactory and that it is alright to proceed (e.g. go ahead, ready, acceptance).

e. White: Used to advise the operator of a system condition without implication of a right or wrong situation. Has no meaning other than that derived from its legend or label (e.g. power on, location of available components, action or test in progress, warm up).

Indicator Lights. These are commonly used qualitative displays. The following is a summary of the general design considerations for indicator lights and their use (Woodson, 1981; CPTG, 1980):

a. Legend lights are used primarily to display qualitative information that requires an immediate reaction by the operator.

b. Simple Indicator Lights: An indicator light containing no legend plate. (ex. pilot lights)

c. Legend Plate: The surface on which the legend (words, symbols or numbers) is inscribed.

d. Word warning lights are used where possible to specify an emergency condition.

e. Every indicator light is labeled according to function.

f. In almost all cases, labeling is located above the indicator.

g. When transilluminated indicators are used under varied ambient illumination, a dimming control is provided.

h. If panel is to be used outdoors, provision is made to prevent reflected sunlight from making the indicators appear illuminated.

i. Provision is made for bulb removal from the front of the display panel without the use of tools, or other rapid and convenient means.

j. A master light test control is incorporated if there are many indicator lights on a control panel.

k. Lights which must attract immediate attention should be at least twice as bright as the immediate background. The background should be dark in contrast to the light and should be in a dull finish.

l. Critical lights should be isolated from less important lights whenever possible.

m. Master warning and caution lights are located near the line-of-sight to attract attention to a subsystem failure.

n. Use flashing lights only to call attention to situations requiring immediate action. Flash rates of 3-5/sec with equal "ON" and "OFF" time.

o. A light going off should never be used to indicate either green or red light functions. Must have positive feedback to attract attention.

Design recommendations for visual displays. The following are general recommendations and guidelines for the design of visual displays:

i. For many uses, digital displays (also called counters) are generally superior to analog displays, especially under the following conditions:

    a. A precise numeric value is required.

    b. The values displayed remain visible long enough to be read.

ii. Analog displays are preferred when it is important to observe the direction or rate of change of the values being presented.

iii. In general, of the two analog displays, a fixed scale - moving pointer is preferred.

iv. If numerical increase is related to some interpretation such as more or less or up or down, it is easier to interpret a straight-line or thermometer type scale with a moving pointer.

v. If slight variable changes in quantity are important to the user, a moving pointer will enhance readability.

vi. When the range of values is too wide to be presented on the face of a relatively small scale, then a fixed pointer - moving scale is advantageous.

vii. Certain moving-scale, fixed pointer designs have the practical advantage of occupying a small panel space, since the scale can be laid out behind the panel face, and exposing only the relevant portion of the scale.

viii. Sometimes, considerations other than human factors may dictate the choice of a display. For example, space limitation may lead to the choice of a moving scale display, even though a moving pointer display is preferable.

Table 1 presents a summary guideline on the recommended display types for different types of information.

TABLE 1
Types of information displayed and recommended displays for each [a]

Information Type	Preferred Display	Feature	Examples in Industry
Quantitative Reading	Digital readout or counter	Minimum reading time Minimum error potential	Number of units produced on a production machine
Qualitative Reading	Moving pointer or graph	Position easy to detect, trends apparent	Temperature changes in a work area
Check Reading	Moving pointer	Deviation from normal easily detected	Pressure gauges on a utilities console
Adjustment	Moving pointer or digital readout	Direct relation between pointer movement and motion of control, accuracy	Calibration charts on test equipment
Status Reading	Lights	Color-coded, indication on status (e.g., "ON")	Consoles in production lines
Operating Instructions	Annunciator Lights	Engraved with action required, blinking for warnings	Manufacturing lines in major production systems

[a]Adapted from Grether and Baker, 1972

Auditory displays

Auditory displays are used to present information to the auditory senses. Table 2 presents the characteristics and features of certain types of audio alarms.

TABLE 2
Characteristics and features of certain types of audio alarms[a]

ALARM	INTENSITY	FREQUENCY	ATTENTION-GETTING ABILITY	NOISE-PENETRATION ABILITY
Diaphone (foghorn)	Very High	Very low	Good	Poor in low-frequency noise
Horn	High	Low to high	Good	Good
Whistle	High	Low to High	Good if intermittent	Good if frequency is properly chosen
Siren	High	Low to high	Very good if pitch rises and falls	Very good with rising and falling frequency
Bell	Medium	Medium to high	Good	Good in low-frequency noise
Buzzer	Low to medium	Good	Good	Fair if spectrum is suited to background noise
Chimes and gong	Low to medium	Low to medium	Fair	Fair if spectrum is suited to background noise
Oscillator	Low to high	Medium to high	Good if intermittent	Good if frequency is properly chosen

[a]Adapted from Deatherage, 1972

Auditory displays are usually preferable to visual displays under the following conditions:

i. When the origin of the signal is itself a sound
ii. When the message is simple and short
iii. When the message will not be referred to later
iv. When the message deals with events in time
v. When warnings are sent or when the message calls for immediate action
vi. When continuously changing information of some type is presented, such as aircraft, radio range, or flight-path information
vii. When the visual system is overburdened
viii. When speech channels are fully employed (in which case auditory signals such as tones should be clearly detectable from the speech)

ix. When illumination limits use of vision
x. When the receiver moves from one place to another
xi. When a verbal response is required

Tactual displays

The primary use of tactual displays is to serve as substitutes for hearing, especially aids in the deaf, and as substitutes for seeing, especially as aids to the blind. The most common types of stimuli used for tactual displays have been mechanical vibration or electric impulses.

CONTROLS

A control is any device - a switch, lever, pedal, button, knob, keyboard - used by a person to put information into a system. Controls can be used optimally, if the controls operate as one expects them to (according to the population stereotypes); they are dimensioned to fit the human body; and their operating characteristics are within the strengths and precision capabilities of people. Controls can be classified as follows:

a) Discrete or Continuous, and
b) Linear or Rotary

Discrete vs. continuous control

A discrete control can take on only a limited number of finite states, e.g., a light switch which can be either on or off, up or down. A continuous control on the other hand, can assume a large number of states, e.g., a dimmer light switch can continuously alter light output from none to full brightness. Continuous controls are used when fine adjustments are necessary. Kantowitz and Sorkin (1983) suggest the use of discrete or continuous control as follows:

Use a discrete control when:

i. There is a limited number of control states (<25).
ii. Numerical or alphabetical information must be entered.
iii. On-Off or Yes-No information is required.
iv. Only small mechanical forces are required or available.

Use a continuous control when:

i. There is a large number of control states (>25).
ii. The operator must exert substantial force.
iii. Speed of operation is more important than accuracy of control setting.

An important advantage of discrete controls is the tactual feedback they provide when operated. For example, an automobile driver can feel the control engage as he/she shifts from one gear to the other.

Linear vs. rotary controls

Linear controls such as pushbuttons, toggle switches, and slides move in a straight line. Rotary controls such as a dial telephone, door knob, and radio and TV knobs move in an arc. According to Kantowitz and Sorkin (1983), at present, there is no well established theoretical basis for choosing a linear versus a rotary control.

Coding of controls

Two important features that are common to control are their shapes and sizes. Figure (4) reveals a set of shapes that can be discriminated by touch alone.

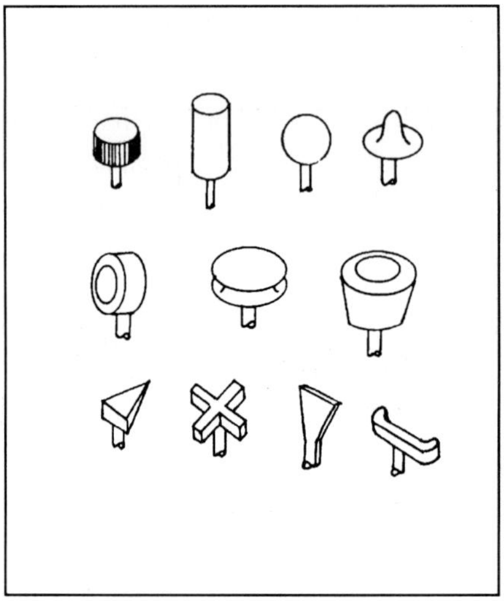

Fig. 4. Examples of shapes that can be distinguished by touch.

As a coding scheme, varying size is not considered as effective as varying shape (Kantowitz and Sorkin 1983). Labels on the other hand, are simple and inexpensive control features. They can be applied to all controls and require little operator training. Some design guidelines for using control labels are as follows (Chapanis & Kinkade, 1972):

i. Locate labels systematically and consistently in relation to controls.
ii. Labels should be brief
iii. Avoid abstract symbols that require special training
iv. Use color coding of labels when a special color is attributable to certain

special occurrences.

v. Color should not be the primary or sole coding method.

vi. Illumination must be adequate to read labels.

Types of controls

Although several types of controls exist, the use of foot pedals and pushbuttons have received increasing interest in industry. Some of the design parameters for foot pedals and pushbutton controls are presented next.

Foot pedals. The design of foot pedals must consider three important features:

i. How much force must be exerted on the pedal.

ii. The speed and accuracy with which the pedal can be reached.

iii. The distance the pedal must travel.

Figure (5) compares different types of pedal designs.

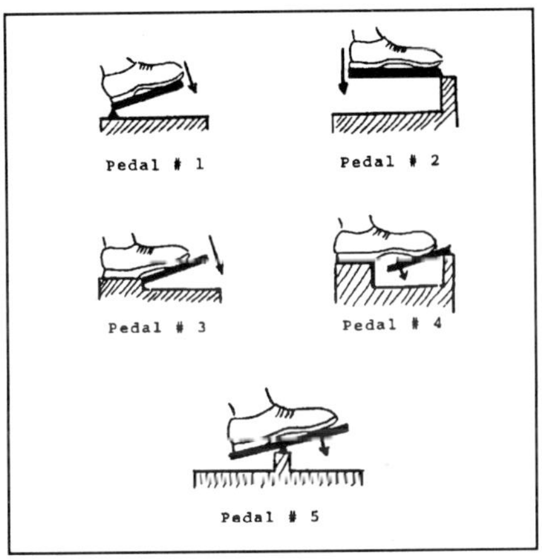

Strokes per minute	187	178	176	140	171
	Pedal No. 1 required least time per stroke	Pedal No. 2 required 5% more time per stroke than No. 1	Pedal No. 3 required 6% more time per stroke than No. 1	Pedal No. 4 required 34% more time per stroke than No. 1	Pedal No. 5 required 9% more time per stroke than No. 1

Fig. 5. Comparison Between Different Types of Pedal Design (Adapted from Barnes, 1980)

Pushbuttons. According to Morgan (1963), there are primarily three types of pushbutton (discrete) control as follows:

i. Momentary contact-push on, release off.

ii. Latching - push on, lock on.

iii. Alternate action - push on, push off.

A summary of the suitability of ten different controls (Column 1) for five different conditions or job requirements (across the top) is given in Table 3. The ratings are based on typical examples of each control type, not on the extremes of performance in each range. The types of controls that should be considered when the design or work situation has certain requirements, such as speed, accuracy, ease of operation, ease of reading, or limited space on a control panel, can be determined from this summary.

TABLE 3
Characteristics of common controls[a]

Control	Suitability Where Speed of Operation is Required	Suitability Where Accuracy of Operation is Required	Space Required to Mount Control	Ease of Operation in Array of Like Controls	Ease of Check Reading in Array of Like Controls
Toggle Switch (On-Off)	Good	Good	Small	Good	Good
Rocker Switch	Good	Good	Small	Good	Fair[1]
Push Button	Good	Unsuitable	Small	Good	Poor[1]
Legend Switch	Good	Good	Small	Good	Good
Rotary Selector Switch (discrete steps)	Good	Good	Medium	Poor	Good
Knob	Unsuitable	Fair	Small-Medium	Poor	Good
Crank	Fair	Poor	Medium-Large	Poor	Poor[2]
Handwheel	Poor	Good	Large	Poor	Poor
Lever	Good	Poor (H) Fair (V)	Medium-Large	Good	Good
Foot Pedal	Good	Poor	Large	Poor	Poor

[a]Adapted from Chapanis and Kinkade, 1972
V = Vertical, H = Horizontal
[1]Except where control lights up for "on"
[2]Assumes control makes more than one revolution.

Direction-of-movement relationships and operator expectancies

One of the major categories of error: reversal errors which are made by moving the control in the wrong direction. Most people expect controls to move in certain ways. In some cases, these expectancies are so universal that they have been termed population stereotypes. Some of the stereotypes related to controls are presented in Table 4.

TABLE 4
United States stereotypes for up and downswitch settings[a]

UP	DOWN
ON	OFF
START	STOP
HIGH	LOW
IN	OUT
FAST	SLOW
RAISE	LOWER
INCREASE	DECREASE
OPEN	CLOSE
ENGAGE	DISENGAGE
AUTOMATIC	MANUAL
FORWARD	REVERSE
ALTERNATING	DIRECT
POSITIVE	NEGATIVE

Some movement stereotypes for toggle switches used in equipment or production system control panels are given. The expected direction of activation of the switch (when mounted vertically) is given for 13 actions or conditions. These expectations are United States stereotypes and may vary in other countries.

[a]Adapted from Alexander, 1976

When controls and control-display arrangements conform to population stereotypes, we find in general that:

1. Reaction time or decision time is shorter.
2. The first control movement the operator makes is more likely to be correct.
3. The operator can use the control faster and make adjustments with greater precision.
4. The operator can learn to use the control (or control-display combination) much faster.

We should not ask the operator to make unnatural movements, especially since it is so easy to design equipment properly in the first place.

Control-display design

Control-display arrangment. It is difficult to devise a set of general rules to cover all possible kinds of control-display arrangements. Some rules seem to be so general that they could be called principles.

i. Displays and controls that are to be used in a fixed order should be arranged in sequence:
   a. from left to right
   b. from top to bottom
   c. in rows from top to bottom and from left to right within the rows.
ii. When a number of displays and controls are each associated with a group of similar components, the arrangement of the controls should correspond with the arrangement of the components.
iii. Controls associated with specific displays should be located so that the operator's hand does not interfere with viewing the display. When a number of associated controls and displays appear on a panel:
   a. put each control directly below its corresponding display or
   b. group all the displays above and all the controls below, but arrange the controls and the displays in exactly the same order.
iv. If concentric or ganged knobs are used to control a set of displays, the uppermost (smallest) knob should operate the farthest - left display, and the lowermost (largest) knob, the farthest - right display.

Control-display ratios. A control-display (C/D) ratio problem exists whenever acontrol is used to change the position of a moveable element in a display and both the control and display are continuously adjustable.

The control-display (C/D) ratio is defined as the ratio of the movement of the control device to the movement of the system response (McCormick and Sanders, 1987). Control movements as a function of the C/D ratio:

1. Normal reaction time: time required for the hand to reach the control and start moving it once the operator has decided to make the movement.
2. Travel time or primary movement time: time required for a gross adjustment, to move the needle to approximately the correct setting.
3. Adjusting-movement time or secondary movement time: time for making fine adjustments, the final precise positioning of the display element.

The best C/D ratio is one that minimizes the sum of these three times.

SUMMARY

Selection of a control or a display depends on the type of job, operation, task, work layout, and the user. Optimal selection should enable compatibility as well as efficiency. This chapter presents general guidelines and recommendations for controls and displays. In a technological society that relies heavily on information exchange, controls and displays become important not only for systems' operation, but also for safety, productivity, and the reduction of human errors.

Optimal design and selection of controls and displays should be based on a thorough understanding of human attributes: dimensions, visual abilities, etc.

It is difficult to devise a set of general rules to cover all possible kinds of control-display arrangements. The most important rules are those that consider the safety and limitations of humans while ensuring the integrity of the system.

REFERENCES

Adams, J.A., 1989. Human Factors Engineering. Macmillan Publishing Co., New York.

Bailey, R., 1982. Human Performance Engineering: A Guide for System Designers. Prentice-Hall, Englewood Cliffs, New Jersey.

Barnes, R. M., 1980. Motion and Time Study Design and Measurement of Work (7th edition). John Wiley and Sons, New York.

Chapanis, A. and Kinkade R., "Design of Controls", in: H.P. VanCott and R.G. Kinkade (Eds.), Human Engineering Guide to Equipment Design, U.S. Government Printing Office, Washington D.C., 1972, Chapter 8.

CPTG, 1980. Consumer Products Technical Group of the Human Factors Society,

CP News, Vol. 5.
Deatherage, B.H., Auditory and Other Sensory Forms of Information Presentation, in: H.P. VanCott and R.G. Kinkade (eds.), Human Engineering Guide to Equipment Design, U.S.Government Printing Office, Washington D.C., 1972.
Diffrient, N., A. Titley, and Bardagly, J., 1974. Humanscale 4/5/6. MIT Press, Cambridge, Massachussets.
Eastman Kodak Co., 1983, Ergonomic Design for People at Work. LifeTime Learning Publishers, California
Huchingson, R.D., 1981, New Horizons for Human Factors in Design. McGraw Hill, New York.
Kantowitz, B. H. and Sorkin, R. D. 1983. Human Factors. John Wiley and Sons, New York.
Konz, S., 1990. Industrial Ergonomics (3rd ed.) Publishing Horizons, Inc.
McCormick, E. J. and Sanders, M. S. 1982. Human Factors in Engineering and Design (5th ed.) McGraw Hill, New York.
Morgan, C.T., and Cook, J. S., Chapanis, A. and Lund, M. W., 1963. Human Engineering Guide to Equipment Design. McGraw Hill, New York.
Woodson, W. E., 1981, Human Factors Design Handbook. McGraw Hill, New York.
Woodson, W. E., and Conover, D. W., 1970. Human Engineering Guide for Equipment Designers (2nd ed.). University of California Press, Berkeley, California.

PART III
GENERAL

Workspace, Equipment and Tool Design
edited by A. Mital & W. Karwowski

ERGONOMIC GUIDELINES FOR DESIGNING AND MAINTAINING UNDERGROUND COAL MINING EQUIPMENT

K. Conway(1) and R. Unger(2)

(1) Monterey Research, Box S-3222, Carmel, CA 93921 (USA)
(2) U.S. Bureau of Mines, Pittsburgh, PA (USA)

ABSTRACT

Thirty-five years ago underground coal mining equipment consisted of simple but rugged machines powered by electric motors and hydraulics. The machines were maintained by maintenance personnel armed with a basic knowledge of hydraulics, electricity, and mechanical design. Over the years the machine has been transformed into powerful complex mining systems. The maintenance burden has also increased. During this period significant advances have been made in maintainability engineering and system design techniques. Many of these innovations and techniques are applicable to the design of underground mining equipment. The objectives of this section to; (1) review recent U.S. Bureau of Mines funded research on mining equipment maintainability, (2) review the impact of equipment design for maintenance on productivity and safety, and (3) present maintainability design guideline to enhance the design of new or rebuilt equipment.

HISTORIC OVERVIEW

Fifty years ago underground coal mining equipment consisted of relatively simple but rugged machines powered by electric motors and hydraulics. These machines were used to cut, dig, load, and transport coal from the mine face to the surface. The machines were maintained by the mine maintenance personnel armed with a basic knowledge of hydraulics, electricity, mechanical design and a few hand tools. These maintainers were expected to repair all of the equipment at the mine site. Because the machines were simple in design, ergonomic design factors, such as component accessibility, did not represent a major problems.

Over the years, however, the basic mining machine has been transformed into powerful, complex mining systems. To boost productivity, the horsepower and size of the original machines have been increased by orders of magnitude. To enhance unit productivity, newer machines were designed that performed multiple functions. To increase throughput, continuous miners, long and short wall systems and continuous haulage were introduced. To improve personnel safety, safety features have been added. To protect the miners' health, environmental control systems have been tacked on. Even with all of the above changes, the same maintenance crew is still expected to maintain these ever more complex mining systems. This must be accomplished in an operational environment providing little in the way of new maintenance tools, procedures, automatic test equipment, or other technology--based aids. Limited ergonomic improvement in the design of equipment for maintainability has been observed. In many cases, equipment maintainability has been sharply decreased as a result of added-on safety systems. The result is a continuing escalation of maintenance costs and equipment downtime.

Significant maintainability engineering (ME) advances have been made in the last 20 years (Conway and Unger, 1989, McBride, 1982). The advances have been necessitated by increased complexity, technical sophistication of newer systems, escalating maintenance costs, growing equipment down time and persistently high levels of maintenance injuries. Coupled with findings from human factors and applied ergonomic research, considerable improvement in maintenance efficiency, cost reduction and personnel safety have be achieved (Dhillon and Reiche, 1985; Sanders and Peay, 1988).

The purpose of this chapter is to integrate ergonomic design guidelines with applicable ME concepts for use in the design and evaluation of underground coal mining equipment. Ergonomic and ME recommendations have been specifically adapt to the underground coal operating environment. The recommendations are based on U. S. Bureau of Mines funded studies.

DESIGN INDUCED MAINTAINABILITY PROBLEMS

An mining equipment ergonomic and maintainability design review was conducted in sixteen operational underground U.S. coal mines (Conway and Unger, 1989). Mining machines in large and small mines operating in high and low seam coal were surveyed. Conventional, continuous, and long wall operations were included. Shuttle cars, scoops, roof bolters, continuous miners, long wall equipment, undercut machines, face drills, utility vehicles, and personnel carriers were reviewed. The study identified the following design induced limitations impacting personnel safety, maintenance time and cost.

1. <u>Accessibility Problems</u> - Inability of maintenance personnel to access failed or suspected components to inspect, remove or replace (R/R) them. Accessibility problems resulted from:

a. Inadequate access opening size,

b. Poor layout of components in a compartment,

necessitating R/R of non-affected parts in order to access

the failed units. In Figure 1, several components under the motor cannot be reached without removing the large motor. The access cover itself had to be removed just to complete the inspection. Removing top mounted access covers in low seam mines is a difficult task.

c. Inability to access mounting bolts or to use tools,

d. Inaccessible of components mounted in interior cavities,

Fig. 1. Inadequate Access - Components Under Motor Cannot Be Reached.

e. Routing of cables inside the frame or chassis where they cannot be reached,

f. Placing components where they cannot be reached unless the machine is partially or completely disassembled.

2. Component Handling - Inadequate component handling capability and component-machine interface design.

3. Inadequate Design For Routine Maintenance - Inability to quickly R/R leaking hydraulic hoses and water lines and to perform routine lubrication.

4. Fault Isolation - Inadequate fault isolation including:

a. difficulty determining the cause and location of a failure,

b. limited or no designed-in fault diagnostic capabilities,

5. Maintenance Inducing Design - Increased maintenance burden resulting from poor design such as placement of components where they are exposed to impact or other damage.

6. Poor Design With Respect To Resources Available - The need for maintenance personnel to "jerry-rig" tools, to handle 100 lb to 1,000 lb components, to substitute brute human strength to overcome poor component interface design or lack of tools.

7. Equipment Layout Complexity - The crowding of components into compartments without regards to the need to maintain them.

8. Design Conveniences - Multiplying the number of valves, connectors, and other high frequency replacement components as a design convenience.

The time required to change hydraulic hoses on miners and shuttle cars at two mines ranged from 15 minutes to over 4 hours. The average time for R/R was 35 minutes. Examination of these machines revealed that the time differences were directly linked to accessibility. In several cases, two or more non-failed components had to be removed to access a failed hose connection. By relocating several components or re-routing hoses, maintenance personnel could directly access over 90% of all hydraulic connections on the surveyed machines. This would reduce average hydraulic line R/R time to under 15 minutes. Other research suggests that similar maintenance task time reductions of 40% to 70% are reported (Foster et. al., 1981; Dhillon, 1986).

MAINTENANCE ACCIDENTS

Mine Safety and Health Administration (MSHA) accident statistics for 1986 reveal that maintenance related injuries account for 33% of all mining lost time accidents. These accidents impact mine operating costs in the form of lost productivity, increased benefits costs, and increased insurance rates. Many injury accidents can be directly traced to equipment design (Conway and Unger, 1989; Conway and Sanders, 1982; Barry and Associates, 1975). Inadequate accessibility, lack of means to lift and maneuver heavy components, inability to visually observe the maintenance task being performed, inadequate maintenance safeguards, and other design induced problems are cited for a significant percentage of all mine maintenance accidents. A summary of the underground coal maintenance related accident statistics for 1986 is presented in Table 1.

TABLE 1

Maintenance Related Mining Injuries (MSHA, 1987)

		Mine Maintenance		Machine Maintenance	
Code	Type of Accident	Number	% Total	Number	% Total
1	Stationary Object	185	5.6%	272	8.9%
2	Moving Object	2		6	
3	Concussion	--		1	
4	Falling Object	611	18.4%	511	16.8%
5	Flying Object	62	1.9%	62	2.0%
6	Rolling Object	62	1.9%	18	
7	--	--		--	
8	Struck by, NEC	231	6.9%	265	8.7%
9	--	--		--	
10	--	--		--	
17	Fall - Walk Away	8		7	
18	Fall on Object	3		5	
20	Caught - Meshing Object	--		2	
21	Caught - Move/Stationary	169	5.1%	190	6.3%
22	Caught - 2 Moving Objects	6		9	
23	Caught - Collapse	2		--	
24	Caught - NEC	261	7.9%	292	9.6%
25	Rub, Abrade	3		1	
26	Bodily Reaction, NEC	2		2	
27	Over-Exert: Lifting	1,132	34.1%	793	26.1%
28	Over Exert: Pull/Push	78	2.3%	147	4.8%
29	Over-Ecxert: Welding	112	3.4%	13	
30	Over-Exert: NEC	360	10.8%	373	12.3%
33	Contact Hot Object	3		26	
36	Inhale Noxious Fumes	1		8	
38	Absorb Noxious Fluid	26		25	
39	Flash Burns - Electrical	--		2	
42	NEC	1		2	
43	Insufficient Data	2		6	
	Total Within Category	3,322		3,038	

1. Perception or detection of an event or clue
2. Operator/maintainer decision making
3. Operator/maintainer physical action

A majority of the maintenance injuries involve strains/sprains, low back injuries and crushing injuries. These injuries typically occur during R/R of components ranging in weigh from 35 pounds to over 1,000 pounds (Conway and Elliott, 1988). In many instances two or more men with crow bars, 4"x4" boards, or other makeshift tools must manhandle the component to remove it from the mining machine or to lift it into place so that it can be secured. In most cases, provision had not been made during component-machine interface design to provide for mechanical assist.

HUMAN ERROR AND DESIGN FOR MAINTENANCE

Errors occur in operating mining machines, performing maintenance tasks, or in making management decisions. Fortunately, most human errors result in limited negative consequences (e.g.; lost time, production waste, etc.). In a smaller percentage of cases, people are injured or killed and equipment destroyed. Dramatic evidence of the impact of maintenance errors was the 1979 American Airlines DC10 crash that killed 272 people. This crash was directly attributed to maintenance error. The probability of recurrence of this type of error was reduced by means of a simple component design change. Maintenance error is a problem that must be addressed in during equipment design. (Dhillon, 1986 and Taylor, 1982).

Christensen et. al. (1981) reports that a significant percentage of all equipment failures are operator error induced. In fact, human error accounted for:

1. 50% - 70% of all electronics failures.
2. 60% - 70% of all aircraft and missile failures.
3. 20% - 30% of all mechanical failures.

Christensen et. al. also reports that 20%-25% of all failures are directly traceable to maintenance errors. Maintenance requirements could be reduced through improved equipment design. A Government Accounting Office report (GAO, 1981) found that 25% of all maintenance problems to be human error induced during maintenance operations. Rasmussen (1981) reports, however, that not all maintenance errors are controlled through equipment design. In a review of underground mining equipment maintenance errors, Conway and Unger (1988) identified several categories of maintainer induced errors. The errors rates are summarized in Table 2. During this study, it was possible to identify factors contributing to maintenance related human error.

TABLE 2

Mining Equipment Maintenance Errors (Conway and Unger, 1989)

FREQUENCY	TYPE OF ERROR
I	Install Incorrect Component
I	Omitting A Component
S	Parts Installed Backwards
S	Failure to Properly Torque
S	Failure To Align, Check, Calibrate
S	Use of Incorrect Fluids, Lubricants, Greases
O	Reassemble Error
O	Failure To Seal Or Close
O	Error Resulting From Failure To Complete Task Due To Shift Change
O	Failure To Detect While Inspecting
O	Failure To Lubricate
O	Failure To Act On Indicators Of Problems Due To Workload, Priorities, or Time Constraints
O	Failure To Follow Prescribed Instructions

I = Infrequent (Less than once per year)
S = Somewhat Frequently (2 to 5 times per year)
O = Often (Over 5 times per year)

The factors include:

1. Confined work spaces - crowded equipment bays
2. Inability to make visual inspections
3. Inaccessible of components:
   a. Lube and adjustment points that could not be reached
   b. Major components that could nor be reached
5. Poor layout of components in a compartment
6. Inappropriate placement of components on machine
7. Poor/No provision for hose and cable management
8. Lack of troubleshooting guides and tools
9. Lack of installation guide pins
10. Insufficient task inspection and checkout time
11. Excessive weight of components handled

ERGONOMIC DESIGN RECOMMENDATIONS

Anthropometric Standards

Proper design for maintenance begins with proper sizing of compartments and equipment openings. This section summarizes minimal accepted anthropometric standards for the 5th through the 95th percentile user. Since the dimensions shown are for lightly clothed individuals with no miner's hat, self-rescue device, battery pack, or hand tools hanging from the belt, the values presented represent minimum dimensions. It is noted that if an access opening is in an awkward location (e.g., near the bottom or behind another machine feature or component), minimum dimension may be inadequate. Table 3 summarizes the anthropometric size standards applicable to the design of underground mining machines.

TABLE 3

Basic Anthropometric Measurements (Conway, 1986)

Prone

Prone height	A	480 mm
Prone length	B	2740 mm

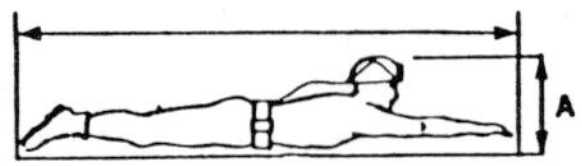

Kneeling

Dimension		
Kneeling length	C	1410mm
Kneeling height	D	1000mm

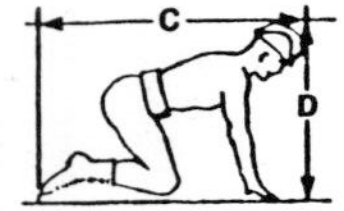

Seated (Confined)

Dimension		Men	Women
Seated height	R	973mm	8541mm
Seated length	S	900mm	[illegible]751mm

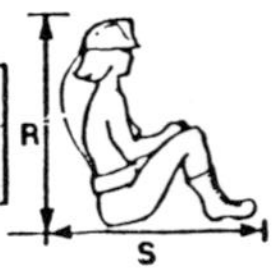

Standing

Dimension	Men			Women		
	5th	50th	95th	5th	50th	95th
Stature B	1640mm	1751mm	1861mm	1519mm	1825mm	1731mm

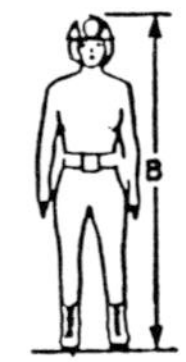

Reaching

Dimension		Men			Women		
		5th	50th	95th	5th	50th	95th
Shoulder height	D	1323mm	1427mm	1530mm	1219mm	1324mm	1429mm
Eye height	E	1526mm	1639mm	1751mm	1400mm	1508mm	1616mm
	F						
Elbow height	R	1006mm	1089mm	1171mm	912mm	993mm	1073mm
Fingertip height	S	601mm	660mm	719mm	542mm	599mm	657mm

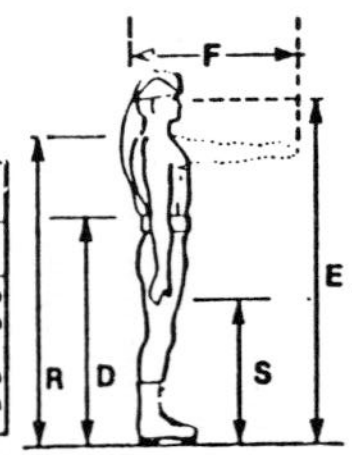

Seated (Side View)

Dimension		Men			Women		
		5th	50th	95th	5th	50th	95th
Elbow rest height	G	195mm	239mm	284mm	188mm	299mm	270mm
Shoulder height	H	549mm	604mm	658mm	516mm	570mm	624mm
Eye height	I	745mm	805mm	865mm	691mm	746mm	800mm
Sitting height	J	857mm	915mm	973mm	796mm	854mm	913mm
Thigh clearance	K	127mm	152mm	177mm	104mm	139mm	174mm
Buttock-knee length	L	548mm	597mm	645mm	525mm	577mm	628mm
Knee height	M	500mm	546mm	591mm	460mm	502mm	543mm
Stool height	N	371mm	415mm	459mm	341mm	382mm	423mm

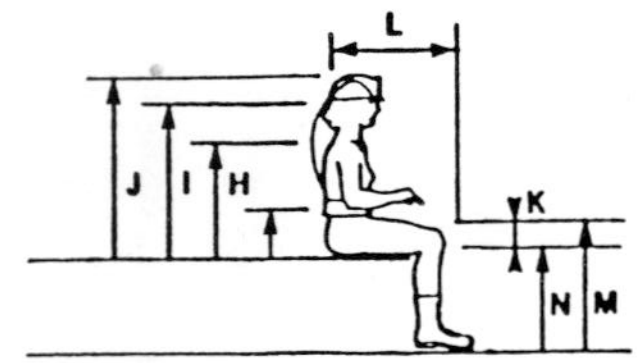

Seated (Front view)

Dimension		Men			Women		
		5th	50th	95th	5th	50th	95th
Shoulder breadth	P	420mm	482mm	504mm	374mm	419mm	464mm
Hip breadth	Q	321mm	360mm	400mm	319mm	377mm	434mm

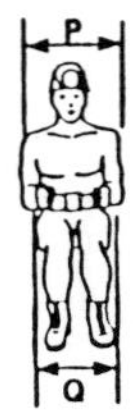

Personal Protective Equipment

Table 4 specifies supplemental space requirements for various items of personal protective equipment typically found in the underground coal mining operations. These values should be added to the basic dimensions listed in Table 3.

TABLE 4

Supplemental Access Opening Dimensions Required For Personnel Protective Equipment (Conway, 1986).

ITEM DESCRIPTION	ADD TO	HEIGHT	WIDTH	LENGTH
Miner's Hard Hat	Head	50mm (2")	50mm (2")	50mm (2")
Miner's Boots (Leather)	Foot	25-50mm	25mm (1")	25mm (1")
Miner's Boots (1/2") (Rubber)	Foot	12mm (1/2")	12mm (1/2")	12mm
Gloves (Light)	Hand	-	-	8mm (1/4")
Gloves (Heavy) (1/2")	Hand	12mm (1/2")	12mm (1/2")	12mm
Battery Pack	Waist	-	50mm (2")	-
Self Rescue Unit	Waist	-	87mm(3 1/2")	-
Flame Safety Lamp	Waist	-	50mm (2")	-
Breathing Apparatus Self-Contained	Waist/ Back	-	75-125mm (3-5")	-
Hand Tools	Waist	-	50mm (2")	-

Units in millimeter/inches

Seated And Prone Body Dimensions

Table 5 presents a summary of body lengths as a function of seated and prone postures. These dimensions are useful in the design of operator compartments, drill stations, tram stations, and protected workstations. It is noted that although the above locations are designed primarily for equipment operators, maintenance personnel must also access these locations in order to complete maintenance tasks or to tram the equipment.

Compartment Dimensions

Table 6 presents a summary of minimum and preferred compartment widths for the population of underground coal miners. It is noted that the data presented are for persons in light clothing and carrying a battery pack.

TABLE 5

Coal miner body height and length dimensions (Conway,1984).

Seated Position	Body Length (L)		Body Height (H)	
	Minimum	Preferred	Minimum	Preferred
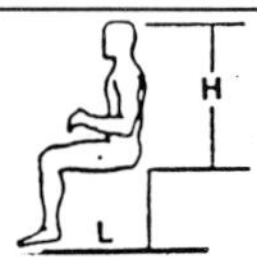	940mm 37in	1000mm 39.4in	1025mm 40.4in	1100mm 43.3in
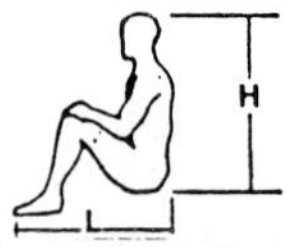	940mm 37in	1010mm 39.7in	980mm 38.6in	1100mm 43.3in
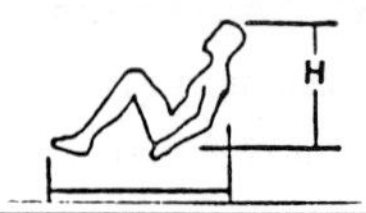	1250mm 49.2in	1400mm 55.1in	640mm 25.2in	650mm 25.6in
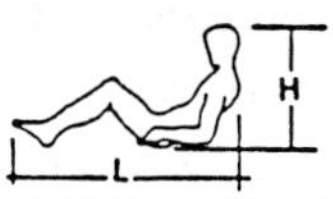	1500mm 59.1in	1600mm 63in	600mm 23.6in	620mm 24.4in
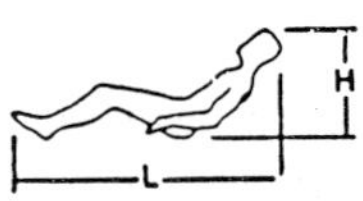	1700mm 66.9in	1800mm 70.9in	500mm 19.7in	540mm 21.3in
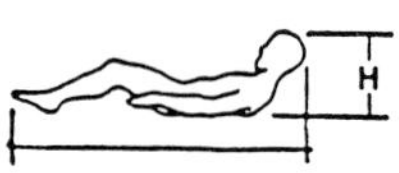	1875mm 73.8in	1950mm 76.8in	480mm 18.9in	520mm 24.5in
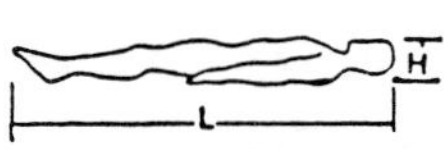	2000mm 78.7in	2100mm 82.7in	380mm 14.2in	460mm 18.1in
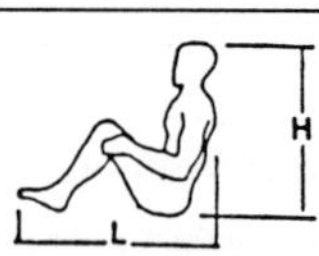	875mm 34.4in	900mm 35.4in	980mm 38.5in	1100mm 43.3in

Table 6

Recommended compartment width and height (Conway, 1984)

SEATED POSITION	MINIMUM 5th Percentile	RECOMMENDED 95th Percentile
Normal Seated	433mm (17.0")	526mm (20.7")
Crossed Leg	650mm (25.6")	750mm (29.5")
Knee To Chest	600mm (23.6")	625mm (24.6")
Supine	450mm (18.0")	560mm (22.0")

Ingress-Egress Dimensions

For machines equipped with enclosed workstations, the ingress-egress opening should permit rapid and unobstructed personnel entry and exit. Table 7 recommended entry widths and heights for three seam heights. The following recommendations are presented.

1. Operator Controls - Operating controls should be designed and located so that they cannot be kicked, bumped, or inadvertently activated by maintainers or operators during entry or exit.

2. Hand Grips - Hand grips and foot holds should be provided to facilitate entry and exit for lower seam mines.

3. Low Seam Entries - On machines designed for 48" or lower mines, the entry width should be increased by 50%, where possible.

4. Control Access - Where possible, controls, control valves, hydraulic lines, and other operator activated devices should be designed and installed to be accessible from outside the operator's compartment.

TABLE 7

Recommended entry widths and heights (Conway, 1984)

	Entry Width		Entry Height	
Height	Minimum	Preferred	Minimum	Preferred
1219mm (48.0")	590mm (22.8")	1200mm (47.2")	1180mm (46.5")	1500mm (59.1")
914 To 1219mm (36.0 - 48.0")	660mm (26.0)	1200mm (47.2")	860mm (33.9")	1050mm (41.3")
914mm And Under	660mm (26.0")	1500mm (59.1")	580mm (22.8")	860mm (33.8")

Anthropometric Safety Distances

Figure 2 illustrates recommended safe reaching distances for hand and arm access. The minimums listed are based on the assumption that the joints or body parts are in fixed contact with the machine surface and the hand or arm cannot be extended.

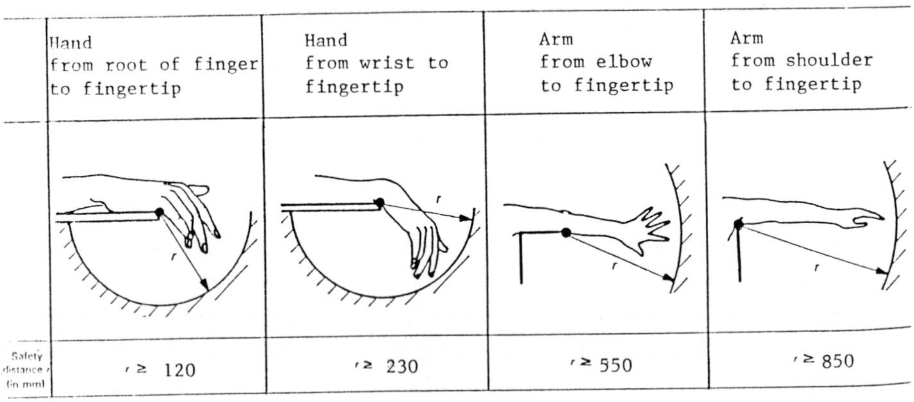

Fig. 2. Hand and arm safe reaching distances (Clark and Corlett, 1984).

It is equally important to design potential pinch points so that if the machine is inadvertently activated or maneuvered, maintenance personnel will not be trapped between moving components. Figure 3 summarized the minimum safety dimensions at potential squeeze points. These minimum dimensions can be achieved by designing the equipment so that moving components will not encroached upon or exceed these minimums or by installing permanent blocks to prevent it from doing so. For a detailed discussion, see the chapter by A. Mital on workspace design guidelines for clearances and access dimensions.

Noise Standards And Protection

CFR 30, Part 70, Subpart F (noise) outlines acceptable noise standards for underground mining operations. These standards are summarized in Table 8. At times it may be necessary for maintenance personnel to work on equipment while it is operating. The following recommendations are to assist in protecting maintenance personnel from excessive noise exposure.

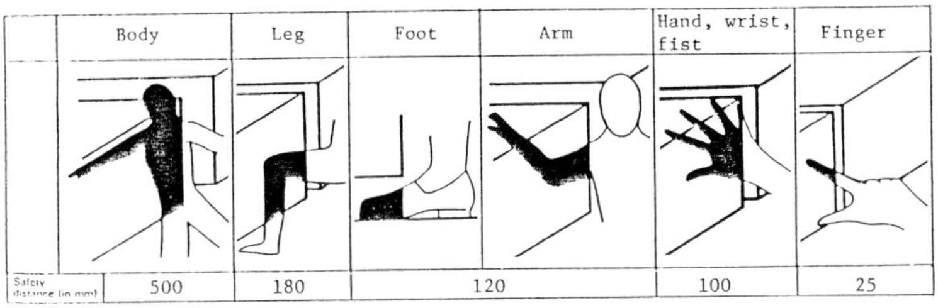

Fig. 3. Minimum safety dimensions at squeeze points (Clark and Corlett, 1984).

1. Service point location - Where possible, locate maintenance points where personnel will not be exposed to high noise sources.
2. Isolate noise source - Attempt to isolate maintenance points from direct exposure to high noise sources by positioning them behind other large components or machine parts, installing deflector shields, or otherwise protecting the maintainer.
3. Identify And Label High Noise Sources - Where isolation or relocation are not possible and noise cannot be reduced to permissible levels, the source should be identified and labeled with a warning sign.

Table 8

Permissible Noise Exposure Levels (CFR, Section 30, Part 70)

Duration Per Day (Hours)	Noise Level (dBA)	Duration Per Day (Hours)	Noise Level (dBA)
8	90	1 to 1 1/2	102
6	92	1	105
4	95	3/4	107
3	97	1/2	110
2	100	1/4 - less	

Manual Component Handling

Considerable research has been focused on the question of how much weight can a person lift (Halliday, 1977; Chaffin et al, 1977; Ayoub et. al, 1979; Snook et. al., 1974; and Ayoub et. al., 1981). Much of this research has been summarized in the National Institute Of Occupational Safety And Health (NIOSH) publication "Work Practices Guide For Manual Lifting" (NIOSH, 1981). These studies focused, for the most part, on physical strength determinations and lumbar area stress analysis and were based on the following lifting assumptions.

1. Two handed lifting.
2. Sagittal plane lifting in front of body.
3. No torso twisting.
4. Use of proper handholds.
5. Balanced loads.
6. Load less than 76cm (30") in width.
7. Person has solid footing.
8. Lifting in good environment with unrestricted posture.

The maximum load suggested by NIOSH is approximately 50 pounds being lifted not more than seven inches off the ground. In the underground equipment maintenance environment, however, it reasonable to assume that most of the nine lifting assumptions that form the foundation of this NIOSH recommendation may not be valid (Conway and Elliott, 1986). Two handed lifting in the sagittal plane, for example, is not always possible. Likewise, handles or grips are not available on mining machine components. Good footing and unrestricted posture cannot be assumed.

Labels

Function labels are used to identify specific functions or modes of operation of a component. Representative functions include:

1. Direction of Flow - Arrows or wording providing positive identification of the direction of flow of hydraulic fluids, water or other fluids through a system.

2. Amperage - Numerics indicating the level of amperage in a circuit or electrical component.

3. Pressure - Numerics indicating the operating pressure or pressure range in hydraulic or water lines, pump outputs, etc.

4. Direction Of Motion - Arrows or words indicating the direction of motion of machine components or moving components.

5. Sweep Area - symbols indicating the sweep area of pivoting or rotating components or appendages.

6. Operating Mode - signs indicating the operating mode of a machine component such as high speed-low speed.

Warning Labels

Warning Labels should be installed at locations where operator or maintainer injury may occur or where inappropriate human action may result in equipment damage. The warning labels should identify:

1. The nature of the hazard.
2. Location of the hazard.
3. Action(s) that will trigger the danger.
4. Consequences.
5. Required corrective actions.

Warning labels should be prominently visible from the position the maintainer (or operator) is likely to be in at the time of the maintenance action. It should be permanently affixed to the machine and should be visible in spite of coal dust, oil accumulation or other visibility degrading factors.

LOCATION OF MAINTENANCE POINTS

A maintenance point is defined as any location where two components are joined or where a component is mounted on the mining machine. Placement of components on a machine has a direct impact on its accessibility and serviceability. This section presents recommendations for component placement.

Primary And Secondary Maintenance Zone

The Primary Maintenance Zone (PMZ) is defined as the area from the outer edge of a machine inward a distance of 18" (Conway and Unger, 1989). The PMZ extends around the circumference of the machine and is illustrated in Figure 4. The Secondary Maintenance Zone (SMZ) is the area extending from the inner edge of the PMZ to the machine centerline.

It is recommended that all routine maintenance and service points be located within the PMZ. All service and maintenance points, including all frequently performed maintenance, should also be located in the PMZ. Only components or service points requiring infrequent maintenance should be located in the SMZ. Visual and physical access must be provided for all components.

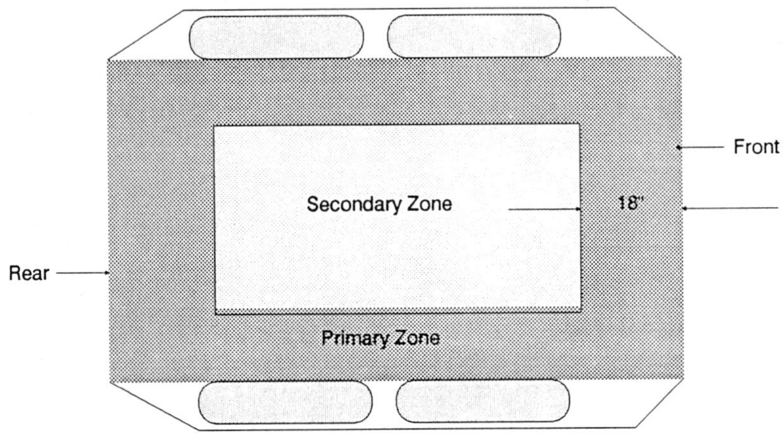

Fig. 4. Primary and secondary maintenance zones.

Tertiary Maintenance Locations

A Tertiary Maintenance Locations (TML) is any service or maintenance point located outside the PMZ or the SMZ. TMLs may include; lubrication points on a roof bolter boom, adjustments on a conveyer boom, or other remotely located service points. The Tertiary Maintenance Location (TML) should be designed to permit full visual and physical access by the maintainer.

Preferred Maintenance Point Location

Table 9 presents a summary of preferred maintenance and service point locations for optimal visual and physical accessibility (Conway and Unger, 1989). Preference is given to locating components and maintenance points in the PMZ.

ACCESSIBILITY

Visual inspection of components or systems is the most frequently performed equipment maintenance task. Accessing a suspected component to visually inspect it often accounts for a substantial percentage of maintenance task time. In other cases, visual inspections may be readily completed, but physical access to the component is often needlessly time consuming due to equipment design and layout. Visual and physical access time can be reduced if minimal design recommendations are adhered to.

TABLE 9

Recommended Equipment Accessibility Locations

Location	For Physical Access To Maintenance Points	For Visual Access To Maintenance Points
Recommended	Machine Exterior-PMZ	Machine Exterior-PMZ
	Machine Exterior-Under Fender	Machine Exterior-Under Fender
	Machine Exterior-Open Compartment	Machine Exterior-Open Compartment
Acceptable	Machine Exterior-End of Vehicle	Machine Exterior-End of Vehicle
	Machine Exterior-Closed Compartment	Machine Exterior Closed Compartment
		Machine Cavity - Opening
Least	Machine Cavity-Top	Machine Cavity-Top
Preferred	Opening	Opening
	Inside Operator's Compartment	Inside Operator's Compartment
	Machine Exterior-Behind Other Components	Machine Exterior Behind Components
Undesirable	Machine Top-Under Access Cover	Machine Top-Under Access Cover
	Inside Machine Feature (E.g.; Boom Frame)	
	Under Machine-Exterior	Under Machine - Exterior
	Interior Machine Cavity-No Direct Access	Interior Machine Cavity-No Direct Access
	Interior Machine Chassis Wall	Interior Machine Chassis Wall

Accessibility Requirements

Accessibility requirements vary with the maintenance task being performed. For instance, if the task is to visually inspect a component, a minimal visual access opening may suffice. If the task is to R/R a component, whole body access may be required.

Less rugged or hazardous components may be located under protective covers, in enclosed compartments or in more protected positions on the machine. Components should not be located behind other components that have to be removed to perform inspection.

Visual Accessibility

Visual inspection is the first step in any maintenance task. Hence, all Maintenance and Service (M&S) points on a mining machine must be visually accessible to the maintainer. Where practical, M&S points should be located in the PMZ. All non--hazardous components located in the PMZ should be directly visible to the maintainer without having to unbolt access covers or remove other components. Hazardous components, such as energized or rapidly moving parts, may be placed behind doors, protective guards or access opening covers. Likewise, brittle, dust or moisture sensitive or other components requiring physical isolation may be located behind protective covering.

All M&S points should provide immediate, direct visual access for the maintainer after protective covers have been removed. Visual access should provide line of sight inspection of all locations on the component or connection where failure may occur or where signs of performance degradation would be visually obvious. When a component cannot be located in the PMZ, it should be located on the exterior (outboard) of the machine frame where it will be directly visible to the maintainer. If a maintenance point must be located on the inboard side of the machine frame, it should be located to permit full and direct visual access from the side or end of the machine. This visual access may be through covered or uncovered visual access openings.

Visual Inspection Openings

Visual inspection openings should be installed on solid access opening covers, cabinet doors or other protective enclosures housing components requiring continuous, daily, or periodic visual inspections. The openings should be designed to permit the maintainer to directly view the M&S point from the side or end of the machine. The maintainer may be in a stooped, kneeling or crouched position while viewing, but his/her head should be in an upright position.

Visual access openings should be located at the top of side of the machine to permit ease of viewing by maintenance personnel.

Openings that must be located near the bottom of the machine side should be sufficiently large to permit viewing of the maintenance point from a kneeling or prone position.

Visual access openings should not be located on the top of machines designed for use in 60 inch or lower seams. If the visual access opening must be placed on the top of the machine, it should be located as close to the side of the machine as possible. Visual access openings should not be located under the machine. Openings should not be located behind components or machine features that restrict visibility or that have to be removed in order to visually inspect the maintenance point.

Maintenance points for routine (e.g., after shift, daily or weekly) maintenance and service tasks should be directly accessible visually. M&S points should not be located behind access plates, enclosure doors, or other components. The maintainer should be able to walk up to the mining machine and complete a visual inspection of all routine maintenance points.

Infrequently Accessed Maintenance Points

For less frequently performed maintenance tasks, maintenance points may be located under an access plate or protective cover, or behind a normally closed cabinet door. The maintenance point should, however, be directly visible when the protective cover is removed. A maximum of five minutes should be required to remove the cover or open the closed door.

Visual Access To Service Points

All service points on mining machines should be immediately visible to the maintainer as he/she approaches the machine. Routine service points should not be located in interior cavities of the machine, under bolted access covers, or behind other components that have to be removed in order to perform the servicing. Routine service points should not be located in a position that requires the maintainer to crawl under the machine or on top of the machine to complete the visual inspection or servicing.

PHYSICAL ACCESSIBILITY

Physical accessibility is the second step in any maintenance task. The degree of physical access required to perform maintenance tasks varies as a function of the task being performed.

Single or two handed access may be required if:

1. The maintenance point is located within eighteen inches of the periphery of the machine and,

2. The maintainer can observe the component being worked on.

Partial or whole body access is required for maintenance tasks if:

1. The maintenance point is located more than 18" from the outside edge of the machine or,

2. The maintenance point is located in an interior cavity.

Routine M&S points should be directly accessible by the maintainer. Accessibility can be achieved by:

1. Locating the maintenance or service point in the primary maintenance zone or by:

2. Providing access openings sufficiently large to permit the maintainer to access the M&S point.

Hand And Arm Access

If a maintenance point is not located on the exterior of a machine but it is within the PMZ, hand and arm access will be required. Visual access is required to the maintenance point while the hand and/or arm are inserted in the opening.

Whole Body Access

If the task requires the use of one or both hands to complete the maintenance action and the maintenance point is in the SMZ, whole body access may not be required. Whole body access should be from the side of the machine.

ACCESSIBILITY WITH HANDTOOLS

Components are secured to the mining machine by means of bolts secured with nuts. The bolt heads or the nuts must be tightened by hand tools. To effectively use these tools, space must be provided to manipulate them. Screw type fasteners are also used to secure components or covers to mining machines. Space must be provided to access the screw head and to turn the screwdriver itself.

Offset screwdrivers are sometimes used to access confined spaces. If this type of screwdriver is to be used, a 180 degree swing radius must also be provided for the user. Bolt heads and nuts should be offset from the side of the component or the machine wall by a minimum of 1/2 inch.

MECHANICAL OR POWERED LIFTING DEVICES

Where practical, components weighing more than 50 pounds should be installed in the primary maintenance zone. Installation in PMZ represents the most effective location with respect to removal and repair of the component. It reduces access time, permits the use of lifting and handling devices, and the use of powered hand tools. It also reduces exposure to maintenance related accidents and injuries.

For components weighing more than 50 pounds and located in the SMZ, provisions should be made to permit the use of mechanical or powered lifting devices to install or remove the component. Mechanical assistance may be in the form of a manually powered swinging arm boom, a mechanical pulley system or lifting devices.

CLEANING

Underground machines are subject to the continual build-up of coal, mud, rock, and other materials on upper surfaces, in crevices, in body cavities, and between exposed components and machine features. This accumulated debris must be removed before inspection and maintenance tasks can be completed. Proper design of this equipment can minimize the accumulation of materials at maintenance points and thereby reduce maintenance time. The following recommendations are presented.

1. Minimize Collection Points - Through design, minimize the number of locations where debris can accumulate on the surface of the machine

2. Optimize Component Location - Minimize the number of components and maintenance points on the upper surface of the machine.

3. Debris Disposal - Where practical, use no decking or flooring materials on the underside of machine cavities to permit debris to fall through the area.

4. Self-Cleaning Materials - Where protection is required on the underside, use expanded metal grating, parallel bars or other design techniques to permit debris to pass.

5. Other Features - Incorporate other self-cleaning features and designs where practical. Designs may include; surfaces inclined at an angle to pass material off, clean-out doors under areas to permit passage of materials down and out or the maintenance point area, locating mounting bolts high up on the component to minimize the need to remove all debris, and other design innovations to minimize the accumulation of waste materials.

REMOVAL AND REPLACEMENT OF PARTS

At some point in the service life of a mining machine, all components and parts will have to be repaired or replaced. It is important, therefore, that provisions be made during the design of the machine for expeditious removal and replacement. The following recommendations are presented to facilitate component replacement.

1. Accessibility - Design the component layout so that no other component must be removed to facilitate the removal of a failed component.

2. Direct Line Access - All components should be removed and replaced in a line perpendicular to it's mounting point. Heavy components should not have to be maneuvered around other components or machine features in order to remove or install it.

3. Cables And Lines - Cables, hydraulic lines, water hoses or other items should not have to be removed from non-failed components in order to remove a failed component. Likewise, machine features should not have to be disassembled in order to remove a failed component.

4. Non-Failed Components - A failed part in a major component or in a movable machine feature (for example, a bearing) should be accessible without having to remove other components or to disassemble non-failed machine features.

5. Frequently Repaired Parts - Parts subject to high stress loading or usage should be directly accessible without having to disassemble the component or major parts of the machine. In the worse case, no more than three or four steps should be required to replace a failed part.

REFERENCES

Ayoub, M. M., R. Dryden, J. McDaniel, R. Knifer, and D. Dixon. Predicting Lifting Capacity. American Industrial Hygiene Association Journal, 1979, Volume 40, pp 1075-1084.

Ayoub, M. M., N. J. Bethea, M. Bobo, C. L. Buford, D. K. Caddel, K. Intaranont, S Morrissey, and J. L. Selan. Mining In Low Seam Coal. Volume I: Biomechanics and Work Physiology (contract H0387022, Texas Tech Univ.). BuMines OFR 162(1)-83, 1981, 175pp.; NTIS PB 83-258160.

Barry, T. A. and Associates. Analysis Of Men, Materials, And Supplies Handling Systems In Underground Metal And Non-Metal Mines. (contract number S0241085, T. Barry And Associates). 1975, 163 pp.; PB-249-132.

Christensen, J. M., J. H. Howard, and B. S. Stevens. Field Experience In Maintenance, in Human Detection And Diagnosis Of System Failures. (edited by J. Rasmussen and W. B. Rouse). Plenum Press, 1981, 363 pp.

Chaffin, D. B., G. Herrin, W. M. Keyserling, and A. Garg. A Method For Evaluating The Biomechanical Stresses From Manual Material Handling Jobs. American Industrial Hygiene Association Journal, 1977, Volume 38, pp. 662-675.

Clark, T. S. and E. N. Corlett. The Ergonomics Of Workspaces and Machines - A Design Guide. Taylor & Francis, 1984, 95 pp.

Code Of Federal Regulations, Section 30, Part 70, Subpart F.

Conway, E. J., and M. Sanders. Recommendations For Human Factors Research And Development Projects In Surface Mining (contract J0395080, Canyon Research Group, Inc.). BuMines OFR211-83, 1982, 86 pp.; NTIS PB 84-143650.

Conway, E. J. Human Factors Design Guidelines For Mobile Underground Mining Equipment - Proposed SAE XJ1314 (contract No H0308110). 1984, 36 pp.

Conway, E., and W. Elliott. Mine Maintenance Material Handling: Volume I - Final Technical Report. (contract H0113018, Canyon Research Group, Inc.). U S Bureau Of Mines, Pittsburgh, Pa, 1988, 44 pp.

Conway, E. J., W. A. Elliott, and R. Unger. Mine Maintenance Material Handling: Volume II - Prototype Device Specification. (BuMines Contract H0113018), Canyon Research Group, Inc, 1988, 51 pp.

Conway, E. J., and R. Unger. Maintainability Design Of Underground Mining Equipment - Final Technical Report. U.S. Bureau of Mines Contract J0145034, U.S. Bureau of Mines, Bruceton Research Center, Pittsburgh, PA., 1988.

Conway, E. J. and R. Unger. Maintainability Design Guidelines For Underground Mining Equipment. U.S. Bureau of Mines contract J0145034, U.S. Bureau of Mines, Bruceton Research Center, Pittsburgh, Pa., 1989.

Dhillon, B. S., and H. Reiche. Reliability And Maintainability Management. Van Nostrand And Reinhold Publishers, 1985, 240 pp.

Dhillon, B. S. Human Reliability. Pergamon Press, 1986, 239 pp.

Ferguson, C. A, et. al. Ergonomics Of The Maintenance Of Mining Equipment. Institute of Occupational Medicine, Edinburgh, 1985, 55 pp.

Foster, J., D. Phillips, and T. Rodgers. Reliability, Availability And Maintainability. M/A Press, 1981, 265 pp.

Halliday, J. (Ed), Industrial Safety Handbook. Mc Graw Hill Company, 1977, 480 pp.

McBride, D. K.(Ed), Design For Maintainers. Proceedings of a Conference Hosted By Naval Air Development Center, Pensacola, Florida, 9-11 March 1982, 205 pp.

Mine Enforcement and Safety Administration. Analysis Of Injuries Associated With Maintenance And Repair In Metal And Non-Metal Mines. U S Department of the Interior, MESA Information Report No 1058, 1985, 34 pp.

National Institute Of Occupational Safety And Health (NIOSH). Work Practices Guide For Manual Lifting. DHHS/NIOSH Publication No 81-122. U S Government Printing Office, 1981, 183 pp.

Rasmussen, J. and W. Rouse (Eds). Diagnosis of System Failures. Plenum Press, 1981, 363 pp.

Sanders, M. S. and J. Peay. Human Factors In Mining. BuMines Information Circular 9182, 1988, 153 pp.

Snook, S. H. and V. M. Ciriello. Maximum Weights and Workloads Acceptable To Female Workers. Journal Of Occupational Medicine, 1974, Volume 16, pp 527-534.

Taylor, R. J., An Introduction To Error Analysis. Oxford University Press, 1982, 270 pp.

U S Government Accounting Office. Effectiveness Of U S Forces Can Be Increased Through Improved Weapon System Design. GAO Report To Congress Number PSAD-81-17. 1981, 23 pp.

Workspace, Equipment and Tool Design
edited by A. Mital & W. Karwowski

WARNING SIGNS AND LABELS IN THE WORKPLACE

MARK R. LEHTO[1] and DAVID R. CLARK[2]

[1]School of Industrial Engineering, Purdue University, West Lafayette, IN 47907, USA.

[2]Department of Industrial & Manufacturing Systems Engineering, GMI Engineering & Management Institute, Flint, MI 48504-4898, USA.

ABSTRACT

This chapter reviews the effectiveness of warning signs and labels, presents standard design guidelines, and discusses the warning design process. Major recommendations are (1) warning signs and labels should focus on providing critical information and not be indiscriminately applied, (2) in many cases, the comprehension of the warning signs and labels should be evaluated, along with the feasibility of the proposed behavior, (3) specific steps to enforce compliance should be considered, and (4) warning labels and signs should be consistent with existing standards.

INTRODUCTION

Warning signs, labels, and other forms of safety markings have long been used in attempts to increase the safety of workplaces. The primary assumption underlying this approach is that the safety-related decisions and actions people make can be improved by providing information that instructs, persuades, informs, or warns. As such, this approach assumes that people will both notice and heed these various forms of safety information.

For this very reason, the provision of safety information must be recognized as a method for supplementing, rather than replacing, design solutions that focus on removing or reducing hazards in the workplace. In general, the accepted hierarchy of hazard control from most to least effective is (1) elimination of hazards, (2) containment of hazards, (3) containment of people, (4) training of people and (5) warning of people. Warnings are at the bottom of this hierarchy because the more fundamental approaches can eliminate the need for warning.

With these caveats aside, this chapter addresses the topic of warning sign and label design. The chapter begins by addressing issues related to the provision of safety information. The following section focuses attention upon the critical issue of warning effectiveness. Sources of help and design guidelines found in safety standards are then provided in the final sections.

Issues Related to the Provision of Safety Information

From an engineering viewpoint, the primary reason to provide safety information is to reduce accidents by informing people of risks. From this point of view, the effectiveness of safety information becomes the primary and most critical issue. Several more tangential issues have, however, become prominent over the past 20 or so years.

In particular, safety information has been heavily emphasized in products liability litigation. From the view of consumer plaintiffs, the "failure to adequately warn or instruct" issue provides easier and less expensive grounds for litigation than do design defects (Twerski, et al., 1976). The warnings issue also gives employers and insurers who provide worker's compensation insurance an opportunity to shift the costs of accidents to manufacturers (Schwartz and Driver, 1983). Inspired by past experience with litigation, many manufacturers have both increased and made more explicit the warning labels they provide with their products, often with the conviction that the hazards being warned against are minor.

The dangers associated with the indiscriminate provision of warnings have been previously noted by Lehto and Miller (1986) and several other authors. One concern is that warning labels and signs may lose effectiveness, if they "cry wolf" too many times. Another, even more troublesome, possibility is that manufacturers may use warnings as a replacement for careful design, because of the present tendency to litigate on the grounds of inadequate warnings rather than design defects.

EFFECTIVENESS OF WARNING SIGNS AND LABELS

We know of only two studies demonstrating any influence on behavior due to safety signs in industrial settings and of no such studies for warning labels. In the first such study, Laner and Sell (1960) evaluated the influence of safety posters which told coal miners to hook slings. An average increase of 7.8% (a change from 37.6% to 45.4%) in hook-slinging behavior was observed when the safety posters were present. It was also observed that in shops with low ceilings, where the value of hooking slings was more easily perceived, the measured increase was 13.5% (a change from 42.2% to 55.7%). A follow-up measure taken 6 weeks later showed that the hooking behavior increased by an additional 4.5%. Interestingly, the observed change in behavior was lowest in shops where hook-slinging behavior was previously either very high or low.

A second study, summarized in **National Safety News** (1966), evaluated the effect of placing posters on the steps entering an aircraft. The three posters considered were a picture of a man holding the rail, a picture of a man stumbling, and a picture of a man sprawled at the bottom of the stairs. During the course of the experiment, 2000 passengers were observed while entering an aircraft. A 6%, 13%, and 21% increase in railing-holding behavior was associated with the above posters, respectively.

Due to the lack of research directly evaluating the behavioral effects of either warning signs or labels, it becomes necessary to take less direct approaches which focus on nonbehavioral measures often obtained in laboratory settings. The following discussion describes such an approach and organizes relevant research within this framework.

The Need for Valid Measures of Effectiveness

Just as safety communication campaigns in the 1960's were often assumed without foundation to be effective (Haskins, 1969; 1970), warning signs and labels are often alleged to be effective or ineffective on the basis of little or no research. Part of the problem is that little research has measured the effectiveness of warnings in terms of safety-related behavior. The measurement of effectiveness is also complicated by the differing functions of warning signs and labels. Many signs and labels often perform persuasive or educational functions long in advance of when the task is performed, rather than performing alerting functions during the task. It is logical to assume that signs and labels which effectively perform alerting functions might not be particularly effective as educational or persuasive tools.

To evaluate the effectiveness of warning signs and labels, it is necessary to define in a organized way the complicated process that takes place when an effective sign or label prevents accidents. An approach we have found useful is to describe the warning process in terms of (1) attention, (2) comprehension, and (3) decisions and behavior. Notably, this description implies that the effectiveness of a warning sign or label can never be greater than the probability of successfully completing any single step in the sequence.

The significance of this result becomes more clear from a simple example. Consider the hypothetical situation where (1) 50% of the population will read the warning, (2) 50% of those individuals understand the warning after reading it, (3) 100% of those individuals, who read and understood the warning, will retain and retrieve the warning from memory when they need to, (4) 90% of those individuals will act in accordance with the warning after they retrieve it, and (5) that action is sufficient to avoid the accident 90% of the time. When effectiveness is defined as the probability of successfully completing the entire sequence, it becomes the product formed by multiplying the conditional probabilities of successfully completing each separate step. This results in an effectiveness of 0.20 for this example.

Recent research has confirmed the implications of this model. For example, Otsubo (1988) observed in an experimental study of warning labels designed to convince people to wear gloves while using a circular saw, that 74% of the subjects noticed the warning, 52% said they read it, and 38% were observed to comply. For a jigsaw, 54% of the subjects noticed the warning, 25% read it, and 13% complied.

The following discussion will separately evaluate the degree to which warning signs and labels have been shown to attract attention, be comprehended, and modify decisions and behavior.

Attention to Warning Signs and Labels

It might initially seem obvious that warning signs and labels effectively expose people to warning messages. However, difficulties may occur that are related to legibility and location. Furthermore, people may filter out or consciously decide not to read a warning sign or label.

(i) Legibility. The primary focus of traditional research pertaining to the design of warning signs and labels has been on legibility. Numerous reference books (McCormick, 1976; Van Cott and Kinkade, 1972; Woodson, 1981; Westinghouse, 1981; FMC, 1980) and safety standards specify legibility requirements (as discussed later in this chapter). Although modern labels and signs will generally meet or exceed these specifications, exceptions do occur, as found in regard to the poor nighttime legibility of guide signs (Hahn et al., 1977), or the low conspicuity of exit signs in smoky buildings (Lerner and Collins, 1983). Difficult problems are also found in certain dirty environments where signs or labels become covered with oil, mud, or other contaminants and where space constraints limit the size of signs or labels.

(ii). Location. The location of signs or labels poses a greater problem. The fundamental question is if the sign or label is likely to be seen at critical times while performing a task. Ideally, the sign or label will appear prominently when the hazard is present, as an integral part of the task, at a stage in performance where the "cost" of compliance is low and the "benefit" of compliance is high. This need for selectivity is essential if the sign or label is to perform an alerting function, since there is evidence that continuous exposure can result in habituation or ignoring of the sign or label (Miller and Lehto, 1990). Additionally, if the safety information becomes available at an inopportune stage in performance (i.e., when taking the safety measure becomes inconvenient), the likelihood of compliance may be lowered. Along these lines, Wogalter et al. (1985) showed that the location of warnings in instructions can be a major determinant of effectiveness.

Numerous examples can be found where warning signs or labels are well integrated into a task, as, for example, when a lockout tag is placed next to an activated lockout, when switches are labeled on a control panel, or when a sign placed on a door warns against entering. On the other hand, many signs and labels fail this criteria, as, for example, when signs requiring use of personal protective equipment are posted in locations where the equipment is not conveniently accessed.

(iii). Filtering. People selectively attend to information, ignoring what they consider to be irrelevant. A major difficulty in designing an effective warning sign or label is to design it so that it will be attended to. Two important questions relevant to this topic are: (1) Will people read warning messages? and (2) When and why do people filter out warning messages?

Will people read warning messages? In a classic experiment reported on by Dorris and Purswell (1977), none of 100 subjects noticed the explicit warning labels placed on hammers, one of which was the warning label supplied by the manufacturer. Wright et al. (1982) studied 52 subjects using 60 different consumer products. It was found that, 34% of the time, subjects stated they would not read any of the instructions that came with a product; 53% of the time the subjects said they would read all the instructions. Fischer et al. (1989) found the average viewing time of warning information in tobacco advertisements by adolescents was only 8% of the total viewing time. In 43.6% of the cases the warning was not viewed at all. Godfrey and Laughery (1984) found that only 42% of women surveyed noticed a tampon warning when changing from one brand to another.

Several additional experiments indicate that even traffic signs are frequently filtered. Among these studies, Ruchel and Folkman (1965) found that 15% to 30% of motorists did not recall seeing forest fire safety signs. Shinar and Drory (1983) found that the average recall by motorists of the last two road signs they passed (they were stopped 200 meters away from the signs) was 4.5% and 16.5% during the day and night, respectively. Johansson and Backlund (1970) found that sign recall levels varied from 21% to 79% for motorists stopped 710 meters after passing a traffic sign.

When and why do people filter out warning messages? The above research indicates that warnings will be filtered out in many situations. The studies of traffic signs provide the clearest evidence of filtering, since Summula and Naatanen (1974) showed that motorists failed to notice only 2.95% of the passed signs when they were explicitly asked to look and then report the signs to an investigator in the back seat of the car. It should be emphasized that very significant filtering occurred in every other study where the behavior of people was observed.

These studies do not, however, determine when warnings will attract attention rather than be filtered out. Many factors can theoretically influence filtering. Among such factors are the perceived risk and importance, information overload, noise, conspicuity, and message tone. Some of the available research addressing these factors is summarized below.

Perceived Risk and Importance. Warnings may be ignored if they seem to be irrelevant to task performance. Experienced workers in particular might be more prone to ignore warning-related information because of past, benign experience, in which accidents rarely occur. Along these lines, Slovic et al. (1978) hypothesize that people may rationally ignore safety-related advice when the probability of an accident is perceived to be low.

When Wright et al. (1982) investigated the reading of instructions, they found that the tendency to read instructions increased when people were unfamiliar with a product or when a product was perceived to be complex, unsafe, or expensive. Complexity and frequency of use correlated significantly with the propensity to read instructions (r=.47 for the former and r=-.24 for the latter). Similar results were obtained by Otsubo (1988) for warning labels. Johansson and Backlund (1970), found drivers were most likely to recall seeing traffic signs perceived as being important.

These findings indicate that people are more likely to attend to warning signs and labels when they perceive the risk to be high or are unfamiliar with a product. Unfortunately, people's risk perceptions are often seriously biased (Dunn, 1972; Lichtenstein et al. 1978), leading to the conclusion that important warnings may be filtered out in some situations.

Information Overload. The number of items included in a warning sign or label and the extent of concurrent information processing activity may also influence filtering. It remains unclear how much information will result in information overload. However, there is evidence that increasing the number of items on a label or sign can cause a division of processing time among the items presented. For example, Scammon (1977) found that subjects remembered important product-related information better, when fewer items were listed on labels. It is consequently logical to selectively provide **critical** information in warning signs or labels.

Noise. Several studies have shown adverse effects of visual clutter (or noise) on the perception of signs. In particular, roadside advertisements (or other non-traffic-related signs) have been shown to influence the perception of traffic signs (Holahan, 1977; Boersema and Zwaga 1985). Holahan was able to show that traffic accidents at a "stop sign" increased with the presence of commercial signs. Boersema and Zwaga, in an experimental study, showed that the presence of advertisements interfered with the perception of routing signs.

The literature also shows that in certain instances warnings become noise that must be filtered out while performing a task. For example, Loomis and Porter (1982) discuss some of the negative experiences of pilots with false alarms provided by ground proximity warning systems (GPWS). It has also been shown that voice warnings in airplanes have distracting effects (Wheale, 1983) that may degrade pilot performance. No studies have yet explored whether warning signs or labels can act as noise, but again we have reason for recommending the selective provision of warnings.

Conspicuity. Several studies have shown that conspicuity related factors can influence the ability of people to correctly perceive safety signs. For example, nonverbal symbols can be perceived at greater distances than written text (Jacobs et al. 1975), and are also less sensitive to degradation. Such research has, however, focused on legibility rather than filtering. Among the few studies that have evaluated filtering, Strawbridge (1986) found that highlighting warning messages in instructions increased reading from 83% to 71%. Zlotnik (1982), on the other hand, found no significant differences in a similar experiment. Godfrey, et al. (1985) found that a large, very noticeable, and explicit sign warning of water contamination caused people to avoid using a water fountain to a much greater extent than did a small sign.

It is quite possible that filtering is more a function of people's search processes than conspicuity. (They attend to what they are looking for.) On the other hand, numerous studies of attention have shown that highly conspicuous signals are more likely to attract attention. For this reason, warning signs and labels should be designed to be conspicuous (numerous guidelines have been developed and are summarized later in this chapter). However, it should also be realized that there is a tradeoff between the costs and benefits of increasing conspicuity beyond certain reasonable bounds. One reason for avoiding over-conspicuous warning signs or labels is that they can be aesthetically displeasing. As Coates (1973) notes, very large deviations from expected values (as in severe color or brightness contrasts) can be unpleasant. Hypothetically, over-conspicuous warning signs or labels might be removed by workers for aesthetic reasons or might induce resistance. Although no studies have demonstrated such effects for warning signs or labels, Robertson (1975) notes that many drivers disabled annoying seat belt interlock systems rather than wearing their belts.

Warning Message Tone. Another factor which hypothetically affects the filtering of warnings is the tone of the message. More direct, explicit, or frightening warning labels for lawnmowers were selected as most effective by consumers (McGuinness, 1977). On the other hand, Sell (1977) concludes that safety posters should not involve horror, or be negative or general. The rationale for his conclusions was that (1) horror incites shock, but has no lasting effects, (2) that negative information simply shows incorrect ways of behaving, and (3) that general information is of little use because all people think they behave safely.

The Comprehension of Warning Signs and Labels

To be effective, the meaning of a warning sign or label must be correctly understood. Designers must realize that people often have difficulty comprehending both written text and symbols. Writing text that poor readers can comprehend is not an easy task. Nonverbal symbols have consequently been justified as a replacement for written text on the assumption that their meanings can be inferred by illiterate people across cultures. However, more recent research has shown that the new symbols introduced in attempts to improve comprehension may require significant learning. Furthermore, their comprehension may vary greatly between different individuals and cultures. On the other hand, relatively obscure information received in the proper context is often well understood. For these reasons, it is difficult to predict how well warning signs and labels will be comprehended. Testing of comprehension in realistic settings therefore becomes important.

The following discussion will separately consider research regarding the comprehension of (1) safety-related words and text, and (2) safety symbols.

(i). Safety-Related Words and Text. Relatively few studies have measured worker's comprehension of the words and text within warning signs or labels. Some perspective is gained by considering the estimates given by Collins et al. (1982): (1) about 5 million individuals in the United States (as of 1976) reported difficulties in speaking or understanding English; and (2) between 2 and 64 million adults in the United States are functionally illiterate (this great variation, of course, reflects differing definitions of literacy).

Other research has demonstrated more specific problems regarding the readability of safety related words and text. For example, Pyrczak and Roth (1976) evaluated the statements of "warning" and "caution" that appeared in the directions for ten aspirin-type drugs. Using the Dale-Chall readability formula they found that many samples could be read only by individuals with reading skills at or above the eleventh- or twelfth-grade level. Some of the words identified as giving difficulty are commonly used in warning labels (words such as accidental, contact, immediately, persists, conditions, consult, affecting). More serious problems were uncovered by Mrvos et al. (1986), who found that only 18.2% of the products they evaluated displayed correct and appropriate precautions and first aid information.

In attempting to resolve such problems, standard terminology has been developed by several different standard making organizations (as discussed in more detail later in this chapter). The following paragraphs will briefly address research regarding the understanding of signal words and safety phrases used in these systems.

Signal Words. An issue that has received recent attention is the use of very particular signal words to signify levels of hazard. It is assumed that using such terminology will minimize possible misperception of the significance of warned-against hazards. Particular words which have been proposed, listed in decreasing order of their signified severity, are DANGER, WARNING, and CAUTION. Several commonly referenced sources including the FMC labeling system, use this terminology, while the ANSI Z35.1 standard uses the terms DANGER and CAUTION.

One study (Bresnahan and Bryk, 1975) has addressed the perception of the terms "danger" and "caution" by industrial workers. Here it was found that greater levels of hazard were associated with the term "danger" than with "caution." It is unclear, however, if the term "warning" has a stereotypical association with levels of hazard that fall between "danger" and "caution." It is also unclear if members of the general population, who have not been exposed to industrial safety signs, view the term "danger" as implying greater hazard than the term "caution." A simpler system might use a numerical risk scale, or the phrases "extreme-danger", "serious-danger", and "moderate-danger."

Safety-Related Phrases. Safety standards (ANSI Z35.1, ANSI D6.1 , ANSI Z129.1) and other standard sources (Westinghouse, 1981; FMC, 1980) list examples of messages to be used in warning signs. The messages prescribed in such sources are often short fragments of sentences (phrases) that describe actions or conditions. It appears that such phrases should generally be easily understood.

To summarize the available findings regarding the most preferable construction of such phrases, one extensive survey of symbol comprehension (Easterby and Hakiel, 1981) indicated that descriptive signs (those that describe conditions) tend to be more well-understood than proscriptive signs (those that recommend against certain actions). Sell (1977) also recommends against proscriptive statements. Another researcher (Dixon, 1982) found that messages which recommended actions prior to describing the condition for which the action should be taken were preferable when few actions were feasible. Messages which first described the condition were preferable when many actions were feasible.

The use of concrete, rather than abstract, words within warning signs and labels (Wright, 1979) and short simple sentences constructed in the standard subject-verb-object form also have been shown to improve comprehension. Wright (1981) notes that negations and complex conditional sentences frequently create comprehension problems. More pessimistically, Kammann (1975) cites what he calls the "two-thirds rule." This rule implies that only two-thirds of written verbal material will ever be comprehended if the material is at all complex. Such conclusions imply that written safety information will often not be meaningful to a large group of individuals.

(ii). Safety Symbols. Safety symbols have been developed as an alternative means of communicating safety messages to both literate and illiterate populations, often with the conviction that such symbols are more easily perceived and understood than written text. Recently, several substantial and well-known research efforts have evaluated the comprehension of safety symbols by industrial workers. Standard making groups such as the International Standards Organization (ISO) and the Occupational Safety and Health Administration (OSHA) have commissioned such work in the hope of developing easily comprehended symbols. In these studies, it has been found that very few symbols are universally comprehended. Consequently, effective symbols have been arbitrarily defined as those which are correctly interpreted by 85% or more of the evaluated population. Several of these studies are briefly summarized below. More detailed information, rather than being reproduced here is available in Miller and Lehto (1990).

The discussion will separately consider symbol semantics, syntax, and context and learning as influences on comprehension. Symbol semantics refers to the meaning implicit within individual symbols. Symbol syntax refers to the meaning encoded by the arrangement or pattern of grouped symbols. Symbol context refers to the meaning dependent upon the environment or content within which the symbol appears.

Symbol Semantics. Several studies have evaluated the comprehension of individual symbols. The following discussion will first present some general findings.

Research on symbol associations or stereotypes will then be presented, followed by a comparison of abstract and pictographic symbol comprehension.

Several of these studies have evaluated the comprehension of symbols commonly found in consumer and public settings. Among such studies, Easterby and Hakiel (1981) tested all known symbols pertaining to fire, poison, caustic, electrical, and general hazard. Approximately 4000 consumers participated in the survey. The comprehension of the best signs was only about 20%, when the criterion of correctness was stringent. When the criterion was lax, comprehension of the best signs increased to 50%. Markedly worse performance was observed for certain signs (5% or worse with the lax criterion). Collins and Lerner (1982) investigated 25 fire-safety signs for a sample of 91 subjects. Comprehension of the symbols varied from nearly zero to nearly 100%. Green and Pew (1978) studied the comprehension of 19 pictographs used in automobiles. Only 6 of the 19 symbols met the criteria of 75% recognition and 5% errors. Easterby and Zwaga (1976) surveyed public information signs, finding large variations in understandability.

Among those studies most focused on industrial settings, Collins, et al. (1982) studied the comprehension of symbols used to convey 33 messages related to hazards, protective gear, first aid and emergency equipment, prohibited actions, and egress. The surveyed individuals consisted of 222 employees. Substantial variation was found for the evaluated symbols. For example, between 18% and 58% of the subjects correctly identified the meaning of at least some "no exit" symbols. In contrast, between 90% and 100% of the subjects correctly identified the meaning of at least some "eye protection" symbols. In a subsequent study, Collins (1983) studied 72 mine safety symbols conveying a total of 40 messages. The surveyed subjects were 267 miners located at 10 different mine sites. The results showed that 34 of the 40 messages were correctly interpreted by 85% or more of the subjects.

A variety of other experiments have evaluated the so-called natural (or stereotypical) associations between abstract symbols and concepts. Such studies seem to provide an initial basis for constructing safety messages from abstract symbols. However, it must be recognized that there much variation between individuals in their safety-related stereotypes (Easterby and Hakiel, 1977). One fairly consistent finding is that pointed shapes, such as diamonds, triangles pointing downward, or other regular figures with a vertex pointing downward, have greater hazard association values than shapes like rectangles oriented parallel to the ground or circles (Jones, 1978; Riley, et al., 1982; Collins, 1983; Cochran et al., 1981). These effects might reflect stereotypes people develop from observing traffic signs. Another commonly cited stereotype is the association between different colors and the perceived degree of danger. Bresnahan and Bryk (1975) found that industrial workers appeared to associate the colors red and yellow with greater degrees of hazard than they did for the colors green and blue. Jones (1978) further tested the importance of color cues in the comprehension of

European road signs. Removal of the red color cue associated with signs that indicated hazard had no significant effect. A negative effect was noted, however, when the blue color cue for information signs was removed. The study by Jones also indicated that (for European traffic signs) the shape cues were more important than the color cues.

In all of the above studies, there was a general trend toward poorer comprehension of abstract, as opposed to pictographic symbols. The understanding of abstract symbols was always very low for naive subjects, while pictographic symbols occasionally met a criterion of 85% correct comprehension. Although such results indicate that pictographs are easier to comprehend, research is needed in which the use and structure of pictographs, as opposed to abstract symbols, is carefully compared. Pictographs might be currently used to convey messages that are innately simpler than those conveyed with abstract symbols. This would force greater dependence upon syntax when abstract rather than pictographic symbols are used. Obviously, if abstract symbols are used to convey more complex concepts, their comprehension can be expected to be lower.

A final point is that pictographs (having a more detailed design) may be less legible than abstract symbols (Lerner and Collins, 1980). This indicates that their comprehension related advantages may be counterbalanced by poorer perception.

Symbol Syntax. In many signs, much of the information is encoded by the arrangement of symbols or equivalently by a pattern, instead of by individual symbols. There are few rules which define how to combine safety symbols into messages. There are examples along the lines of linguistics, as in the use of slashes to imply negation. Evaluation of what principles there are and how they influence comprehension corresponds to the study of symbol syntax.

Flow charts, logical trees or decision tables are one method often used to organize symbols (Wright and Reid, 1973; Kammann, 1975; Green, 1982). In certain instances, these graphical approaches have been useful for describing complex processes or procedures. For example, Kammann (1975) compared a standard telephone directory to a flowchart. Employees using the flowchart were significantly better able to understand the dialing procedures. Another study showing the successful use of flow diagrams was reported by Booher (1975). Here, it was found that simple action sequences, describing the use of a task simulator to naval personnel, were well conveyed by sequences of nonverbal symbols.

The use of pictographs is another method for graphically combining multiple symbols into messages. However, this approach often results in comprehension problems associated with determining the referents of symbols that describe actions or combine multiple meanings. In particular, Johnson (1980) discusses research where his group was unable to develop an easily comprehended pictorial method of describing to passengers (making emergency exits from planes) that they had to open either a door or a hatch. Positive conclusions were given in regard to more simple sequences. Cahill (1976) found that certain pictographs used to indicate machine elements were poorly understood. The most poorly understood symbols were used to convey actions or combine different elements of meaning. Cairney and Sless (1982b) found similar comprehension-related problems for pictographs. In particular, a symbol that indicated the availability of both gas and service for cars was commonly misunderstood. Jones (1978) provides an example of how even the prohibitive slash interacted with the type of symbol it was expected to negate. Specifically, when the symbol to be negated was abstract (shape and/or color coding only), the message was correctly identified 50% of the time. When the symbol to be negated was concrete (an arrow, pedestrian, vehicle, etc.), the proportion of correct identifications rose to 89.9%.

Symbol Context. In ordinary tasks, workers understand and react to a multitude of messages often presented in very incomplete form. Without such ability, skilled performance of most if not all tasks would be impossible. Much of this phenomenon is due to the large amount of knowledge people possess regarding everyday events and activity. In other words, people are able to accurately interpret incomplete and vague information in a familiar context. On the other hand, even very explicit information may be quite incomprehensible in an unfamiliar context. Cahill (1975) emphasizes the role of such effects on symbol comprehension. She notes the paradoxical situation where "symbols are almost superfluous for the highly experienced operator" while in-experienced subjects are unable to use almost any symbol.

The context within which information is given can generally be defined in terms of the task performed, or equivalently in terms of the product used. Past research clearly demonstrates that the context can influence the comprehension of symbols. For example, Cahill (1976) performed an experiment in which one group of subjects received a drawing of cab within which the symbols were said to be used. A control group of subjects was given the symbols alone. The subjects who received contextual information correctly identified 62% of the symbols, while the subjects in the control group correctly identified 44% of the symbols. Galer (1980) evaluated the influence of a task-related context on the comprehension of signs (by truck drivers). A comprehension rate of 71% was found when no contextual information was given. Of those drivers who did not understand the sign, 37% were able to understand it when contextual information was given.

Such results show that measures of the comprehension of symbols, as normally obtained in surveys or tests, might be artificially low, since such approaches do not usually provide the contextual information found within real tasks. On the other hand, such results also imply that explicit warning signs and labels may be adding little to the information obtained from context specific signals emitted by many products.

(iii). Learning of Symbols. Many problems in comprehension may be due to a lack of standardization of design and a lack of opportunity for people to learn correct interpretations (Collins, et al., 1982). In any case, the learning of symbols is an important issue. Among related studies, Green and Pew (1978) found that the ease of learning the meaning of pictographs was not related to initial measures of comprehension. Cairney and Sless (1982a, 1982b) found that most of the pictographs which gave trouble to their subjects during initial testing were readily learned. They also found high levels of retention (85% or better) for the majority of symbols when the subjects were retested one week later. Further research is needed to determine whether people get adequate opportunities to learn the meaning of symbols, not to mention remember them over long time intervals.

Modifying Decisions and Behavior

It is often assumed that a comprehended warning message will be heeded. This assumption is unfortunate because many additional factors complicate the transition between understanding what should be done and actually behaving safely. Among these factors, aspects of memory and decision making play an important role. Memory is important because it may be necessary for people to remember the warning message and retrieve it at the appropriate time. Decision making plays an even more central role because people must always decide whether or not to behave consistently with the warning message. Numerous questions related to memory and decision making become important when evaluating the effectiveness of warning signs or labels. Among these questions are "How well can people remember messages within such media?", "How likely are people to be educated or persuaded to behave more safely by a warning sign or label?", "What factors affect decisions by people to behave safely?" The following discussion addresses such questions.

(i). The Role of Short Term Memory. As a general rule, only 7 ± 2 items can be stored in short term memory, or consciousness. This means that no person can always be conscious of the information within a warning sign or label, no matter how motivated they are. Other items will always enter consciousness, either from long term memory or because objects or events are perceived. Consequently, to be effective a warning sign or label must cause critical information to be placed into short term memory at the appropriate time. Those warning messages that are read from a sign or label will be eliminated from short term memory as soon as task performance begins, unless the reading is an integral element of the task (as when an airplane pilot reads a checklist before leaving the ground).

In accordance with these above points, the task-specific context is a PRIMARY determinant of warning effectiveness. In other words, warnings that are well-integrated into a task-specific context are the most likely to usefully exploit the human's knowledge, since such warnings can act as cues that trigger the retrieval of additional, and hopefully relevant, information from long-term memory.

(ii). Recall of Warning Information. People appear to be only marginally able to recall the warnings or instructions given to them. As summarized below, such results have been documented for warning labels, propaganda posters, traffic signs, and to a lesser degree for symbols.

Warning Labels. Very little research documents the ability of people to remember the information given in warning labels. Among such studies, Wright (1979) discovered that less than 10% of the purchasers of antacids remembered even a portion of an in-store warning label (placed next to the antacid display) when they were questioned while leaving the store. The level of recall stayed the same even if the shoppers were observed to read the warning, and the time spent reading the label also had no effect. On the other hand, Strawbridge (1986) found that shortly after performing a task, subjects were able to recall most of the information off a product's warning label. In a more specific study, Ursic (1984) found that the use of a pictograph, the strength of a signal word, and the presence of capital letters all had no significant impact on memorability of the content of the warnings.

Propaganda Posters. Several studies have evaluated the ability of people to remember propaganda posters. Among such studies, Harper and Kalton (1966) placed two posters in a coal mine and then measured the extent to which the employees recalled and recognized the posters. One of the posters was humorous, the other serious. The findings were as follows: 18.2% of the subjects recalled and 49% recognized both posters; 52.3% recalled and 26.7% recognized neither poster; 18.3% recalled and 14.6% recognized only the humorous poster; 11.2% recalled and 9.7% recognized only the serious poster.

Belbin (1956b) tested the recall of six traffic safety posters presented to subjects who were exposed to the posters in a room for approximately three minutes. An average of about 42% of the posters were recalled. No significant differences occurred as a result of whether the type of posters was pictorial, horror, verbal positive, or verbal negative. When the subjects were retested one day after viewing the posters, recall dropped to about 28%, after one week the recall was 15%, and after two weeks the recall was about 3%.

Traffic Signs. A number of studies have measured the extent to which people remembered the traffic signs they drove past. In summary, Johansson and Backlund (1970) reported that the percentage of drivers who recalled a road sign after passing it varied from 21% to 79%, depending upon the particular sign; Shinar and Drory (1983) found percentages of 4.5% and 16.5% during the day and night, respectively.

Safety Symbol Meaning. The ability of people to remember the meaning of safety-related symbols has seen much less study than this topic deserves. The most relevant studies we found that directly evaluated the ability of people to remember the meaning of safety symbols were performed by Cairney and Sless (1982a; 1982b). During their initial testing each subject was told the meaning of the considered symbols. One week after this initial testing, their subjects still comprehended most of the symbols 85% or more of the time.

(iii). The General Influence of Knowledge, Education, and Persuasion. The knowledge people possess is frequently assumed to influence their safety-related decisions and behavior. A reasonable conclusion is that people will take safety precautions when they have a strong belief in danger and the cost of compliance is viewed to be low. Otherwise, safety precautions become less likely. Safety education and persuasion are obviously oriented toward modifying behavior by increasing the belief in danger. Research regarding each of these items is summarized below.

Belief in Danger. One of the most obvious principles is that people are more likely to take safety precautions if they believe the danger is large. This principle was confirmed in the experiment by Otsubo, (1988) where subjects were more likely to behave in accordance with a warning label when the product was perceived as being more dangerous. Perry (1983) summarizes a number of findings regarding the response of people to volcano, flood, and nuclear power plant-related warnings. The belief that real situational danger was present, as when officials or police warned them, was a very major determinant of behavior. If people didn't believe the warning (as when newspapers reported problems), they were much less likely to behave in accordance with the warning. Consistently, McGuire (1980) notes that credible sources are more likely to be persuasive. Craig (1978) also found such effects for public utility customers. Messages stated to come from the public service commission were more likely to elicit effects than those from the electrical utility.

Cost of Compliance. Several studies have shown that the likelihood of following a warning is influenced by the cost of compliance. Godfrey et al., (1985) found that only 3 out of 51 people obeyed a sign telling them to use an alternate exit fifty feet away, while 61 of 64 people obeyed a sign telling them to use the right instead of left door. Wogalter et al., (1988) found similar effects regarding the use of gloves and masks in a chemistry lab task. Fhaner and Hane (1974) found that perception of discomfort tended to outweigh people's knowledge of the effectiveness of seatbelts as a predictor of use.

From an economic view, people may weight the cost of compliance against the perceived benefit of compliance. Slovic (1978) notes that people may perceive the probability of an accident to be very low, causing other considerations to become relatively more important. If the value of behaving safely is made explicit, as when people are paid to behave safely, or penalized when they do not, safe behavior becomes more prevalent. This implies that, in the industrial setting, enforcement of compliance with warning signs and labels is critical.

Modifying Behavior. As noted by Belbin (1956a) even if people understand and are able to recall safety precautions, they may behave in conflicting ways. A particularly prevalent finding is that people's preconceived impressions regarding product safety are expressed by behavior patterns that are very difficult to modify. For example, Robertson and Haddon (1974) found that buzzers and warning lights in automobiles did not influence the use of seatbelts. Similarly, it is doubtful that warnings on cigarette packages significantly modify the decisions of smokers. Lehto and Miller (1986) summarize several additional studies evaluating educational and persuasive programs. Perhaps the most notable conclusion which can be drawn from that and other research is that educational and persuasive campaigns are often ineffective (Sell, 1977; Haskins 1969; 1970).

(iv). Skill, Ability and Training. An adequate response is the final test of a warning's effectiveness, it is also the least reliably evoked of any measure of effectiveness and most difficult to measure in a realistic setting. Warnings, to be effective, must prescribe actions that do not exceed the abilities of the receiver. To attain this goal, the abilities of the user must be carefully matched to the requirements of the action prescribed by the warning. This can be difficult because the conditions under which the recommended actions prescribed by a warning become relevant may occur infrequently. Furthermore, under such conditions the response itself may become much more difficult.

In summary, the research on effectiveness clearly supports the need for (1) selectively applying warning signs and labels, (2) focusing on critical information, and (3) enforcing compliance. Furthermore, in certain instances there may be need for evaluating conspicuity, comprehension, existing behavior patterns, and the feasibility of the proposed behavior.

GUIDELINES FOR WARNING DESIGN

During the design of warnings, it is helpful to have specific criteria which can be applied during their development. This section summarizes many such guidelines found in safety standards or the human factors literature. It begins by listing standard sources and then describes selected design guidelines in more detail. The following section describes the application of these guidelines during the development of warning signs or labels.

Sources of Design Guidelines

Warning design guidelines are primarily provided by standard making organizations that provide voluntary recommendations. Regulatory agencies also develop and mandate certain types of warnings as explicit requirements, while the courts through their decisions provide implicit requirements for warnings. Other less commonly used sources include human factors handbooks and academic research.

Voluntary Recommendations. Voluntary recommendations are developed in the private sector by several different types of organizations. This approach has several advantages in that it promotes consistency in warning designs and allows designers to make use of the substantial experience other organizations have in developing warnings.

Consensus organizations develop voluntary requirements that represent a consensus, or agreement, among its membership and the affected industries or public. This process is somewhat similar to that which the federal government follows under the rules of the APA discussed below. Notable among consensus-standard-writing organizations are: (1) those of international importance, such as the United Nations, the European Economic Community (EURONORM), the International Organization for Standardization (ISO), and the International Electrotechnical Commission (IEC), (2) national organizations, such as the American National Standards Institute (ANSI, also formerly known as the American Standards Association and the United States of America Standards Institute), the British Standards Institute, the Canadian Standards Association, the German Institute for Normalization (DIN), and the Japanese Industrial Standards Committee, and (3) independent agencies which often submit their standards to ANSI for approval as consensus standards, such as Underwriters Laboratories (UL) and the National Fire Protection Association (NFPA).

Professional organizations such as the Society of Automotive Engineers (SAE), the American Society of Mechanical Engineers (ASME), the National Safety Council (NSC), and the American Society of Safety Engineers (ASSE) as well as *Industrial trade associations* such as the National Electrical Manufacturers Association (NEMA), the Chemical Manufacturers Association (CMA), and the Material Handling Institute, (MHI) also develop recommendations for warning signs and often are the initial source of standards developed by consensus organizations.

Corporations also make extensive efforts to develop warning guidelines, and often make them generally available. Two notable examples of this type are the **Product Safety Sign and Label System** developed by the FMC Corporation (1980) and the **Product Safety Label Handbook** developed by the Westinghouse Electric Corporation (1981) Elements of the FMC system have become the basis of the new ANSI labeling standards (ANSI Z535).

Additionally, several companies provide warning signs and labels that usually conform to standards set by federal authorities and consensus organizations.

Explicit Requirements. Regulatory agencies develop explicit warning requirements, in a process governed by the Administrative Procedures Act (APA), which requires advance notification of the public of the proposed rulemaking, sufficient time for comment by the public, and consideration of these comments by the regulatory agency involved. The developed requirements can: (1) be promulgated by regulatory agencies wherein the language of concern is essentially new and included as part of the regulation itself, (2) adopt the language contained in another source, typically a consensus standard (for example: 29 CFR 145 adopted ANSI Z35.1), (3) give the language of other standards the force of law in specific situations by reference (for example: 16 CFR 1402.4(a)(1)(i)(E)(1) "The colors...shall conform to ANSI Standard

Z53.1-1971, Safety Color Code for Marking Physical Hazards,..."), or (4) refer to the source upon which the present regulation is based.

Implicit Requirements. The Common Law of Torts and the theories of Negligence and Strict Liability describe implicit requirements warnings must meet, by holding manufacturers and others liable when the failure to adequately warn is considered to be the cause of injury or loss. The *Theory of Negligence* specifically requires that the failure to adequately warn be unreasonable conduct based on (1) the foreseeability of the danger by the manufacturer, (2) the reasonableness of the assumption that a user would realize the danger and, (3) the degree of care that the manufacturer took to inform the user of the danger. The *Theory of Strict Liability* only requires that the failure to warn caused the injury or loss. These implicit requirements are, of course, highly related to the issue of effectiveness discussed earlier in this chapter.

Other Sources. Human factors handbooks are available that summarize research findings, many of which are applicable to the design of warning signs and labels. Among such handbooks, McCormick (1976), Van Cott and Kincade (1972), and Woodson (1981), all provide useful information. The technical reports of Collins et al. (1982) and Easterby and Hakiel (1977) provide much recent information concerning warning symbols. Miller and Lehto (1990) provide an annotated bibliography summarizing academic and applied research relevant to the issues of warning design and effectiveness.

Standard Systems for Sign and Label Design

Several standards-making organizations and industrial groups have developed systems for designing warning labels. These systems provide recommendations for signal words, symbols, and colors to convey hazards in a rather stereotypical and explicit way. However, among the various systems, a wide diversity of design recommendations exists (Table 1) as briefly discussed below. A more detailed discussion will follow for selected standards.

Among the most popular signal words recommended are: DANGER, to indicate the highest level of hazard; WARNING, to represent an intermediate hazard; and CAUTION, to indicate the lowest level of hazard. Color coding methods, also referred to as a "color system," consistently associate colors with particular levels of hazard. For example, red is used in all of the standards in Table 1 to represent the highest level of danger. Explicit recommendations regarding typography are given in nearly all the systems. The most general commonality between the systems is the recommended use of san serif typefaces. Varied recommendations are given regarding the use of symbols and pictographs. The FMC and the Westinghouse systems advocate the use of symbols to define the hazard and to convey the level of hazard. Conversely, the ANSI Z35.1 standard recommends symbols only as a supplement to words. Another area of

TABLE 1. Selected Warning Systems (adapted from Lehto & Miller, 1986; Westinghouse, 1981)

System	Signal Words	Color Coding	Typography	Symbols	Arrangement	Hazard ID
ANSI Z35.1 Specifications for Accident Prevention Signs (to be replaced by ANSI Z535.2)	Danger Caution	Red Yellow	Sans serif typeface. All upper case or upper and lower case.	Symbols only as supplement to words	Defines Signal word, Message, Symbol panels (optional, attached to side of label)	Not specified
ANSI Z129.1 Precautionary Labeling of Hazardous Chemicals	Danger Warning Caution Poison optional words for "delayed" hazards	Not specified	Not specified	Skull-and-crossbones as supplement to words. Acceptable symbols for 3 other hazards types.	Label arrangement not specified; examples given	Provides guidance about how to select signal words
ANSI Z535.2 Environmental and Facility Safety Signs (to replace ANSI Z35.1)	Danger Warning Caution Notice [general safety] [arrows]	Red Orange Yellow Blue Green as above; B&W otherwise per ANSI Z535.1	Sans serif, upper case, acceptable typefaces, letter heights	Symbols and pictographs per ANSI Z535.3	Defines Signal word, Word message, Symbol panels in 1-3 panel designs. 4 shapes for special use. Can use ANSI Z535.4 for uniformity.	Provides guidance
ANSI Z535.4 Product Safety Signs and Labels	Danger Warning Caution	Red Orange Yellow per ANSI Z535.1	Sans serif, upper case, suggested typefaces, letter heights	Symbols and pictographs per ANSI Z535.3; also SAE J284 Safety Alert Symbol	Defines Signal word, Message, Pictorial panels in order of general to specific. Can use ANSI Z535.2 for uniformity. Use ANSI Z129.1 for chemical hazards.	Provides guidance
NEMA Guidelines	Danger Warning	Red Red	Not specified	Electric shock symbol	Defines Signal word, Hazard, Consequences, Instructions, Symbol. Does not specify order.	Not specified

TABLE 1. Selected Warning Systems (continued)

System	Signal Words	Color Coding	Typography	Symbols	Arrangement	Hazard ID
SAE J115 Safety Signs	Danger Warning Caution	Red Yellow Yellow	Sans serif typeface, upper case	Layout to accommodate symbols; specific symbols/ pictographs not prescribed	Defines 3 areas: Signal word panel, Pictorial panel, Message panel. Arrange in order of general to specific.	Provides guidance
ISO Standard	None. 3 kinds of labels: Stop/prohibition Mandatory action Warning	 Red Blue Yellow	Message panel is added below if necessary	Symbols and pictographs	Pictograph or symbol is placed inside appropriate shape with message panel below if necessary	Not specified
OSHA 1910.145 Specification for Accident Prevention Signs and Tags	Danger Warning (tags only) Caution Biological Hazard, BIOHAZARD, or symbol [safety instruction] [slow-moving Vehicle]	Red Yellow Yellow Fluorescent Orange/Orange-Red Green Fluorescent Yellow-Orange & Dark Red per ANSI Z53.1	Readable at 5 feet or as required by task	Biological Hazard Symbol. Major message can be supplied by pictograph (Tags only). Slow-Moving Vehicle (ASAE S276/ANSI B114.1)	Signal Word and Major Message (tags only)	Provides guidance
OSHA 1910.1200 [Chemical] Hazard Communication			In English		Only as Material Safety Data Sheet	Provides guidance
	Per applicable requirements of EPA, FDA, BATF, and CPSC; not otherwise specified.					
Westinghouse Handbook; FMC Guidelines	Danger Warning Caution Notice	Red Orange Yellow Blue	Helvetica bold and regular weights, upper/lower case	Symbols and pictographs	Recommends 5 components: Signal word, Symbol/pictograph, Hazard, Result of ignoring warning, Avoiding hazard	Provides guidance about how to select signal words

substantial variation shown in Table 1 pertains to the recommended label arrangements. The proposed arrangements generally include elements from the above discussion and specify the: Image - graphic content, color; Background - shape, color; Enclosure - shape, color; and Surround - shape, color. Many of the systems also precisely describe the content and arrangement of the written text. Certain systems also provide guidance regarding methods of hazard identification.

(i). Consensus Standards. Both consensus and governmental safety standards specify ways of designing warnings, and are available to anyone who desires to consult them. These standardized designs have usually been developed for very particular products, and also tend to distinguish between industrial and consumer settings. Many consensus standards relevant to the design and application of warnings exist (see Table 2).

The major consensus standards for warning labels have come from the American National Standards Institute (ANSI). Over the last decade or so, ANSI has made a concerted effort to rethink the standards on hazard labeling. The primary standards in this area since World War II have been the ANSI Z35 series of standards on accident prevention and informational signs and tags and ANSI Z53.1, the color code. In 1979, these separate committees responsible were combined into the Z535 Committee on Safety Signs and Colors with the purpose to develop standards for the design, application, and use of signs, colors, and symbols intended to identify and warn against specific hazards and for other accident prevention purposes. The five standards developed include: (1) ANSI Z535.1 Safety Color Code (replaces Z53.1), (2) ANSI Z535.2 Environmental and Facility Safety Signs (replaces Z35.1), (3) ANSI Z535.3 Criteria for Safety Symbols, (4) ANSI Z535.4 Product Safety Signs and Labels, and (5) ANSI Z535.5 Accident Prevention Tags (replaces Z35.2). These standards specifically exclude from coverage all chemicals and chemical compounds covered under ANSI Z129.1-1988, Hazardous Industrial Chemicals – Precautionary Labeling. To illustrate the current state-of-the-art in warning systems, a summary of the provisions of ANSI Z535.1-.4 and Z129.1-1988 are included in the following sections.

ANSI Z535.1 Safety Color Code. The goal of this standard is to increase the uniformity of color coding and the spontaneity of action in times of emergency by associating specific colors with particular meanings (Table 3). The standard makes specific reference to other ANSI standards that have color specifications and provides colorimetric specifications for each of the proposed colors along with methods for accommodating color blind people. It is interesting to note that the writers of this standard intend that it be used to supplement the proper guarding or warning of hazardous conditions and never as a substitute for the reduction or elimination of a hazard when the latter approach is feasible. This prioritization is consistent with the hierarchy of hazard control discussed earlier in this chapter.

TABLE 2. Commonly-Available U.S. Consensual Standards

Designation	Topic
ANSI	
A13.1	Piping
C95.2	Radio Frequency Radiation
D6.1, 6.1b, 10.1	Traffic Control
N2.1	Radiation
N12.1	Fissile Material
Z35.1	Accident Prevention Signs
Z35.2, 535.5	Accident Prevention Tags
Z35.4	Informational Signs
Z35.5	Biological Hazard
Z53.1, 535.1	Color Code
Z129.1	Industrial Chemicals
Z138.2	Color
Z241.1	Sand Foundry Industry
Z244.1	Lock Out/Tag Out
Z535.2	Environmental and Facility
Z535.3	Symbols
Z535.4	Consumer Product
JOINT WITH ANSI	
ASAE S338.1	Towed Equipment
ASTM D56, 93	Flash Point
ASTM D1535	Color
ISA S5.5	Process Display Symbols
MH11.3	Powered Industrial Trucks
NEMA ICS1, 6	Industrial Control & Systems
NFPA 30	Flammable and Combustible Liquids
NFPA 70	Electrical
NFPA 101	Safety to Life from Fire
NFPA 178	Fire Fighting Operations
NFPA 1901	Automotive Fire Apparatus
SAE J99	Industrial Equipment on Highways
SAE J115	Safety Signs
SAE J137	Agricultural Equipment on Highways
SAE J208, 389, 841, 1170	Agricultural Equipment
SAE J284	Agricultural, Construction, and Industrial Equipment
SAE J298	Industrial Equipment
JOINT WITH ANSI (continued)	
SAE J575	Motor Vehicle Lighting
SAE J594	Reflex Reflectors
SAE J674	Motor Vehicle Glazing
SAE J725, 943	Slow-Moving Vehicles
SAE J1116	Off-Road Work Machines
SAE J1164	ROPS and FOPS
SAE J1500	Operator Controls
OTHER	
ASAE S441	Safety Signs
ASTM C1023, ES6	Ceramic Art Material
ASTM D1014	Paints on Steel
ASTM D1729, 2244, E308	Color
ASTM D1788	ABS Plastic
ASTM D2794	Organic Coatings
ASTM D3278	Flash Point
ASTM D4086	Metamerism
ASTM D4257	Coatings & Lining Industry
ASTM D4267	Parenteral Drug Containers
ASTM E42, 188, 822, G23, 26	Nonmetallic Material
ASTM E239	Paint, Varnish, Lacquer, and Related Products
ASTM E991	Fluorescent Color
ASTM ES9, F926	Kerosine Containers
ASTM F406	Play Yards
ASTM F839	Gasoline Containers
EIA RS257	Mercury
NEMA 260	Switchgear & Transformers
NEMA EW6	Arc-Welding and Cutting
NEMA IB1	Lead-Acid Batteries
NFPA 291	Fire Hydrants
NFPA 704	Fire Hazards
SAE J107	Motorcycles
SAE J179	Truck Wheel Rims
SAE J1048	Motor Vehicles
SAE J1150	Agricultural Equipment
TAPPI UM586	Label & Tape Aging Testing

TABLE 3. ANSI Z535.1 Safety Colors and Intended Meanings

COLOR	INTENDED MEANING
Red	identification of DANGER or STOP
Orange	identification of hazardous parts of machines
Yellow	identification of CAUTION; used for traffic and housekeeping markings
Green	identification of SAFETY, emergency egress, and first aid and safety equipment
Blue	identification of SAFETY INFORMATION on informational signs & bulletin boards
Brown	unassigned individual color meaning
Purple	unassigned individual color meaning
White	unassigned individual color meaning; used for traffic and housekeeping markings
Gray	unassigned individual color meaning
Black	unassigned individual color meaning; used for traffic and housekeeping markings (alone or in combination with While or Yellow)

ANSI Z535.2 Environmental and Facility Safety Signs. Considered a revision of Z35.1-1972 as well as a replacement of Z35.4-1973, the standard establishes requirements for a uniform visual system of identification related to potential hazards in the environment and provides requirements for the design, application, and use of signs and placards. This standard recognizes that to communicate information about environmental hazards and safety, it is necessary to minimize the number of layouts and design formats while still providing the flexibility to meet individual situations. It is applicable to the design of fixed and larger, movable signs as opposed to product or chemical labels, although the use of either Z535.2 or Z535.4 for both environmental and product signs/labels is not precluded if desired for the sake of consistency. An interesting provision of this standard allows the use of components of the standard, including color, size, legend, or symbols, when the standard as a whole is in conflict with requirements of a higher authority. Specifically excluded from coverage are chemicals covered by ANSI Z129.1, vehicular movement both in-plant and public, and safety bulletins and posters.

The format of a Z535.2 sign includes from 1 to 3 panels containing (1) a signal word designating the degree or level of safety alerting (see Table 4 and Figure 1), (2) a word message identifying the hazard, how to avoid it, and/or the consequences of failing to do so, and/or (3) a symbol or pictorial intended to represent a hazardous situation, precaution, and/or consequence. Standard layouts are illustrated in Figure 2. Provision is made for the use of one of 10 specific surround shapes around a symbol to convey added information about the level of the hazard; some of these surround shapes are illustrated in Figure 3.

TABLE 4. ANSI Z535.2 Hazard Classifications, Signal Words, and Indication

LEVEL	SIGNAL WORD	INDICATION
Danger	DANGER	Imminently hazardous situation, if not avoided, will result in death or serious injury
Warning	WARNING	Potentially hazardous situation which, if not avoided, could result in death or serious injury
Caution	CAUTION	Potentially hazardous situation which, if not avoided, may result in minor or moderate injury
Notice	NOTICE	Statement of company policy related directly or indirectly to the safety of personnel or protection of property
General Safety	SAFETY FIRST, BE CAREFUL, THINK, etc.	General instructions relative to safe work practices, reminders of proper safety procedures, and the location of safety equipment

Fig. 1. ANSI Z535.2 Signal Word Panels (from Clark, 1988)

ANSI Z535.3 Criteria for Safety Symbols. This completely new standard addresses the difficulty, in part, of developing easily understood written signs for a U.S. population that is multi-ethnic, highly mobile, and derived from a multiplicity of social and educational backgrounds. The standard presents an initial set of selected symbols and, more importantly, methods for designing and evaluating safety symbols to be used in applications other than those specifically addressed therein. The purpose of the standard is to promote the adoption and use of uniform and effective safety symbols. As in the case of Z535.2, this standard does not address chemical hazards. A grace period of 5 years is suggested for the conversion of existing, conflicting symbols to this new system.

Recommendations for color usage on single-panel signs include: (1) white on red for Danger, (2) black on orange for Warning, (3) black on yellow for Caution, (4) white on blue for Mandatory Action, (5) white on green, or black on white with green border for General Safety or Egress, (6) white on red, or black on white with red border for Fire, and (7) black on white with red slashed circle for Hazard Alerting Prohibition. For multi-panel signs, it is recommended that ANSI Z535.2, Z535.4, and Z535.5 be followed.

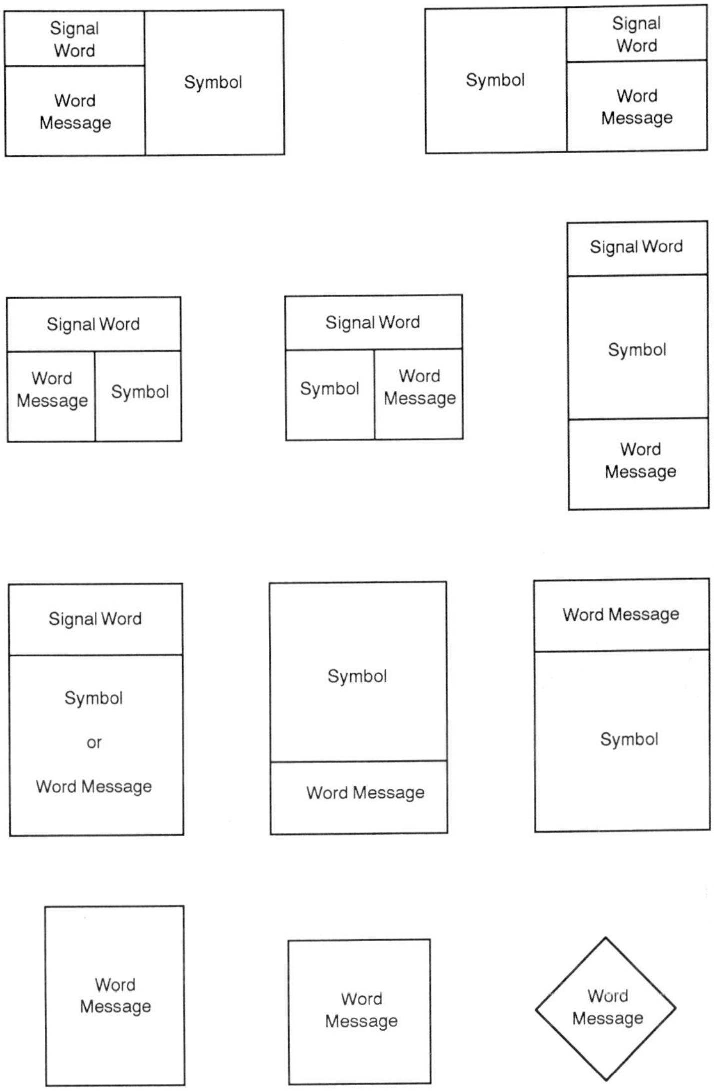

Fig. 2. Permitted Layouts of ANSI Z535.2 Signs (from Clark, 1988)

DANGER: White on Red Diamond or Tr iangle
WARNING: Black on Orange Diamond or Triangle
CAUTION: Black on Yellow Diamond or Triangle

NOTICE: White Circle on Blue Square
GENERAL SAFETY: White on Green Squ are

Prohibition At Any Level: Slashed Circle of Appropriate Color

Fig. 3. Examples of Permitted Surround Shapes in ANSI Z535.2 Signs (Clark,1988)

Example symbols are given for the categories shown in Table 5. In addition to the symbols, surround shapes are also addressed (see the discussion under ANSI Z535.2 above). Suggested surround shapes and their designated meanings are shown in Table 6. Surround shapes are intended for application on both environmental signs (Z535.2) and product signs (Z535.4).

The recommendations given in this standard for symbol selection and design are potentially more important than the symbols themselves since they are the basis for the creation of warnings with new symbols. The guidelines include that (1) symbols should have demonstrated understandability as verified by acceptable selection procedures involving appropriate test groups, (2) symbols without this demonstrated understandability should only be used in conjunction with word messages in a multi-panel signs, (3) training/recognition procedures should be conducted by employers and product manufacturers to familiarize users with the intended meaning of the symbols,

TABLE 5. ANSI Z535.3 Symbol Referents and Image Contexts

MESSAGE	REFERENT	IMAGE CONTEXT
Hazard Alerting	Flammable	Flames
	Explosion	Object Exploding
	Electrical	Lightning Bolt
	Electrical	Hand (or other parts of body) with Wire
	Corrosion	Test Tube, Hand, Drops
	Poison	Skull and Crossbones
	Slip	Person (or other parts of body) Falling on Surface
	Trip	Person (or other parts of body) Tripping over Object
	Fall	Person (or other parts of body) Falling from Surface Backward or Forward
	Entanglement	Hand (or other parts of body) and Rollers
	Pinch	Hand (or other parts of body) and Rollers
	Cut/Sever	Hand (or other parts of body) and Sharp Object
	Hot Surface	Hand (or other parts of body) and Radiating Surfaces
	Radiation*	Abstract 3-Bladed Object
	Laser*	Radiating Sunburst and Line
	Biohazard*	Abstraction without surround shape (from OSHA)
	General Alert*	Exclamation Point and Triangle (used only in signal word panel of multi-panel signs, from SAE J284)
	Radio Frequency*	Abstraction
Mandatory Action	Eye Protection	Head with Eye Glasses or Goggles
	Ear Protection	Head with Ear Protection
	Head Protection	Head with Hard Hat
	Respiratory Protection	Head with Respirator, or Head with Mask, or Airline Respirator
	Foot Protection	Safety Shoe
	Hand Protection	Two Gloves
General Safety	First Aid	Cross
	Safety Shower	Person, Shower Head, Drops
	Eye Wash	Head, Drops, Basin
	Fire Hose and Reel	Hose and Reel
	Extinguisher	Extinguisher
	Exit	Person, Open Door (consistent with arrow and true direction of exit
	Direction	Arrow (used only with another symbol, consistent with that symbol and true direction of travel)
Prohibition	No Smoking	Lighted Cigarette
	No Open Flame	Lighted Match, or Flame
	Do Not Touch	Hand, Surface
	No Exit or No Way Out	Person, Open Door
	No Entrance or No Admittance for People	Standing Person
	No Entry for Vehicle	Circle, Horizontal Bar

* Should only be used with supplementary word messages

TABLE 6. ANSI Z535.3 Surround Shapes and Designated Meanings

SHAPE	MEANING
Triangle or Diamond	Hazard Alerting
Circle	Mandatory Action
Square	Information
Slashed Circle	Prohibition

and (4) new symbols developed under these guidelines can be considered for inclusion in future revisions of the standard.

In developing new symbols, stress is placed on using (1) representational, rather than abstract symbols, (2) the U.S. Dept. of Transportation human figure symbol when a full human figure is needed, (3) views, such as front or side, shown to be most easily recognized, (4) symbols components consistent with those already included in the standard, (5) symbols depicting action, and (6) symbols involving components depicting both the human and the hazard. Additional design considerations for such things as proportion, symmetry, form, size, and placement are given.

The recommended symbol evaluation procedure includes considerations for (1) a representative target audience, including subgroups with unique problems, (2) a sample size (minimum of 50) chosen for statistical reliability and ease of testing, (3) collecting and objectively assessing meanings as perceived by the test audience, and (4) the criteria for accepting the proposed symbol. This last item specifies a minimum of 85% correct responses and a maximum of 5% critical confusions (defined as the opposite of correct). Because of the extremely variable situations that this standard must address, specific test procedures are not given for such things as age, education, experience, and visual capabilities of subjects, uniformity of test instructions, manner of presentation, size of symbols, viewing time, and specific number of subjects. Provisions are made, however, for utilizing validated alternative testing procedures.

ANSI Z535.4 Product Safety Signs and Labels. Although the design of product safety signs and labels has been influenced for many years by standards such as the ANSI Z35.1 standard for signs in industrial environments, basic differences have evolved. Product signs usually (1) are smaller, (2) contain more information, (3) are viewed from closer range, and (4) must contend with a larger variety of product backgrounds than environmental signs such as those specified by ANSI Z35.1/Z535.2.

The Z535.4 standard is designed to alert users, operators, service people, and those in proximity to the product labeled to specific hazards, of the level of hazardousness, the consequences of exposure, and how the hazards can be avoided. The purpose is to establish a uniform system across the U.S. and across different products. Other standard-writers are encouraged to adopt the standard in future revisions of their standards. Once again, chemicals covered by ANSI Z129.1 are specifically exempted as are hazards covered by mandatory regulations. Either ANSI Z535.2 or Z535.4 can be used if only one system for both environmental/facility and product labels is preferred. The effective date of this standard is left to the adopting authorities having jurisdiction.

Under ANSI Z535.4, sign classifications (level of hazard), signal words, word message, and symbols are treated the same as under ANSI Z535.2. The main differences between the two standards are the inclusion in ANSI Z535.4 of (1) permission to address more than one hazard in a given sign (the hazard level of the signal word must correspond to the worst case), (2) the mandatory use of the safety alert symbol in the signal word panel (Figure 4), and (3) different panel layouts including the use of a border contrasting with the background (the basic 3-panel layout is shown in Figure 5).

The ANSI Z535.4 standard also recommends that the symbols used in signs conform to ANSI Z535.3 and provides detailed specifications for lettering. These include (1) lettering in sans serif only, (2) signal words in upper case only, (3) word message in either upper or mixed case, and (4) signal word letter size at least 50 percent greater than word message letter size and legible at a viewing distance that allows for reasonable avoidance reaction time. The standard further specifies that sign or label must be located where it will both alert the viewer in time to take appropriate action and be readily visible.

In cases of extended expected life of the product on which signs or labels are located, the ANSI Z535.4 standard recommends that the user contact the manufacturer or some other source for replacements. Ironically, no recommendation is made that manufacturers maintain replacement stock.

Fig. 4. ANSI Z535.4 Signal Word Panels (from Clark, 1988)

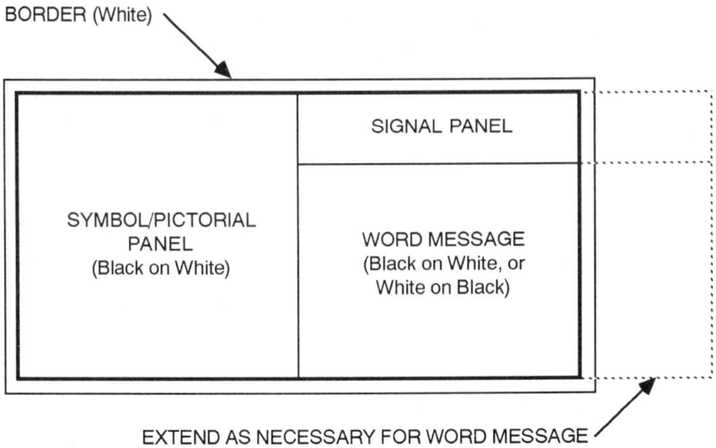

Fig. 5. ANSI Z535.4 Layout of 3-Panel Horizontal Sign (from Clark, 1988)

ANSI Z129.1-1988 Hazardous Chemical Labeling–Precautionary Labeling. An outgrowth of the Labels and Precautionary Information (LAPI) Committee of the Manufacturing Chemists Association (MCA), this guideline first appeared as Manual L-1 in 1946. It became an ANSI standard in 1976, was revised in 1982 and 1988, and is sponsored by the Chemical Manufacturers Association (CMA). This standard addresses the need to furnish appropriate precautionary information for newly developed chemicals, during the reevaluation of existing chemicals, and for the ever-widening use of chemicals and chemical processes. The intent is to provide guidelines for chemical container labels, however, it may be applicable to other forms of hazard communication. These labels are not intended to cover all information on chemical properties and handling since other means, such as Material Safety Data Sheets (MSDS), are required by OSHA for these purposes. ANSI Z129.1 labels are intended to warn of appropriate hazards, meaning those that result from reasonably foreseeable occupational use, misuse, handling, and storage. [There may be a gap in the standards here. If a chemical is primarily used in the home, the manufacturer may claim exclusion from both ANSI Z535.4 and Z129.1] A chemical is to be labeled for its reasonably foreseeable, immediate and delayed, physical and health hazards.

The components of a Z129.1 label are (1) identification of the chemical product or its hazardous component(s), (2) signal word, (3) statement of hazard(s), (4) precautionary measures, (5) instructions in case of contact or exposure, (6) antidotes, (7) notes to physicians, (8) instructions in case of fire and spill or leak, and (9) instructions for container handling and storage. To identify the chemical product, stress is placed on using common names and avoiding esoteric designations. The signal words are the same as under ANSI Z535.2 and Z535.4, but specified meanings of each are not given. The word POISON and the skull and crossbones are used in addition to the signal word when labeling highly toxic chemicals. The statement of hazard is required to give the hazard(s) present with customary or reasonably foreseeable use. It is recommended that statements identifying immediate hazards precede those for delayed hazards. Precautionary measures are defined to be those actions to be taken to avoid exposure and injury. Instructions in case of contact or exposure are generally descriptions of remedial measures that can be taken before professional medical assistance (for example, antidotes that can be administered by an untrained person) and are required to be labelled as First Aid. If antidotes must be administered by a physician, they are to be labeled as Notes to Physicians. It is further recommended that instructions in case of fire and spill or leak give information for controlling fires, confining spills and leaks, and reducing contamination. The instructions for container handling and storage are to address special or unusual handling and storage procedures. A general format for an ANSI Z129.1 label incorporating these components is shown in Figure 6. The standard also provides specific recommended texts for most of the prescribed components of the label.

While this standard recognizes the use of symbols, they are included only as an appendix. The appendix contains symbols tested in a 1986 CMA-sponsored study and shown to be acceptably valid for four hazard categories: (1) toxic, (2) corrosive, (3) flammable, and (4) explosive. Symbols for other hazards (sensitizer, oxidizer, reactive, and irritant) did not achieve acceptable scores. Usage guidelines for symbols include (1) using only recognition-validated symbols, (2) using symbols only in conjunction with

PRODUCT NAME OR IDENTIFICATION
(Identity of Hazardous Component(s), Where Appropriate)

SIGNAL WORD
STATEMENT OF IMMEDIATE HAZARD(S)
DELAYED HAZARD(S) LABEL STATEMENT

Precautionary Measures

Instruction in Case of Contact or Exposure
(First Aid Statements and Antidotes Where Appropriate)

Instructions in Case of Fire and Spill or Leak

Instructions for Container Handling and Storage

Additional Useful Statements

Net Contents

Name and Address of Company

Fig. 6. General Example of ANSI Z129.1 Label Format (from Clark, 1988)

precautionary text, and (3) limiting the symbol use to only the most serious immediate hazard(s).

(ii). Governmental Standards. There also are, of course, many governmental standards which include labeling requirements. In the U.S., a number of these are related to the FDA labeling requirements for food and drugs. The EPA has developed several labeling requirements for toxic chemicals, that are somewhat similar to those of the DOT regarding the labeling of transported hazardous materials. The CPSC provides several specific labeling requirements for consumer products, with a particular emphasis on products for children. The OSHA has incorporated new requirements for the labeling of hazardous materials in the workplace (29 CFR 1901.1200), and has recently cooperated with the NBS to conduct a study on workplace symbols (Collins, et al., 1982). Clark (1988) reviewed almost 500 warning-related regulations of 10 agencies; some of the results are summarized in Table 7.

From these findings, one can see the emergence of a set of guidelines for the design of new warnings. The effective use of the guidelines will depend upon warning designers' knowledge of particular hazards and the context in which they will be found including characteristics of the people expected to be exposed to the hazards.

The Warning Design Process

Given that a warning is to be developed, without the benefit of using existing warning systems, the steps necessary to develop a good warning might include: (1) research basic population characteristics, (2) assess the ability of warnings to affect population behavior, (3) choose the best probable or required design, (4) implement the design, (5) educate the population where necessary, i.e., where the purpose or value of the information contained in the warning is not obvious or inherent, (6) expose the population to the warning *in situ*, and (7) evaluate the effectiveness of the specific warning design.

In many instances, however, warning design is more a process of selecting an appropriate warning system and then developing a prototype consistent with the system's requirements or recommendations. This latter approach has several advantages in that it promotes consistency in warning designs and allows designers to make use of the substantial experience other organizations have in developing warnings. As noted by Clark (1988), it is extremely important that consistency be maintained unless objective research shows that effectiveness could be improved. Those warnings as they now exist in the field provide useful insights into what people's expectations are and how consistency can be achieved. Additionally, the substantial research and experience embodied in existing warning standards helps organizations who do not have the necessary resources to develop effective warnings on their own. Hence, we will focus on this aspect of warning design in this chapter. When existing

TABLE 7. Characteristics of U.S. Federal Warning Regulations (from Clark, 1988)

CHARACTERISTIC	FINDING	IMPLICATION
Type of Warning	56% health, 27% safety, 0% both, 17% neither (systems)	Individual requirements; incompatibilities not a problem
Structure vs. Content	18% structure only, 69% content only, 13% both	Can add attention-getting without violating content
Types of Information Likely to be found in Health vs. Safety Hazard Warnings	Health, degree of danger and how to remedy exposure; Safety, nature of hazard	Assumes correct response knowledge for safety hazards.
Types of Safety Hazards Covered	Fire and explosion (59%), explosion (15%), electrical (11%), mechanical (7%)	Fire warnings may provide guidelines for other safety hazard warnings
Information Grouping	Tends to be either (1) the nature of the hazard, or (2) everything else	Far-reaching assumptions about the knowledge of the reader
Types of Information Likely to be Prescribed	Nature of hazard (83%), degree of danger (57%)	This information should always be present
Instructions (what not to do to expose or what to do to avoid)	46% give instructions. More likely in warnings for public than for workers	Assume public is not knowledgeable. Workers can be trained.
Consequences or Remedies	4% on what will happen if improper action taken; 11% on how to recover	Emphasis is on avoidance, not recovery
Symbols	Prescribed by 11%. Tend to be in public environments and for short time exposures.	Use symbols where reader knowledge is not known and reading time is limited
Attention-Getting Features	Significant in transportation of hazardous substances; usually symbols.	Opportunity to broaden application
Structure, Layout, and Arrangement	Commonly found for worker warnings, not otherwise	Public likely to be exposed to inconsistent designs
Non-English Readers	<1% consider	Should be considered
Testing for Effectiveness	<1% consider	Effectiveness assumed.

systems are less appropriate, we recommend design methods such as those described in Lehto and Miller (1986).

More specifically, the approach to warning design emphasized in this chapter involves the following stages: (1) determining whether and what type of warning is required or recommended, (2) prioritization and conflict resolution between applicable warning standards or systems, and (3) development. Note that the general effectiveness based guidelines presented earlier still remain appropriate, but may be preempted in certain cases by the requirements of an existing system.

(i). <u>Determining Whether and What Type of Warning is Required</u>. In certain situations, the provision of a warning is required or mandated by law. Warnings may also be recommended in voluntary standards for which compliance is not legally required. The requirements within regulations requiring warning signs or labels are either specification or performance oriented. Those within voluntary standards are generally specifications.

Specifications set out in more or less exact terms what is to be done, including combinations of words, symbols, shapes, colors, contrasts, reflectances, sizes, fonts, locations and/or borders, within a specific context (for example: "shall have a sign with letters at least 2 inches tall saying NO SMOKING"). Since they are so explicit, they are often used by defendants in litigation, with varying degrees of success, as a defense against allegations of inadequacy or ineffectiveness.

Performance requirements, on the other hand, spell out a level of effectiveness that is to be achieved and leave varying amounts of flexibility in how it is to be achieved. A particularly unique example can be found in the Consumer Product Safety Commission labeling requirements for antenna supporting structures, which states that "no particular wording is required ..., but it shall be in legible English and understandable to a user with a sixth grade reading ability (other languages may be included as appropriate)" (16 CFR 1402.4 (a)(2)(ii)(B)). The most general example of a performance standard is the duty of a manufacturer under the Common Law (for example: products "shall not be unreasonably dangerous"). This latter guideline becomes a default criteria for selecting a warning when no existing standards provide requirements or recommendations.

(ii). Selection, Priority and Conflict Resolution. An engineer, having decided that a hazard warning is an appropriate hazard control measure, must complete at least three tasks. First, potentially relevant standards must be selected. Second, this set of standards may be reduced by assigning priority. Third, conflicting requirements and recommendations must be resolved.

Selection. The ANSI Z535 series and horizontal standards that focus on particular hazards are nearly always relevant. Determining which other standards are potentially relevant can be difficult. Over 900 safety related standards have been developed by ANSI alone, not to mention the comparable number of warning requirements in governmental regulations. Fortunately, the number of standards developed for any specific product is generally small. However, isolation of the warning provisions within such standards will often require that they be exhaustively examined.

Prioritization. Once potentially relevant standards have been selected, they should be prioritized in terms of their relative importance. In order to be consistent with the widest base of user expectation, the following prioritization scheme is suggested (1) requirements over recommendations, (2) international over national standards, (3) consensus over non-consensus standards, (4) horizontal over vertical standards, and (5) specification over performance standards.

Conflict Resolution. Assuming that several standards result in equal priority, and that their provisions are not wholly compatible, some objective method is needed to resolve the conflicts. This process may result in one or more selections being discarded or in the development of a composite design which meets the intents of the underlying standards. The process of conflict resolution requires (1) isolation of each component of a warning sign or label proposed in the selected set of standards, (2) assignment to each component its intended function, (3) selection of the best components on a function by function basis.

The components of a warning sign or label have, of course, been referred to earlier and include signal words, symbols, color, arrangement, text, and physical implementation details such as size, typography, contrast, and reflectance. Each of these components has functions devoted to either attracting attention or encoding information. Attention-related functions can be subdivided into the categories of (1) initiating attention, (2) focusing attention, and (3) maintaining attention. Information-related functions can be subdivided into either explicitly or implicitly describing (1) the degree of hazard, (2) the type of hazard, (3) the proscription or inappropriate action, (4) the prescription or appropriate action, (5) the consequences of not heeding the proscription or prescription, and (6) the remedial action to counter the consequences. Implicit descriptions assume the recipient is able to infer certain meanings from an abstract message (such as inferring the consequence "possible electrocution" from a component labeled with the message HIGH VOLTAGE). Explicit descriptions make no such assumptions for the meaning they intend to convey.

The process of selecting the best components from among the conflicting standards can be guided by (1) separately considering each of the intended functions of the warning sign and label and (2) eliminating redundant or incompatible components on a function by function basis. During this process it is important not to eliminate the required components of any mandatory regulations; recommendations within standards of long tradition or high exposure may also be justifiably given priority. On the other hand, certain choices may require research and testing concerning effectiveness.

In regard to the attention related functions of warning sign or label components, it will be necessary to choose between physical implementation details such as size, typeface, color, and shape. Decisions regarding the warning size and typeface can be guided by the required visual angle for the given application or proximity as expanded upon later. Choices of color and shape may be guided by the expected contrast such features will provide in the intended application environment. In regard to the information-related functions of warning sign or label components, it may be desirable to keep more than one component, as when text and a symbol provide the same information. On the other hand, when more than one version of text (or more than one symbol) convey the same information, it will be necessary to choose between them. Such problems may occur when one version is prescriptive and the other proscriptive, when one version uses the active voice and the other passive, when one version provides a positive statement and the other a double negative, or when one version is more explicit. The preferred choice in such situations is often the prescriptive, active, positive, or more explicit version.

(iii). Development. After completing conflict resolution, the designer will have a set of specified components. The final step is to resolve the issues of (1) unspecified components, (2) temporal and physical proximity, and (3) the medium used to provide the warning information. During the development process, it may also become necessary to perform applied research and testing to establish effectiveness, especially if the existing standards leave some of the development issues unresolved.

Unspecified Components. Attention-related components of a warning sign or label are almost always well specified by existing standards. Information-related components, on the other hand, are quite likely to remain partially unspecified after completing conflict resolution, especially when symbols, rather than text, are recommended. Perhaps the easiest solution is to explicitly specify all of the potentially relevant information. Such an approach, however, fails to recognize the need for conciseness (which becomes especially pertinent when multiple hazards are present or when users have limited viewing time) and does not take advantage of the target population's knowledge. Determining whether the target population is capable of inferring each of the intended meanings from the implicit specifications therefore may become necessary. This can be done by surveying comprehension of the warning prototypes or in some cases by consulting previous research.

Temporal and Physical Proximity. Determining when and how people contact a warning sign or label is also important, as mentioned earlier in the section on effectiveness. The concern regarding temporal proximity arises from the need to reduce filtering and forgetting of the message, by providing warnings when they will be relevant. Furthermore, if people are to respond before the hazard becomes uncontrollable, it is necessary to give adequate time for the warning to be perceived. The concern regarding physical proximity arises from the need to ensure visibility at a safe and intended viewing distance.

Addressing the issues of temporal and physical proximity may require simulation of when and how people will contact the proposed warning sign or label. Few, if any, standards describe how to do this. A limited set of studies illustrating such efforts are summarized in Miller and Lehto (1990).

Medium. The medium used to convey warning information is not necessarily limited to free standing signs and labels. Possible mediums include: (1) the product itself, (2) the permanent or semi-permanent attachments to the product, such as tags, (3) one or more of the possibly-nested containers in which the product is packaged, shipped, or stored before and/or after use, (4) unattached documents such as owners' manuals or instructions, and (5) the environment into and out of which the product moves.

Of particularly recent concern is the effective placement of warnings in instruction manuals. Clark (1988) recommends: (1) The warnings that appear on a product or in the packaging should also appear in the manual in the same form and content. (2) A summary of provided warnings should be included near the beginning of the manual.

(3) Additional instances of the warnings should appear where appropriate in the manual. (4) The order of the material within the manual should be based on the frequency of need. Although additional research and testing is necessary to validate any proposed instruction manual guidelines, a proposed order is given in Table 8. The format for completing each of the above steps should follow the **PHIPO** format given in Table 9.

TABLE 8. Proposed Organization of Instruction Manuals (adapted from Clark, 1988)

1	Product identification (therefore, manual identification)
2	Manufacturer identification and means of contact
3	Purchaser's rights
4	Use indications and contraindications including general availability of options/modifications to extend/specialize use, compatibilities and incompatibilities with previous/current products
5	Summary of warnings
6	Manual organization (table of contents)
7	Key drawing of product/major subsystems with nomenclature consistent with remainder of manual
8	Normal operation instructions
9	Normal or periodic adjustment/calibration instructions
10	Troubleshooting instructions
11	Service/repair instructions
12	Re-installation instructions
13	First-time installation instructions (note: separate unpacking instructions)
14	Repacking instructions for return for credit/repair/disposal
15	Parts lists with explored drawings using nomenclature consistent with remainder of manual

TABLE 9. PHIPO Format for Instruction Manual Components (adapted from Clark, 1988)

1	**P**urpose of step; what is to be accomplished.
2	**H**azards that may be present during step
3	**I**nput to process including product state (from previous step), operator skills and information, manufacturer-supplied parts and tools, and operator-supplied parts and materials
4	**P**rocedure including use of "you are here" drawing, detail drawing, and consistent nomenclature.
5	**O**utput of step including product state, skills and information acquired by operator, leftover parts, tools and material, and waste or by-products of procedure

CONCLUSIONS

This chapter reviews the effectiveness of warning signs and labels and presents design guidelines found in safety standards. The review of the effectiveness of warning signs and labels for industrial settings revealed:

(1) A need for selectively providing warning signs and labels because (a) continuous exposure can result in habituation and (b) people ignore warnings when they perceive the risk to be low or are familiar with a product.

(2) Warning signs and labels should be conspicuous, but not beyond certain reasonable bounds.

(3) People often have difficulty comprehending both written text and symbols. Effective symbols have been defined as those which are correctly interpreted by 85% or more of the evaluated population.

(4) Abstract symbols are often not comprehended as well as pictographs, but determining the referents of symbols within pictographs that describe actions or combine multiple meanings can be difficult. Pictographs also may be less legible than abstract symbols.

(5) The ease of learning pictographs and abstract symbols is not related to initial measures of comprehension. However, most pictographs are readily learned.

(6) User knowledge significantly influences the capability of people to comprehend symbols, as does the context within which the symbol is given. Furthermore, symbols or signals that are well-integrated into a task-specific context are most likely to usefully exploit people's knowledge. Consequently, the real-world comprehension of symbols or signals may be higher than estimated in traditional surveys.

(7) People's preconceived impressions regarding product safety are expressed by behavior patterns that are very difficult to modify. They are most likely to take safety precautions when they have a strong belief in danger and the cost of compliance is viewed to be low. When either condition is not met, a warning alone is unlikely to ensure compliance with prescribed safety precautions; compliance will have to be mandated and enforced.

(8) An adequate response is the final test of a warning's effectiveness, but least reliably evoked of any measure of effectiveness and most difficult to measure in a realistic setting. To be effective, the response prescribed by a warning must be within the abilities of the receiver.

The survey of design guidelines revealed that several consensus organizations, industrial groups, and governmental agencies have developed systems for designing warning labels. The major consensus standards have come from the American National Standards Institute (ANSI). Particularly important ANSI standards include ANSI Z535.1 Safety Color Code, ANSI Z535.2 Environmental and Facility Safety Signs, ANSI

Z535.3 Criteria for Safety Symbols, ANSI Z535.4 Product Safety Signs and Labels, ANSI Z129.1-1988 Hazardous Chemical Labeling–Precautionary Labeling.

It is further noted that the warning design process generally involves the following stages: (1) determining whether and what type of warning is required or recommended, (2) prioritization and conflict resolution between applicable warning standards or systems, (3) specification of the warning design, and (4) development. The process of conflict resolution is particularly important and requires (1) isolation of each component of a warning sign or label proposed in the selected set of standards, (2) assignment to each component its intended function, and (3) selection of the best components on a function by function basis. This approach is advantageous because it promotes consistency in warning designs.

REFERENCES

American National Standards Institute, "American National Standard Manual on Uniform Traffic Control Devices for Streets and Highways," **ANSI D6.1**, 1978.

American National Standards Institute, "American National Standard Specifications for Accident Prevention Signs," **ANSI Z35.1**, 1972.

Belbin, E., "The Effects of Propaganda on Recall, Recognition and Behaviour I. The Relationship Between the Different Measures of Propaganda Effectiveness," **British Journal of Psychology**, 47:3, August, 1956a, p. 163- 174.

Belbin, E., "The Effects of Propaganda on Recall, Recognition and Behaviour II. The Conditions Which Determine the Response to Propaganda," **British Journal of Psychology**, 47:4, 1956b, p. 259-270.

Boersema, T., and Zwaga, H., "The Influence of Advertisements on the Conspicuity of Routing Information," **Applied Ergonomics**, 16:4, December, 1985, p. 267-273.

Booher, H.R., "Relative Comprehensibility of Pictorial Information and Printed Words in Procedural Instructions," **Human Factors**, 17:3, 1975, p. 266-277.

Bresnahan, T.F., and Bryk, J., "The Hazard Association Values of Accident- Prevention Signs," **Professional Safety**, 1975, p. 17-25.

Cahill, M.C., "Interpretability of Graphic Symbols as a Function of Context and Experience Factors," **Journal of Applied Psychology**, 60.3, 1975, p. 376- 380.

Cahill, M.C., "Design Features of Graphic Symbols Varying in Interpretability," **Perceptual and Motor Skills**, 42:2, 1976, p. 647-653.

Cairney, P.T., and Sless, D., "Evaluating the Understanding of Symbolic Roadside Information Signs," **Australian Road Research**, 12:2, 1982a, p. 97-102.

Cairney, P., and Sless, D., "Communication Effectiveness of Symbolic Safety Signs with Different User Groups," **Applied Ergonomics**, 13:2, 1982b, p. 91-97.

Clark, D.R., **Model for the Analysis, Representation, and Synthesis of Hazard Warning Communication** (Ph.D. dissertation), University of Michigan, 1988.

Coates, F.D., "Human Factors and the Consumer: The Aesthetic Factor," **Proceedings of the Human Factors Society**, 17th Annual Meeting, 1973, p. 440- 443.

Cochran, D.J., Riley, M.W., and Douglass, E.I., "An Investigation of Shapes for Warning Labels," Rochester, New York, **Proceedings of the Human Factors Society, 25th Annual Meeting**, 1981, p. 395-399.

Collins, B.L., "Evaluation of Mine-Safety Symbols," **Proceedings of the Human Factors Society - 27th Annual Meeting**, 1983, p. 947-949.

Collins, B.L. and Lerner, N.D., "Assessment of Fire-Safety Symbols," **Human Factors**, 24:1, 1982, p. 75-84.

Collins, B.L., Lerner, N.D., and Pierman, B.C., "Symbols for Industrial Safety," NBSIR 82-2485, Washington, D.C., National Bureau of Standards, June, 1982.

Craig, C.S., and McCann, J.M., "Assessing Communication Effects on Energy Conservation," **Journal of Consumer Research**, 5, 1978, p. 82-88.
Dixon, P., "Plans and Written Directions for Complex Tasks," **Journal of Verbal Learning and Verbal Behavior**, 21, 1982, p. 70-84.
Dorris, A.L., and Purswell, J.L., "Warnings and Human Behavior: Implications for the Design of Product Warnings," **Journal of Products Liability,** 1,1977, p. 255-263.
Dunn, J.G., "Subjective and Objective Risk Distribution," **Occupational Psychology**, 46, 1972, p. 183-187.
Easterby, R.S., and Hakiel, S.R., "Safety Labelling of Consumer Products: Shape and Colour Code Stereotypes in the Design of Signs", A.P. Report No. 75, University of Aston in Birmingham, December, 1977.
Easterby, R.S., and Hakiel, S.R., "Field Testing of Consumer Safety Signs: The Comprehension of Pictorially Presented Messages," **Applied Ergonomics**, 12:3, 1981, p. 143-152.
Easterby, R.S., and Zwaga, H.J.G., "Evaluation of Public Information Symbols ISO Tests: 1975 Series," A.P. Report No. 60, University of Aston in Birmingham, March, 1976.
Fhaner, G., and Hane, M., "Seat Belts: Relations Between Beliefs, Attitude, and Use," **Journal of Applied Psychology**, 59:4, 1974, p. 472-482.
Fischer, P.M., Richards, J.W., Berman, E.J. and Krugman, D.M., "Recall and Eye Tracking Study of Adolescents Viewing Tobacco Advertisements", The Journal of the American Medical Association, 251:1, 1989, 84-89.
FMC, **Product Safety Sign and Label System**, 3rd ed, Santa Clara, California, FMC Corporation, 1980.
Galer, M., "An Ergonomics Approach to the Problem of High Vehicles Striking Low Bridges," **Applied Ergonomics**, 11:1, 1980, p. 43-46.
Godfrey, S.S. and Laughery, K.R., "The Biasing Effects of Product Familiarity on Consumers' Awareness of Hazard", **Proceedings of the Human Factors Society** - 28th Annual Meeting, 1984, Human Factors Society.
Godfrey, S.S., Rothstein, P.R. and Laughery, K.R., "Warnings: Do They Make a Difference", **Proceedings of the Human Factors Society** - 29th Annual Meeting, 1985, Human Factors Society, CA, .
Green, P., and Pew, R.W., "Evaluating Pictographic Symbols: An Automotive Application," **Human Factors**, 20:1, 1978, p. 103-114.
Green, T.R.G., "Pictures of Programs and Other Processes," **Behaviour and Information Technology**, 1:1, 1982, p. 3-36.
Hahn, K.C., McNaught, E.D., and Bryden, J.E., "Nighttime Legibility of Guide Signs," Research Report 50, Albany, N.Y., NY State Department of Transportation, 1977, p. 1-64.
Harper, D.G., and Kalton, G., "A Study of the Effectiveness of the National Coal Board's Safety Propaganda in Section 3 Safety Posters," Unpublished Report, Jan., 1966.
Haskins, J.B., "Effects of Safety Communication Campaigns: A Review of the Research Evidence," **Journal of Safety Research**, 1:2, 1969, p. 58-66.
Haskins, J.B., "Evaluative Research on the Effects of Mass Communication Safety Campaigns: A Methodological Critique," **Journal of Safety Research**, 2:2, 1970, p. 86-96.
Holahan, C.J., "Relationship Between Roadside Signs and Traffic Accidents: A Field Investigation," Research Report 54, Austin, Texas, Council for Advanced Transportation Studies, November, 1977, p. 1-14.
Jacobs, R.J., Johnston, A.W., and Cole, B.L., "The Visibility of Alphabetic and Symbolic Traffic Signs," **Australian Road Research**, 5:7, 1975, p. 68- 86.
Johansson, G., and Backlund, F., "Drivers and Road Signs," **Ergonomics**, 13:6, 1970, p. 749-759.
Johnson, D.A., "The Design of Effective Safety Information Displays," Poydar, H.R. (Ed.), Tufts University, Medford, Mass., **Proceedings of the Symposium: Human Factors and Industrial Design in Consumer Products**, 1980, p. 314-328.

Jones, S., "Symbolic Representation of Abstract Concepts," **Ergonomics**, 21:7, 1978, p. 573-577.

Kammann, R., "The Comprehensibility of Printed Instructions and the Flowchart Alternative," **Human Factors**, 17:2, 1975, p. 183-191.

Laner, S., and Sell, R.G., "An Experiment on the Effect of Specially Designed Safety Posters," **Journal of Occupational Psychology**, 34:3, 1960, p. 153- 169.

Lehto, M. R. and Miller, J.M., **Warnings: Volume I: Fundamentals, Design, and Evaluation Methodologies**, Ann Arbor, MI, Fuller Technical Publications, 1986, p. 286.

Lerner, N.D., and Collins, B.L., "The Assessment of Safety Symbol Understandability by Different Testing Methods," PB81-185647, Washington, D.C., National Bureau of Standards, August, 1980.

Lichtenstein, S., Slovic, P., Fischhoff, B., Layman, M., and Combs, B., "Judged Frequency of Lethal Events," **Journal of Experimental Psychology: Human Learning and Memory**, 4:6, 1978, p. 551-578.

Loomis, J.P., and Porter, R.F., "The Performance of Warning Systems in Avoiding Controlled-Flight-Into-Terrain (CFIT) Accidents," **Aviation, Space, and Environmental Medicine**, 53:11, 1982, p. 1085-1090.

McCormick, E.J., **Human Factors in Engineering and Design**, 4th edition, N.Y., McGraw-Hill, 1976.

McGuinness, J., "Human Factors in Consumer Product Safety," **Proceedings of the Human Factors Society-21st Annual Meeting**, 1977, p. 292-294.

McGuire, W.J., "The Communication-Persuasion Model and Health-Risk Labeling," in **Product Labeling and Health Risks**, Banbury Report 6, Morris, L.A.,Mazis, M.B., and Barofsky, I. (Ed.), Cold Spring Harbor Laboratory, 1980, p. 99-122.

Miller, J.M. and Lehto, M.R., **Instructions and Warnings: The Annotated Bibliography**, Ann Arbor, MI, Fuller Technical Publications, 1990, p. 395.

Mrvos, R., Dean, B.S. and Krenzelok, E.P., "An Extensive Review of Commercial Labels...The Good, Bad and Ugly", **Veterinary and Human Toxicology**, 28:1, 1986, 67-69.

National Safety News, June, 1966.

Otsubo, S.M., "A Behavioral Study of Warning Labels for Consumer Products: Perceived Danger and Use of Pictographs", Proceedings of the Human Factors Society - 32nd Annual Meeting, 1988, Human Factors Society.

Perry, R.W., "Population Evacuation in Volcanic Eruptions, Floods, and Nuclear Power Plant Accidents: Some Elementary Comparisons," **Journal of Community Psychology**, 11, 1983, p. 36-47.

Pyrczak, F., and Roth, D.H., "The Readability of Directions on Non- Prescription Drugs," **Journal of the American Pharmaceutical Association**, NS 16:5, 1976, p. 242-243,267.

Riley, M.W., Cochran, D.J. and Ballard, J.L., "An Investigation of Preferred Shapes for Warning Labels," **Human Factors**, 24:6, 1982, p. 737-742.

Robertson, L.S., "Safety Belt Use in Automobiles with Starter-Interlock and Buzzer-Light Reminder Systems," **American Journal of Public Health**, 65:12, 1975, p. 1319-1325.

Robertson, L.S., and Haddon, W., "The Buzzer-Light Reminder System and Safety Belt Use," **American Journal of Public Health**, 64:8, 1974, p. 814-815.

Ruchel, G.J., and Folkman, W.S., "Roadside Fire Prevention Signs - Standard and New Designs Compared," U.S. Fire Research Note PSW 65, 1965.

Scammon, D.L., "'Information Load' and Consumers," **The Journal of Consumer Research**, 4, 1977, p. 148-155.

Schwartz, V.E., and Driver, R.W., "Warnings in the Workplace: The Need for a Synthesis of Law and Communication Theory," **Cincinnati Law Review**, 52, 1983, p. 38-83.

Sell, R.G., "What does Safety Propaganda do for Safety? A Review," **Applied Ergonomics**, H77, p. 203-214.

Shinar, D., and Drory, A., "Sign Registration in Daytime and Nighttime Driving," **Human Factors**, 25:1, 1983, p. 117-122.

Slovic, P. "The Psychology of Protective Behavior," **Journal of Safety Research**, 10:2, Summer, 1978, p. 58-68.

Strawbridge, J.A., "The Influence of Position, Highlighting, and Imbedding on Warning Effectiveness", **Proceedings of the Human Factors Society - 30th Annual Meeting**, 1986, Human Factors Society.

Summala, H., and Naatanen, R., "Perception of Highway Traffic Signs and Motivation," **Journal of Safety Research**, 6:4, 1974, p. 150-154.

Twerski, A.D., Weinstein, A.S., Donaher, W.A., and Piehler, H.R., "The Use and Abuse of Warnings in Products Liability: Design Defect Litigation Comes of Age," **Cornell Law Review**, 61, 1976, p. 495.

Uršic, M., "The Impact of Safety Warnings on Perception and Memory," **Human Factors**, 26:6, December, 1984, p. 677-682.

Van Cott, H.P., and Kinkade, R.G., **Human Engineering Guide to Equipment Design**, Washington, D.C., U.S. Government Printing Office, 1972.

Westinghouse Electric Corporation, **Product Safety Label Handbook**, Trafford, PA, Westinghouse Printing Division, 1981.

Wheale, J.L., "Evaluation of an Experimental Central Warning System with a Synthesized Voice Component," **Aviation, Space, and Environmental Medicine**, 1983, p. 517-523.

Wogalter, M.S., Fontenelle, G.A. and Laughery, K.R., "Behavioral Effectiveness of Warnings", **Proceedings of the Human Factors Society** - 29th Annual Meeting, 1985, Human Factors Society,

Wogalter, M.S., McKenna, N.A. and Allison, S.T., "Warning Compliance: Behavioral Effects of Cost and Consensus," **Proceedings of the Human Factors Society** - 32th Annual Meeting, 1985.

Woodson, W.E., **Human Factors Design Handbook, Information and Guidelines for the Design of Systems, Facilities, Equipment, and Products for Human Use**, N.Y., McGraw-Hill, 1981.

Wright, P., "Concrete Action Plans in TV Messages to Increase Reading of Drug Warnings," **Journal of Consumer Research**, 6, 1979, p. 256-269.

Wright, P., "'The Instructions Clearly State...' Can't People Read?," **Applied Ergonomics**, 12:3, 1981, p. 131-141.

Wright, P., Creighton, P., and Threlsall, S.M., "Some Factors Determining When Instructions Will Be Read," **Ergonomics**, 25:3, 1982, p. 225-237.

Wright, P., and Reid, F., "Written Information: Some Alternatives to Prose for Expressing the Outcomes of Complex Contingencies," **Journal of Applied Psychology**, 57:2, 1973, p. 160-166.

Zlotnik, M.A., "The Effects of Warning Message Highlighting on Novel Assembly Task Performance," **Proceedings of the Human Factors Society-26th Annual Meeting**, 1982, p.93-97.

Workspace, Equipment and Tool Design
edited by A. Mital & W. Karwowski

WORK DESIGN FOR OLDER ADULTS

SARA J. CZAJA[1,2]

[1]Industrial Engineering Department, State University of New York at Buffalo, Buffalo, New York, 14260
[2]Stein Gerontological Institute, 151 NE 52nd St., Miami, Florida, 33137 (United States)

ABSTRACT

Current demographic trends underscore the need to understand the implications of an aging population for job and workplace design. The aging of the population coupled with the declining birthrate implies that organizations will need to utilize older adults to perform work activities. Also high rates of unemployment among older people would create a significant increase in the economic dependency burden thus it is important to identify strategies which delay early retirement and encourage active labor force participation among older adults. This chapter will present an overview of the aging process and provide specific examples of the potential impacts for job performance. In addition, examples will be given to illustrate how Human Factors Engineering can make contributions towards successfully integrating older workers into organizational settings. These examples, will encompass job and workplace design and retraining.

OVERVIEW/EMPLOYMENT TRENDS

Current trends in the demographic structure of the population underscores the need to understand the implications of an aging population for job and workplace design. In 1900, the number of persons over age 65 in the United States numbered approximately 3.1 million; by 1978 this number increased to 24 million. Recent census estimates suggest that by the year 2000 people in this age group will number at least 32 million and represent 12 percent of the population (OTA, 1985). This change in the composition of the population has direct implications for the economy and the work force.

Considering the employment patterns of older adults, recent trends demonstrate that the number of people who are gainfully employed decreases beyond age 55. The labor force participation rates of older people are declining, especially for older men. It is projected that by 1995 workers aged 55 and older will only represent 11% of the labor force as compared to 14% in 1981. The declining labor force participation rates are expected to continue into the next decade among older people. Also, in 1980 56 percent of all male and 69 percent of all female beneficiaries were receiving benefits for early retirement and this trend is expected to continue.

The decline in work activity among older persons creates several problems. One is an increase in the economic dependency burden created by the large number of unemployed older people. It will be costly to maintain these people in society especially since Social Security benefits have an uncertain future. High rates of unemployment may also have a negative impact on the quality of life of older people. Economic security is important to independence and self-esteem. This is especially true for the upcoming cohorts of elderly who are more active, healthier and better educated than previous generations.

Given these trends one important task is to understand why work activity rates among the elderly are declining. General reasons for the decline in work activity among older people include: early retirement options; poor health; forced retirement, age discrimination, burnout, skill obsolescence and desire for leisure. A second task is to identify strategies which delay early retirement and encourage labor force participation of older people. These strategies may involve job or workplace redesign, retraining or alternative work schedules such as part time work. Several surveys of older people have reported a preference for part time employment (Robinson, Coberly and Paul, 1985).

Another reason that we need to consider the aging process in job and workplace design is that the age composition of the labor force is changing. In the next few decades the labor force will consist largely of middle-aged people. During the 1980's and 1990's the "baby boom" generation will move into middle age and by the year 2000 the number of middle aged workers will increase by 79%, (OTA, 1985). This trend will be accompanied by a decline in the number of younger workers. Estimates indicate that the number of workers aged 16-24 years will drop 8 percent by the year 2000. This implies a labor shortage especially for entry level positions.

One way to address this problem is to encourage older people to remain in the work force or re-join the labor market after retirement. A small number of companies have adopted programs to do so. These programs include: retiree re-employment; older worker recruitment; retraining; job sharing; flexible work hours, part time work and job redesign (Jessup and Greenburg, 1989). However, in order for these programs to be successful, we need to have a clear understanding of the implications of the aging process for job performance. This type of knowledge is needed in order to be able to identify specific strategies for promoting employment opportunities for older people.

When considering aging and work it is important to do so in the context of technology. During the past decade computer and communication technologies and other forms of automation such as robotics have been introduced into most occupational settings. For example, currently, it is estimated that one out of every three white collar workers uses a computer on a regular basis. The prevalence of white collar jobs is rising and will continue to do so during the next decade. Specific occupations which will experience the fastest growth are clerical workers, computer technicians, operators and programmers and other service providers such as bank tellers and clerical workers. Most of these occupations involve the use of computer technology. Computer-interactive tasks are also becoming prevalent within the manufacturing, chemical and nuclear power industries. For example, computer-aided manufacturing (CAM) has become, the most important technology in the manufacturing industry. CAM involves both direct and indirect applications of computers in the manufacturing process. In other words there are dramatic changes in the way traditional production jobs are being performed.

In sum, most people will need to interact with some form of computer technology to perform their job. One important question to consider is how this influx of technology impacts on the work life of the elderly as technology changes the nature of work. In some cases jobs are eliminated and new ones are created and in other cases the context or the demands of the job are altered. These changes need to be evaluated with regards to older adults.

Technology may have both positive and negative impacts on the employment status of older people. For example, in many cases, technology reduces physical demands. associated with jobs and in this sense promotes employment opportunities for older people. Declines in health and physical capabilities are a common reason for early retirement. Computer technology also makes work at home a more likely option. This is beneficial for older people who find it difficult to leave home because of health or transportation problems. Older adults frequently report problems accessing services or attending events because of problems with restricted mobility (Burrus-Bammel and Bammel, 1985). Finally computer tasks, such as data entry tasks, are amenable to part time work which is a preferred option for many older people.

Technology may also have negative impacts on the employment status of older people. One obvious problem is that of skill obsolescence and displacement (Miller, 1983). Older people will need to acquire new skills to both successfully re-enter the labor market and to interact with new equipment on the job. Technology creates a need to learn new work methods as well as computer skills.

The manner in which technology changes jobs may also be detrimental to older people. Although computer technologies reduce the physical demands of jobs, in many cases the information processing demands are increased. Two important ways computers may change jobs is in terms of pacing requirements and cognitive complexity. This may be problematic for older workers given the changes in cognitive functioning such as declines in response speed which occur with age. In addition to creating performance problems, these changes in job demands may also be stress inducing. The potential impact of job stress has been well documented. Job stress has been linked to adverse physiological and psychological outcomes. It is postulated that the impact of job stress may be more consequential for older workers because of age related declines in physiological systems and adaptive capacities (Czaja and Sharit, 1988).

In order to insure that employment opportunities are available for older workers, it is important to understand the new skill requirements of tasks so that jobs can be matched with worker capabilities. Currently, knowledge about the impact of technology on an aging work force is limited. Yet, these issues are critical to consider in order to successfully prepare for an impending older society.

AGING AND PERFORMANCE

There is vast literature regarding developmental changes in functional capabilities. For a more detailed discussion of these issues, refer to: Charness, 1985; Birren and Schaie 1985; or Botwinick, 1984. In this section the process of aging will be briefly reviewed in order to illustrate the importance of considering adult development for job and work place design.

Of the findings regarding aging and performance there are several which are relevant to work activities. These include declines in visual and auditory functioning and strength; a general slowing of behavior and changes in perception and cognition. Before discussing these changes, it is important to point out that the elderly as a group are more heterogeneous than younger people with respect to health, social, economic and other characteristics. Although aging is a continuous process, it is non-uniform, which means that the age at which functional changes occur and the rate of change varies considerably across individuals. Thus chronological age is an over simplistic predictor of a person's functional and behavioral
capacity (Birren and Cunningham, 1985).

Although, there has been a vast amount of research regarding the aging process little has been done with respect to translating this information into guidelines for design. In the sections which follow design guidelines for older adults will be presented where ever possible. It should be recognized, that this type of data is limited and the accuracy of some of the data is unknown. Clearly, this is an area where Human Factors Specialists can make important contributions.

SENSORY PROCESS

Age-related changes in visual functioning have significant implications for the design of jobs and work places. Aging of the visual system involves changes in the optics of the eye as well as alterations in the visual nervous system. Although most older people will not experience severe visual impairment they will encounter decline sufficient to influence task performance.

In fact, a recent survey (Kosnick, Winslow, Kline, Rosinski and Sekuler, 1988) of healthy adults ranging in age from 18 to 100 yrs. found that declines in visual functions increased two to sixfold across the adult lifespan. Specifically, the older adults reported that they had problems with glare, dim illumination and near vision tasks. In addition, they reported trouble tracking moving targets, locating a target in a cluttered visual scene and need more time to perform visual tasks.

In general, age-related losses in visual function include: reduced static and dynamic acuity; constriction of the visual field; a loss of accommodation or focusing ability; loss of contrast and color sensitivity and dark adaptation ability and increased susceptibility to glare. Also, the illumination requirements for satisfactory visual performance increase with advanced age. (Kline and Schieber, 1985).

Obviously, these changes in vision have important implications for work activities as most tasks have a large visual component. For example, older people may require larger characters on screens or different color combinations between the characters and background to successfully interact with computer technologies. Changes in vision may also impact on tasks which require the discrimination of detail such as visual inspection or tasks which involve manipulating vehicles such as piloting or driving. Also, older workers may require different types and levels of lighting to efficiently perform jobs. For example, Hughes and McNelis (1975) found that the performance of older clerical workers improved under high levels of illumination. However, appropriate lighting for the elderly is not as simple as adding more light as older adults have heightened susceptibility to glare.

Unfortunately, our knowledge of the practical implications of vision losses with age is limited. The National Research Council (1987) points out that there is a need for research directed towards understanding the relationship between age related changes in vision and job performance. Research is also needed to identify and evaluate the suitability of intervention strategies to compensate for visual declines.

Some existing guidelines for designing visual environments to accommodate the "aging eye" include: increased levels of illumination, avoidance of sudden shifts in illumination, avoidance of glare, elimination of difficult color discrimination (e.g., blue, green range or colors of the same hue), increased letter size (font size of 10 or 12) and increased contrast for targets (Charness and Bosman, 1989). Table 1 presents some guidelines for designing the visual and auditory environments to compensate for age related declines in acuity.

Changes in auditory functioning also have relevance to work activities. Age-related changes in hearing ability, commonly referred to as presbycusis, are characterized as follows: 1) impairment of pure tone thresholds especially for tones in the high frequency range; 2) impaired frequency discrimination; 3) impaired speech localization ability; 4) impaired speed discrimination ability especially when speech is distorted and 5) decreased ability to recall long sentences (Olsho, Harkins and Lenhardt, 1985). This does not imply that the majority of older people suffer serious hearing losses. Most people experience moderate reductions in hearing level and some loss of speech perception. However, these declines are sufficient to influence performance on auditory tasks especially under noisy listening conditions. Older adults are likely to have difficulty communicating or detecting auditory signals if the listening conditions are less than ideal.

Currently, however, little is known about the psychoacoustic performance of older people for complex or real world tasks. More basic research is needed to understand these issues before appropriate environmental and technological interventions can be identified. At the present time, it is appropriate to suggest that older people should work in relatively noise free environments, auditory and speech signals should be distinct and free from distortion; signals in the high frequency range (above 4000 HZ) should be avoided and sufficient time should be allowed for processing auditory signals.

TABLE 1

Design Guidelines for Visual and Auditory Environments

Visual Environment

1. Increase levels of illumination. Use high range of IES guidelines.
2. Provide task lighting where possible. Allow individual control of light.
3. Avoid sudden and pronounced shifts in level of illumination.
4. Control glare
   - use matte finish on all surfaces
   - shield light fixtures
5. Provide maximum contrast on signs and on labels.
6. Avoid difficult visual discriminations - discriminations in the blue-green range or discrimination of the same hue.
7. Increase the size of letters and important visual objects.
   - suggested letter text for size reading is 10-12 point type (IES, 1981)

- Target luminance necessary to achieve sufficient contrast for different age groups can be determined by multiplying the natural log of the background luminance by the following contrast multipliers: (Blackwell and Blackwell, 1971)

Age	Contrast
40-50 yrs.	1.20
50-60 yrs.	1.86
60-70 yrs.	2.51

8. Avoid clutter on visual displays and highlight important visual information.
9. Use redundant cues (visual and auditory) where possible.
10. High color rendering full spectrum lamps should be used to illuminate task areas wherever possible. This will help compensate for the loss of color sensitivity.

Auditory Environment

---

1. Background noise should be kept to a minimum.
   - ceiling floors and walls should be acoustically treated.
   - eliminate reverberation and echos
2. Ensure that auditory signals are of sufficient duration for detection.
3. Avoid auditory signals in the high frequency range (above 4000 H_z)
4. Avoid distorted speech

BODY SIZE AND STRENGTH

Older adults have reduced abilities in strength, force exertable and dexterity. There is, on average, a decline in muscle mass with age which results from a decrease in the number and size of muscle fibers. In addition, the average maximal capacity of the cardiovascular system to deliver oxygen to the muscle is reduced relatively early in the aging process which translates into reduced endurance. Some rough estimates indicate that by age 40 muscle strength is reduced by about 5%, by age 50 about 15% and by age 65 about 25% with further declines thereafter (Stoudt, 1987). There is, however, a great deal of variability among different muscle groups, in types of muscular performance and among individuals. For example, the effects of training and motivation can be considerable in this area.

Declines in strength and endurance have obvious implications for work activities. Heavy work such as lifting and carrying may create more difficulties for older workers than for younger workers. Generally, as age increases there is an increased frequency of transfer to lighter work (Davies, and Sparrow 1985). Older people may require more frequent rest breaks to compensate for declines in stamina or a job rotation schedule which results in reduced physical demands.

There is also some loss of overall joint mobility and dexterity with advanced age. Generally however, physical mobility is not greatly reduced in healthy people until the seventh decade. The incidence of arthritis, which substantially reduces joint mobility is more prevalent among older people. The National Health Interview Survey, (Harris, 1987) indicates that about 45% of the population aged 65-74 years report this condition. Arthritis can play a major role in functional limitations among the elderly. For example, activities such as reaching, grasping with the fingers or bending are difficult for persons afflicted with this condition. This may,
affect one's ability to use hand tools, operate controls or perform small manipulative tasks such as assembly tasks. In addition, a loss in finger dexterity may make it difficult to use a keyboard. Therefore, when possible, alternative input devices should be available.

The aging process also results in changes in body size. Generally, there is lack of data regarding the anthropometric characteristics of older people especially for functional anthropometric measures. The most complete data are those presented by the National Health Examination Survey (Stoudt, Damon, MacFarland, and Roberts, 1965, 1973). Select data from this survey are presented in Table 2. As shown in the table, older people are generally smaller than their younger cohorts. With advanced age, there is a loss stamina on the order of magnitude of 2 inches which results from anatomical and postural alterations. Sitting height, which includes seated eye height also declines by amounts that parallel the loss in stature. Reach abilities are also less for older populations, especially for those afflicted with arthritis.

Changes in body size may have implications for work place design and layout. For example, optimal screen placement may be different for older people especially since the likelihood of wearing bifocals increases with age. Similarly ideal work heights or reach distances may need to be lower to accommodate the loss in stature. More data is needed before specific design recommendations can be generated. It appears that adjustable furnishings (e.g., chair) are especially critical for older workers.

TABLE 2

Selected Anthropometric Measures of Older Adults

	Males			Females		
	55-64 yrs	65-74 yrs	75-79 yrs	65-54 yrs	65-74 yrs	75-79 yrs.
Weight (lbs)[1]	166	160	150	152	146	138
Height (in)[1]	67.4	66.9	65.9	62.4	61.5	61.1
Sitting Height (Normal)[1]	33.8	33.4	33.0	31.9	31.1	30.5
Knee Height[1]	21.1	21.0	20.6	19.4	19.3	19.4
Poplitial Height[1]	17.1	17.0	16.6	15.4	15.3	15.5
Bullock Poplitial Length[1]	19.3	19.2	18.9	18.9	18.8	18.6
Elbow Rist Height[1]	9.3	9.0	8.6	8.9	8.4	8.2
Seat Breadth[1]	14.0	13.9	13.7			

	Males		Females	
	50-59 yrs.	60-69 yrs.	50-59 yrs.	60-69 yrs.
Combined Grip[2] Strength (lbs)	97	88	56	52
Forward Arm[2] Reach (Males, 72-91 years)	34.21			
Forward Arm[2] Reach (Females, 65-80 years)	28.56			

Source:

1. Stoudt, Damon, MacFarland and Roberts (1965, 1973)
2. Charness and Bosman, 1989

Notes: Height and other dimensions in inches, data represent mean values.

INFORMATION PROCESSING

There are a number of age-related changes in cognitive abilities which have important implications for work activities. One of the most reliable findings regarding aging is that there is a general slowing of behavior as age increases. Currently, there is some controversy regarding whether the locus of slowing is primarily due to changes in the perceptual and motor system or in the central processing system. However, despite the underlying causality older people generally take longer to encode information, retrieve information from long term memory, choose among response alternatives, and execute responses (Fozard, 1981). This trend of slower behavior with age has been documented for a variety of tasks including: choice reaction time, digit coding, card sorting, visual search (Salthouse, 1985) and text-editing (Czaja, Hammond, a disadvantage if tasks require quick decisions or rapidly paced activities. Therefore, they should not be placed on tasks which have stringent pacing requirements. In fact, data from industrial studies (e.g., Belbin, 1953) suggest that older people tend not to be employed in these types of jobs as they are stress-inducing.

The size of the age difference in response speed is variable and influenced by factors such as health status, task complexity, response requirements and practice. This implies that factors such as signal or information duration; response interval, and the number of mental operations between stimulus and response, are significant with respect to the magnitude of the age difference in rate of responding and thus to the performance efficiency of older people.

As suggested to some extent, the age decrement in response speed can be compensated for by practice or experience. For example, Plude and Hoyer (1981) found that age differences in visual search time diminished as the task became well-practiced. They concluded that age differences will be minimal on well-practiced tasks where processing becomes automatic and are more likely to be observed in situations where effortful processing is required. It is therefore important to ensure that older workers are provided with sufficient training. Also stimulus response compatibility and consistency are important to rate of responding as inconsistencies in tasks, or equipment make it difficult for tasks to become automatic.

Overall, the data indicate that slowing of response may place older people at a disadvantage in the performance of many tasks. In order to offset or minimize this disadvantage, design inconsistencies and ambiguities should be minimized and sufficient time should be provided for information extraction and response execution (Table 3). Additionally, the pacing requirements of tasks must be evaluated in terms of their suitability for older persons and adequate training must be provided to compensate for declines in processing abilities.

There are several changes in percepual abilities which also have relevance to work activities. These include declines in perceptual flexibility and speed of encoding information. Older adults often have difficulty processing complex, confusing or inconsistent information and are more likely to experience interference from irrelevant or surplus information (Plude and Hoyer, 1985). These findings are pertinent to the design of visual displays. For example, amount of information should be minimized and only necessary and meaningful information should be presented. Also, grouping related information and highlighting critical information is especially important for older persons. Finally, as suggested earlier, incompatibility and inconsistencies should be avoided and adequate time for processing information should be provided.

Age related changes in information processing capacity may also impact on the performance of some tasks. However, this is difficult to predict as the relationship between aging and attentional capacity is unclear. Some researchers (e.g., Hunt and Herzog, 1981) claim that attentional capacities decline with age and tasks that demand attention and are complex with respect to demands on processing resources will show age related declines. Other researchers claim (e.g., Wickens, Braune and Stokes, 1987) that the suggested decline in processing capacity may not exist. Wickens et. al. (1987) paired a tracking task with a memory task and found an age-related decline in processing-speed but age-related difference in time sharing efficiency. These results must be interpreted with some caution as laboratory tasks were used and the age range of the sample was restricted to 65 years (X=59 yrs; older group). However, they suggest that attentional capacity does not change with age. Ponds, Broewer and Van Wolffelear (1988) did find an age-related decrement in dual task performance when a tracking task was paired with a visual choice reaction time task. Their sample was older than that of Wickens, et. al. with the older group ranging in age from 61-80 yrs (X=69 yrs). In fact there was no difference in time-sharing abilities between the middle aged and younger subjects. These results suggest that age decrements in divided attention may not occur until the later decades. Clearly, this is an area where more research is warranted. We need to understand if in fact time sharing ability declines with age and, if so, to what extent this decline is compensated for by experience and practice. McDowd (1986) for example, found that practice improved the divided attention performance of both younger and older adults. This area of research is relevant to many types of tasks such as aviation tasks, computer-interactive tasks and managerial tasks.

The potential for age-related differences in job performance can also be derived from the memory component of information processing. Essentially, there are two age-related effects in memory which are relevant to jobs: shortening of the immediate memory span and difficulty in retrieval of information from the long-term store (Welford, 1985). These changes may make it difficult for an older person to recall the meanings of codes especially if they are "unnatural" or ambiguous or recall a long series of information or the sequence of steps in a complex task. It is also difficult for an older person to maintain large amounts of information in working memory which is important to the performance of some tasks such as piloting or manipulating databases.

In many cases these changes in memory can be offset by practice and experience. There is substantial evidence which shows that the memory performance of older people is less impaired if the task is "automatic" and environmental support such as cues or job aids is provided. Performance is also improved if information is organized and distracting or interfering information is minimized. Table 3 presents some design guidelines relevant to the information processing aspects of jobs.

Learning skills also change with age. This change is often reflected as a decline in learning ability when comparing younger and older learners. For example, Gomez, Egan and Bower (1986) and Czaja, Hammond, Blascovich and Swede (1989) found that older people had more difficulty than younger people learning to use a text-editor. Similar results have been obtained when comparing the performance of younger and older persons for other types of tasks. These results suggest that older people are unable to learn new skills or strategies which compensate for changes in abilities. However, several investigators (Belbin and Shimmin, 1964; Czaja and Drury, 1981) have shown that if appropriate training strategies are employed age differences in learning performance can be eliminated or reduced. Air Lingus mounted a successful campaign to retrain older workers. The airline automated a significant portion of their cargo operation which required that workers learn new skills to operate the new technology. The retraining program was tailored to incorporate the needs of older workers and incorporated elements such as trainee participation and "on-the-job" simulation. The program was successful as workers of all ages were successfully able to use the technology (OTA, 1985).

These results and others indicate that older people learn better when a problem solving approach is used as opposed to the traditional approach of "drill and practice". In addition, allowing extra learning time, spacing training and providing ready access to assistance during learning facilitates the performance of older adults (Hartley, Hartley and Johnson, 1984). Also providing practice on learning skills and drawing analogies between new information and already existing concepts has been shown to aid learning (Hultsch, 1974). In most cases, if appropriate training strategies are employed older people will be able to learn to perform a task at an acceptable level. Also, it is important to insure that older people have equitable access to training opportunities. Training opportunities may be influenced by informal age restrictions or biases and assumptions regarding lowered capabilities of older workers (Sterns and Alexander, 1987).

The preceding discussion summarizes age-related changes in performance which may have relevance to work activities. Much of this discussion is speculative in the sense that it is based largely on findings from the laboratory as opposed to real world settings. This does not mean that laboratory data is not useful with respect to understanding issues related to aging and work. However, caution must be exercised when generalizing from laboratory studies to predict performance in real world situations. Laboratory studies typically involve artificial tasks and do not account for the effects of experiences or compensatory strategies which develop after considerable time is spent performing a task.

Salthouse (1984) examined age differences in performance on two tasks which were structurally similar; a normal typing task and a choice reaction time task which required rapid key stroke responses to rapidly presented alphanumeric characters. He observed different age trends for the two tasks. Age decrements in performance were found for the reaction time task but not for the typing task. He speculates that the older typists employed an anticipatory compensatory strategy to overcome their slower perceptual motor processes. This type of strategy would be unlikely to develop in a lab setting where practice on a task is limited in comparison to real world settings. The role of experience is an especially important consideration when evaluating the potential for age differences in performance. Evidence clearly demonstrates that performance improves with experience and in most activities increased age is positively related to amount of experience.

TABLE 3

Design Guidelines for Information Processing Aspects of Tasks

1. Fast pacing of tasks or rapid presentation of information should be avoided to compensate for decreased rate of processing.
   motor processor cycle time[1]: $Y^{1} = 33.9 + .67$ (age)
   cognitive processor cycle time[1]: $Y^{1} = 18.49 + .66$ (age)
2. Allow sufficient time for responding
   Fitt's Law[1]: $Y = 60.68 + 1.86$ (age), SE=14.49
   Hick's Law[2]: $Y^{1} = 179.92 + 1.64$ (age), SE = 49.95
3. Avoid conflicting, complex or ambiguous coding or instructions.
   - use familiar in instructions
4. Avoid inconsistencies in design of displays and controls.
5. Highlight important information in displays.
6. Group related information where possible.
7. Avoid excessive memory requirements such as having to remember a long series of steps or information items.
8. Provide adequate task training.
9. Provide cues where possible.

Source:
1. Charness and Bosman, 1989

AGING AND WORK PERFORMANCE

The relationship between aging and performance in the workplace is complex. This is because performance is multi-faceted and the effects of age are often confounded with other factors such as education and experience. When examining this issue there are several aspects of performance which need to be considered including: productivity, absenteeism, turnover, accident rate and grievances. Current knowledge regarding how these measures vary with age is limited and unclear. At the outset it is important to recognize that it is over-simplistic to regard chronological age as a valid predictor of an individual's employment potential. As discussed aging is a non-uniform process and thus people within a certain age group vary considerably with respect to their functional abilities. A more important issue than chronological age is understanding how to design jobs and work environments to maximize the potential contributions for a wide age range of workers.

Regarding accidents, the data clearly indicate that older workers have lower accident rates, than younger workers. However, the severity of accidents increases with age. Older workers tend to stay off the job longer if they are injured. Absenteeism and turnover rates are also lower for older people. Martocchio (1989) recently completed a meta-analysis across 34 samples which included data regarding age and employee absenteeism. Results indicated that both voluntary, absence, based on the frequency index, and involuntary absences, based on the time lost index, are inversely related to age. The finding regarding absence frequency is consistent with other investigators whereas the finding regarding lost time was contrary to expectations. Martocchio suggested that this finding may have been due to the fact that older workers have lower accident rates than younger workers. Overall, the results suggest that older workers are able to meet job requirements and are also dependable and have a high degree of job commitment.

Considering productivity or work output, data regarding age related effects is limited and somewhat ambiguous. Examining the current literature in this area reveals it is hard to draw any conclusions on how age affects work. Rhodes (1983) conducted a comprehensive review of the existing data on aging and productivity and concluded that there was equal evidence supporting that job performance increases with age, decreases with age, or remains the same. She concluded that the result depends on the type of performance measures used, the type of job and other factors such as experience.

Waldman and Avolia (1986) conducted a meta-analysis on 40 samples that included data pertaining to the relationship between age and job performance. They classified the samples into three categories: a) supervisory ratings b) peer ratings and c) individual productivity. They found different results according to type of performance measure. For objective indices performance was found to increase with higher ages whereas supervisor ratings showed a tendency to be lower for older employees. They found however that the relationship between ratings and age varied according to job type; ratings were higher for professionals than non-professionals. They concluded that age only accounts for a small amount of the variance in the range of individual differences in work performance and thus is not useful for estimating performance potential.

McEvoy and Cascio (1989) also performed a meta-analysis of over 65 studies and concluded that age and job performance are unrelated. However, they caution that most of the studies they examined were cross-sectional, involved small samples and a restricted age range. They point out that there is a need for more research on the 50-70 year old age cohorts.

Examining the studies that have been done there appears to be little, if any basis for the widespread belief that performance declines with age. Instead, it may be that some tasks are unsuitable for older workers because the demands inherent in these tasks exceed their capabilities. The data that are available suggest that tasks which involve heavy physical demands or those which are externally paced may be unsuitable for older workers (Davies and Sparrow, 1985). However, these conclusions are general and not specific to jobs or occupations.

With respect to other types of jobs the findings are mixed or lacking. For example, although there appears to be age deficits in visual search performance it is unclear as to what extent this affects performance on industrial inspection tasks. Drury and Sheehan (1971), and Czaja and Drury (1981) found age-related declines in inspection performance whereas Evans (1951) reported no age differences in performance on an inspection task. In contrast Jamieson (1966) showed an increase in inspection performance with age. Similarly, Davies and Parasuraman (1982) found that older people performed at about the same level as younger people in a task that required sustained visual attention. Clearly research is needed in this area as inspection is a prevalent and an important task in many industries including the microelectronics industry, garment industry and automobile industry. Visual search is also a component of other tasks such as driving or computer-interactive tasks.

With respect to production tasks the findings are also mixed. Clay (1956) examined output records at two printing plants and found that performance reached a peak in the 30's or 40's and declined gradually thereafter. Greenberg (1960) reports data from a large study conducted by the U.S. Department of Labor which examined output records in a wide range of industries. The results were similar to those reported by Clay (1956); performance tended to improve up to age 35 and decline with increased age. However, the amount of decline was gradual and relatively small. Salvendy (1972) found no performance differences according to age in manufacturing jobs. Geniger, Despenzieri and Eisenberg (1983) also failed to find an age effect on performance for garment industry workers. Their study examined the variables of age and experience in relationship to productivity, absenteeism and turnover for jobs requiring speed and those requiring skill. They found that for both categories of jobs older workers were more productive, had lower absenteeism and turnover rates and less turnover than younger workers. Their findings for the speed jobs were contrary to expectations given the response slowing which occurs with age. They postulated that the jobs were in the light manufacturing industry and were not very physically demanding and also that the competence and experience of the older worker may have compensated for their reduced physical stamina. They found, in fact, that experience rather than age was the primary determinate of job performance.

The results are similar for white collar workers. Kutscher and Walker (1960) reported a decline in productivity with age for clerical workers where the U.S. Department of Labor study reported no age differences in output per work hour for office workers. Considering age and performance in professional and managerial jobs the data is again limited and inconclusive. Different results are obtained among fields and within fields depending on the output measure used in the analysis. For example, Stumpf and Rabinowitz (1981) found a negative relationship between age and research productivity among scientists but a positive relationship between age and peer assessment of performance. Dalton and Thompson (1970, 1971) investigated the relationship between age and performance among a sample of professional engineers in the aerospace industry. They found that performance peaked in the mid-30's and gradually declined with advances in age. However, they found that some of these effects were due to skill obsolescence due to rapid developments of new technologies.

The importance of training and skill maintenance has been found by other investigators. For example, Sparrow and Davies (1988) examined the effects on performance of age, tenure, job complexity and training among a sample of service engineers. They found an inverted-U relation between age, quality and speed of performance. However, three other important findings emerge from their study. One is that job complexity did not moderate the effect of age and performance and in fact for all age groups performance was higher for the more complex tasks. Secondly, age only accounted for a small portion of the variance. Finally, training is an important mediating variable between age and performance. Older engineers who had received recent task training performed better than those who had not received recent training. This suggests that training can mediate age declines in performance. This is a critical finding given the rapid influx of new technologies into the workplace.

As illustrated, the relationship between age and job performance is complex and far from understood. One problem is that the data which is available is sometimes unreliable because of methodological shortcomings in data collection procedures. For example, studies which rely on supervisory ratings of performance may be unreliable if the rater has negative attitudes about older workers. Also, many studies involve small samples, restricted age ranges or are cross-sectional which may confound age effects with factors such as experience, education or exposure to technology. Finally, the results vary according to type of task or type of measure. What's lacking is a systematic body of knowledge that links age related changes in skills and abilities to skill demands of jobs. Clearly,
more research is needed in this area especially for information processing tasks such as computer-interactive tasks or supervisory control tasks which are becoming increasingly prevalent within today's work environment.

Strategies to Promote Employment Opportunities for Older Adults

The discipline of Human Factors Engineering with its focus on fitting the designed environment to user capabilities, needs and limitations has direct conceptual and methodological relevance to the study of work activities and older adults. The major potential contributions of Human Factors Engineering with respect to promoting employment opportunities for older people are: 1) conducting research which furthers our understanding of the relationship between aging and work activities and 2) designing and evaluating intervention strategies.

One important area of research is the study of aging and work performance. As discussed there is a paucity of data regarding how the functional changes which occur with age impact on the performance of specific jobs. This type of knowledge is needed to better design workplaces to accommodate older people and also to dispel myths that suggest that aging is associated with performance declines. Discriminatory practices towards older people are still prevalent in many industries and many of these age barriers stem from the belief that older people are unproductive (Sterns and Alexander, 1987).

Research efforts should be directed towards developing a systematic body of knowledge that ties age-related changes in skills and abilities to skill requirements of jobs. This type of information is especially important for information processing tasks as these types of tasks are prevalent in most work settings given the increased use of technologies. For example, given the slowed response rate of older people one task aspect which needs to be evaluated is the task pacing dimension. In addition, issues surrounding the appropriateness of software applications are important given age-related changes in cognition. For example, age differences in memory may make it difficult for older people to perform tasks which require manipulation of varied databases unless some type of memory aid is provided.

Human Factors research techniques such as task analysis can be used to yield information regarding these issues. Task analysis can be used to specify the skill requirements of jobs and this information can be compared to databases regarding capabilities of older adult populations. This will allow us to identify which tasks are suitable for older people to perform and to target potential changes in job design. In addition to task analysis, experiments that examine age changes in performance on specific tasks should be conducted.

As a cautionary note if these studies are cross-sectional the sample size should be adequate and factors such as experience and education should be controlled. Also, there is a need for longitudinal data.

Another area in which insufficient research has been conducted is in the training of older workers. Documented declines in cognition have often been based on situations where older people have limited opportunity for learning. Several investigators (e.g., Belbin) have shown that, if appropriate training strategies are employed, age differences in performance can be reduced or eliminated. To date, there are only limited data available on task specific training strategies appropriate for older learners. It will also be useful to determine the extent to which training on component skills, such as visual search, benefits performance on tasks such as inspection. We also need to develop methods to ensure that older people have equitable access to retraining programs. Negative stereotypes about aging among employers may create obstacles for older people attempting to enter retraining programs. Perhaps, the development of awareness seminars regarding the aging process may alleviate some of these negative attitudes.

Human Factors efforts can also be directed towards identifying and evaluating design interventions to remove formal age barriers to employment. These interventions can take the form of equipment redesign, job redesign and environmental redesign. This would involve applying data on older adult populations to the design process. Equipment design efforts might include redesign of displays, controls or equipment dimensions. For example, efforts might be directed towards designing alternative input devices to keyboards such as speech recognition systems. Also, as indicated more research is needed in the area of vision and design of displays and environments. Job redesign efforts might include determining suitable work/rest schedules for older people. We know very little about how energy expenditure for various tasks changes with age.

This type of data is needed before guidelines for job design can be developed. Similar questions arise regarding the work environment. We only have limited quantitative guidelines on older adults. Also, the accuracy of some of the existing guidelines are unknown. Finally, as indicated by Charness (Charness and Bosman, 1989) many of the guidelines are prescriptive rather than quantitative in nature. Prescriptive data does not provide a strong basis for making design decisions.

Part of the problem, resides in the fact that we have limited normative data on certain human characteristics. For example, both static and dynamic anthropometric data is needed for older populations. Without this data, it is difficult to arrive at design solutions. It should be pointed out that in some cases it is not necessary to collect new data but to apply what we know to work situations. Much is already known in the Human Factors and gerontological communities about the performance problems of older adults, however, such knowledge rarely makes its way into real work environments.

In general, research attention to those aspects of work such as performance, equipment, job and environmental design could make worthwhile contributions towards enhancing employment opportunities for older people. To a large extent they represent a largely untapped resource who can make worthwhile contributions to our nation's economy.

CONCLUSIONS

It is likely the observed age differences in performance can be accommodated in the workplace through changes in system design and the provision of adequate training. Designing work environments to accommodate older people will increase employment opportunities for this age group. Also, it is likely that people from all ages will benefit from many of the design changes.

However, before this goal can be fully realized there is a need for more research. Ultimately, we need answers to specific scientific and practical questions about identified problems of aging and the workplace. This includes task analysis data as well as more narrowly focused research to fill in gaps in our existing knowledge base. There is a need for a better engineering response directed towards the problems encountered by older adults in the whole spectrum of work activities.

REFERENCES

Belbin, R.M. (1953) Difficulties of older people in industry. Occupational Psychology, 27: 177-190.

Belbin, E., and Shimmin, S., 1964. Training the middle aged for inspection work. Occupational Psychology, 38: 49-57.

Birren J.E., and Cunningham, W., 1985. Research on the psychology of aging: Principles, concepts and theory. In J.E. Birren and K.W. Schaie (Eds.) Handbook of the Psychology of Aging, (2nd Ed.), New York: Van Nostrand Reinhold Co., 3-30.

Birren, J.E., and Schaie, K.W., 1985. Handbook of the Psychology of Aging, (2nd Ed.) New York: Van Nostrand Reinhold Co.
Blackwell, D.M., and Blackwell, H.R., 1971. Visual performance data for 156 observers of normal age. Journal of the Illuminating Engineering Society, 1: 3-13.
Burrus-Bammel, L.L. and Bammel, G., 1985 Leisure and recreation. In J.E. Birren and K.W. Schaie (Eds.), Handbook of the Psychology of Aging Second Edition, pp. 848-863. New York: Van Nostrand Reinhold Company.
Charness, W., 1985. Aging and human performance, New York: John Wiley and Sons.
Charness, W., and Bosman, E.A., 1989. Human factors and design for older adults. In J.E. Birren and K.W. Schaie (Eds.) Handbook of the Psychology of Aging (3rd Ed). New York: Van Nostrand Reinhold Co. 446-463.
Clay, H. M., 1956. A study of performance in relation to age at two printing works. Journal of Gerontology., 4: 417-442.
Czaja, S. J. and Drury, C.G., 1981. Age and pretraining in industrial inspection. Human Factors, 23: 485-494.
Czaja, S.J., Hammond K., Blascovich, J., and Swede, H., 1989. Age related differences in learning to use a text-editing system. Behavior and Information Technology, 8: 309-319.
Waldman, D.A. and Avolio, B.J., 1986. A meta-analysis of age differences in job performance. Journal of Applied Psychology, 71: 33-38.
Czaja, J.S. and Sharit, J., 1988. Aging and the performance of computer-interative tasks: Job design and stress potential. Trends in Ergonomics/Human Factors V. 185-190.
Dalton, G.W. and Thompson, P.H., 1970. Age obsolescence and performance. Unpublished report.
Dalton, G.W. and Thompson, D. H., 1971. Accelerating obsolescence of older engineers. Harvard Business Review, 49: 57-67.
Davies, D.R., and Parasuraman, R., 1985. The Psychology of Vigilance. London: Academic Press.
Davies, R.D., and Sparrow, P.R., 1985. Age and work behavior. In N. Charness (Ed.) Aging and human performance, New York: John Wiley and Sons.
Davis, K., 1985 Health care policies and the aged: Observations from the United States. In R.H. Binstock, E. Shanas,(Eds.) Handbook of Aging and the Social Sciences 2nd Edition. pp. 727-744. New York: Academic Press.
Drury, C.G. and Sheehan, J.J., 1971. Ergonomic and economic factors in an industrial inspection task. International Journal of Production Research, 7: 333-341.
Evans, R.N., 1951. Training improves micrometer accuracy. Personnel Psychology, 4: 231-242.
Fozard, J. L., 1981 Person-environment relationships in adulthood: implications for human factors engineering. Human. 23: 7-28.
Geniger, S., Despenzieri, A., and Eisenberg, J., 1983 Age, experience, and performance on speed and skill jobs in an applied setting. Journal of Applied Psychology. 68: 469-475.
Greenberg, L., 1960. Productivity of olderworkers. Gerontologist, 1: 38-41.
Gomez, L.M., Egan, D.E., and Bower, C., 1986. Learning to use a text-editor: some learner characteristics that predict success. Human-Computer Interaction, 2: 1-23.
Harris, L., 1989. Problems facing elderly americans living alone:

A national survey. New York: Harris and Associates.
Hartley, A.A., Hartley, J.J. and Johnson, S.A., 1984. The older adult as a computer user. In P.K. Robinson, J., Livingston and J.E. Burren (Eds.) Aging and Technological Advances, New York: Plenum Press, 347-348.
Hartley, A.A., Hartley, J.T. and Johnson, S.A., 1984. The older adult as a computer user. In P.K. Robinson, J. Livingston and J.E. Birren (Eds.) Aging and Technological Advances New York: Plenum Press, 347-348.
Hughes, P.C. and McNelis, J.F., 1987. Lighting, productivity and the work environment. Paper presented at the annual Illuminating Engineering Technical Society, Denver, August.
Hunt, E., and Herzog, C., 1981. Age related changes in cognition during the working years (final report). Arlington, VA: Personnel and Training Research Programs, Office of Naval Research.
Hultsch, D., 1974. Learning to learn in adulthood. Journal of Gerontology, 29: 302-308.
Jamieson, G.H., 1966. Inspection in the telecommunications industry: A field study of age and other performance variables. Ergonomics, 9: 297-303.
Jessup, D.D., and Greenberg B., 1989. Innovative older worker programs. Generations. Summer, 23-27.
Kline, D.W. and Schieber, F.J., 1985. Vision and aging. In J.E. Birren and K.W. Schaie (Eds.) Handbook of the Psychology of Aging (2nd Ed.), New York: Van Nostrand Reinhold, 296-331.
Kosnick, W., Winslow, L. Kline, D., Rasinski, K., and Sekuler, R., 1988. Visual changes in daily life throughout adulthood. Journal of Gerontology, 43: pp. 63-70.
Kutscher, R.E. and Walker, J.F., 1960. Comparative job performance of office workers by age. Monthly Labor Review, 83: 39-43.
Martocchio, J.J. 1989. Age-related differences in employee absenteeism: A meta-analysis. Psychology and Aging, 4: 409-414.
McDowd, J.M., 1986. The effects of age and extended practice on dividend attention performance. Journal of Gerontology, 41: 764-769.
Miller, K.D., 1983. Psychological factors in retraining. Technical paper, World Congress on the Human Aspects of Automation, August, Ann Arbor, Michigan.
National Research Council, 1987. Work, aging and vision. Washington, D.C.: National Academy Press.
Olsho, L.W., Harkins, S.W., and Lenhardt, M.L., 1985. Aging and the auditiry system. In J.E. Birren and K.W. Schaie (Eds.), Handbook of the Pshychology of Aging. New York: Van Nostrand Reinhold Company.
Office of Technology Assessment, 1985. Technology and Aging in America. Washinton, D.C.: U.S. Government Printing Office
Ponds, R.W., Brouwer, W.H., Van Wolffelar, P.C., 1988. Age differences in divided attention in a simulated driving task. Journal of Gerontology, 43: pp. 151-156.
Plude, D.J., and Hoyer, W.I., 1981. Adult age differences in visual search as a function of stimulus mapping and information load. Journal of Gerontology, 36: 598-604.
Plude. D.J. and Hoyer, W.I., 1985. Attention and performance: identifying and localizing age deficits. In N. Charness (Ed.) Aging and Human Performance, New York: John Wiley and Sons, 48-89.

Robinson, P.K., Coberly, S., and Paul, C.E., 1985. Work and retirement in R. H. Binstock and E. Shannas (Eds.), Handbook of Aging and the Social Science, Second Edition. New York: Van Nostrand Reinhold Comapny.

Rhodes, S.R., 1983. Age-related differences in work attitudes and behavior. A review and conceptual analysis. Psychological. Bulletion. 93: 328-367.

Salthouse, T.A., 1984. Effects of age and skill on typing. Journal of Experimental Psychology: General. 113: 345-371.

Salthouse, T.A., 1985. Speed of behavior and its implication for recognition. In J.E. Birren and K.W. Schaie (Eds.), Handbook of the Psychology of Aging. New York: Van Nostrand Reinhold Comapny, 100-422.

Salvendy, G., 1972. Effects of age on some test scores on production criteria. Studia Psychologica, 14: 186-189.

Sparrow, P.R. and Davies, D.R., 1988. Effects of age, tenure, training and job complexity on technical performance. Psychology and Aging, 3: 307-314.

Sterns, H. L. and Alexander, R. A., 1987. Industrial gerontology: The aging individual and work. In K.W. Schaie and C. Eisdorfer (Eds.) Annual Review of Gerontology and Geriatrics, 243-264.

Stoudt, H.W., 1987. Changes in the physical characteristics and capabilities of the aged: An overview. Paper prepared for the National Research Council Workshop, Human Factors Research Issues for Aging Population, Washington, D.C., August.

Stoudt. H.W., 1987. Changes in the phsycial characteristics and capabilities of the ages--an overview. Paper prepared for the N.R.C. Workshop, Human Factors Research Issues for an Aging Population. Washington, D.C. August, 12-13.

Stoudt, H.W., Damon, A., McFarland, R.A., and Roberts, J., 1965. Weight, height and selected body dimensions of adults. Visual and Health Statistics, Series 11, 8, Rockville, MD: Public Health Service.

Stoudt, H.W., Damon, A., McFarland, R.A., and Roberts, J., 1973. Skinfolds, body girths, biocromial diameter and selected anthropometric indicies of older adults. Vital and Health Statistics, Series 11, 35, Rockville, MD: Public Health Services.

Stumpf, P.A. and Rabinowitz, S., 1981. Career stages as a moderator of performance relationships with facets of job satisfaction and role perceptions. Journal of Vocational Behaviors, 16: 282-298.

Welford, A.J., 1985. Changes of performance with age: an overview. In N. Charness (Ed.), Aging and Human Performance, New York: John Wiley and Sons, 333-365.

Wickens, C.D., Braune, R. and Stokes, A., 1987. Age differences in the speed and capacity of information processing: a dual-task approach. Psychology and Aging, 2: 70-78.